TUNNELS AND UNDERGROUND CITIES: ENGINEERING AND INNOVATION MEET ARCHAEOLOGY, ARCHITECTURE AND ART

T0175068

PROCEEDINGS OF THE WTC2019 ITA-AITES WORLD TUNNEL CONGRESS,
NAPLES, ITALY, 3-9 MAY, 2019

Tunnels and Underground Cities: Engineering and Innovation meet Archaeology, Architecture and Art

Volume 2: Environment sustainability in underground construction

Editors

Daniele Peila
Politecnico di Torino, Italy

Giulia Viggiani
University of Cambridge, UK
Università di Roma "Tor Vergata", Italy

Tarcisio Celestino
University of Sao Paulo, Brasil

CRC Press
Taylor & Francis Group
Boca Raton London New York

CRC Press is an imprint of the
Taylor & Francis Group, an **informa** business

A BALKEMA BOOK

Cover illustration:

View of Naples gulf

CRC Press/Balkema is an imprint of the Taylor & Francis Group, an informa business

Typeset by Integra Software Services Pvt. Ltd., Pondicherry, India

Published by: CRC Press/Balkema
          Schipholweg 107C, 2316XC Leiden, The Netherlands
          e-mail: Pub.NL@taylorandfrancis.com
          www.crcpress.com – www.taylorandfrancis.com

ISBN: 978-0-367-46579-7 (Hbk)
ISBN: 978-1-003-02972-4 (eBook)

Tunnels and Underground Cities: Engineering and Innovation meet Archaeology,
Architecture and Art, Volume 2: Environment sustainability in
underground construction – Peila, Viggiani & Celestino (Eds)
© 2020 Taylor & Francis Group, London, ISBN 978-0-367-46579-7

Table of contents

Preface

The World Tunnel Congress 2019 and the 45th General Assembly of the International Tunnelling and Underground Space Association (ITA), will be held in Naples, Italy next May.

The Italian Tunnelling Society is honored and proud to host this outstanding event of the international tunnelling community.

Hopefully hundreds of experts, engineers, architects, geologists, consultants, contractors, designers, clients, suppliers, manufacturers will come and meet together in Naples to share knowledge, experience and business, enjoying the atmosphere of culture, technology and good living of this historic city, full of marvelous natural, artistic and historical treasures together with new innovative and high standard underground infrastructures.

The city of Naples was the inspirational venue of this conference, starting from the title Tunnels and Underground cities: engineering and innovation meet Archaeology, Architecture and Art.

Naples is a cradle of underground works with an extended network of Greek and Roman tunnels and underground cavities dated to the fourth century BC, but also a vibrant and innovative city boasting a modern and efficient underground transit system, whose stations represent one of the most interesting Italian experiments on the permanent insertion of contemporary artwork in the urban context.

All this has inspired and deeply enriched the scientific contributions received from authors coming from over 50 different countries.

We have entrusted the WTC2019 proceedings to an editorial board of 3 professors skilled in the field of tunneling, engineering, geotechnics and geomechanics of soil and rocks, well known at international level. They have relied on a Scientific Committee made up of 11 Topic Coordinators and more than 100 national and international experts: they have reviewed more than 1.000 abstracts and 750 papers, to end up with the publication of about 670 papers, inserted in this WTC2019 proceedings.

According to the Scientific Board statement we believe these proceedings can be a valuable text in the development of the art and science of engineering and construction of underground works even with reference to the subject matters "Archaeology, Architecture and Art" proposed by the innovative title of the congress, which have "contaminated" and enriched many proceedings' papers.

Andrea Pigorini
SIG President

Renato Casale
Chairman of the Organizing Committee WTC2019

Acknowledgements

REVIEWERS

The Editors wish to express their gratitude to the eleven Topic Coordinators: Lorenzo Brino, Giovanna Cassani, Alessandra De Cesaris, Pietro Jarre, Donato Ludovici, Vittorio Manassero, Matthias Neuenschwander, Moreno Pescara, Enrico Maria Pizzarotti, Tatiana Rotonda, Alessandra Sciotti and all the Scientific Committee members for their effort and valuable time.

SPONSORS

The WTC2019 Organizing Committee and the Editors wish to express their gratitude to the congress sponsors for their help and support.

*Tunnels and Underground Cities: Engineering and Innovation meet Archaeology,
Architecture and Art, Volume 2: Environment sustainability in
underground construction – Peila, Viggiani & Celestino (Eds)
© 2020 Taylor & Francis Group, London, ISBN 978-0-367-46579-7*

WTC 2019 Congress Organization

HONORARY ADVISORY PANEL

Pietro Lunardi, President WTC2001 Milan
Sebastiano Pelizza, ITA Past President 1996-1998
Bruno Pigorini, President WTC1986 Florence

INTERNATIONAL STEERING COMMITTEE

Giuseppe Lunardi, Italy (Coordinator)
Tarcisio Celestino, Brazil (ITA President)
Soren Eskesen, Denmark (ITA Past President)
Alexandre Gomes, Chile (ITA Vice President)
Ruth Haug, Norway (ITA Vice President)
Eric Leca, France (ITA Vice President)
Jenny Yan, China (ITA Vice President)
Felix Amberg, Switzerland
Lars Barbendererder, Germany
Arnold Dix, Australia
Randall Essex, USA
Pekka Nieminen, Finland
Dr Ooi Teik Aun, Malaysia
Chung-Sik Yoo, Korea
Davorin Kolic, Croatia
Olivier Vion, France
Miguel Fernandez-Bollo, Spain (AETOS)
Yann Leblais, France (AFTES)
Johan Mignon, Belgium (ABTUS)
Xavier Roulet, Switzerland (STS)
Joao Bilé Serra, Portugal (CPT)
Martin Bosshard, Switzerland
Luzi R. Gruber, Switzerland

EXECUTIVE COMMITTEE

Renato Casale (Organizing Committee President)
Andrea Pigorini, (SIG President)
Olivier Vion (ITA Executive Director)
Francesco Bellone
Anna Bortolussi
Massimiliano Bringiotti
Ignazio Carbone
Antonello De Risi
Anna Forciniti
Giuseppe M. Gaspari

Giuseppe Lunardi
Daniele Martinelli
Giuseppe Molisso
Daniele Peila
Enrico Maria Pizzarotti
Marco Ranieri

ORGANIZING COMMITTEE

Enrico Luigi Arini
Joseph Attias
Margherita Bellone
Claude Berenguier
Filippo Bonasso
Massimo Concilia
Matteo d'Aloja
Enrico Dal Negro
Gianluca Dati
Giovanni Giacomin
Aniello A. Giamundo
Mario Giovanni Lampiano
Pompeo Levanto
Mario Lodigiani
Maurizio Marchionni
Davide Mardegan
Paolo Mazzalai
Gian Luca Menchini
Alessandro Micheli
Cesare Salvadori
Stelvio Santarelli
Andrea Sciotti
Alberto Selleri
Patrizio Torta
Daniele Vanni

SCIENTIFIC COMMITTEE

Daniele Peila, Italy (Chair)
Giulia Viggiani, Italy (Chair)
Tarcisio Celestino, Brazil (Chair)
Lorenzo Brino, Italy
Giovanna Cassani, Italy
Alessandra De Cesaris, Italy
Pietro Jarre, Italy
Donato Ludovici, Italy
Vittorio Manassero, Italy
Matthias Neuenschwander, Switzerland
Moreno Pescara, Italy
Enrico Maria Pizzarotti, Italy
Tatiana Rotonda, Italy
Alessandra Sciotti, Italy
Han Admiraal, The Netherlands
Luisa Alfieri, Italy
Georgios Anagnostou, Switzerland

Andre Assis, Brazil
Stefano Aversa, Italy
Jonathan Baber, USA
Monica Barbero, Italy
Carlo Bardani, Italy
Mikhail Belenkiy, Russia
Paolo Berry, Italy
Adam Bezuijen, Belgium
Nhu Bilgin, Turkey
Emilio Bilotta, Italy
Nikolai Bobylev, United Kingdom
Romano Borchiellini, Italy
Martin Bosshard, Switzerland
Francesca Bozzano, Italy
Wout Broere, The Netherlands
Domenico Calcaterra, Italy
Carlo Callari, Italy

Luigi Callisto, Italy
Elena Chiriotti, France
Massimo Coli, Italy
Franco Cucchi, Italy
Paolo Cucino, Italy
Stefano De Caro, Italy
Bart De Pauw, Belgium
Michel Deffayet, France
Nicola Della Valle, Spain
Riccardo Dell'Osso, Italy
Claudio Di Prisco, Italy
Arnold Dix, Australia
Amanda Elioff, USA
Carolina Ercolani, Italy
Adriano Fava, Italy
Sebastiano Foti, Italy
Piergiuseppe Froldi, Italy
Brian Fulcher, USA
Stefano Fuoco, Italy
Robert Galler, Austria
Piergiorgio Grasso, Italy
Alessandro Graziani, Italy
Lamberto Griffini, Italy
Eivind Grov, Norway
Zhu Hehua, China
Georgios Kalamaras, Italy
Jurij Karlovsek, Australia
Donald Lamont, United Kingdom
Albino Lembo Fazio, Italy
Roland Leucker, Germany
Stefano Lo Russo, Italy
Sindre Log, USA
Robert Mair, United Kingdom
Alessandro Mandolini, Italy
Francesco Marchese, Italy
Paul Marinos, Greece
Daniele Martinelli, Italy
Antonello Martino, Italy
Alberto Meda, Italy

Davide Merlini, Switzerland
Alessandro Micheli, Italy
Salvatore Miliziano, Italy
Mike Mooney, USA
Alberto Morino, Italy
Martin Muncke, Austria
Nasri Munfah, USA
Bjørn Nilsen, Norway
Fabio Oliva, Italy
Anna Osello, Italy
Alessandro Pagliaroli, Italy
Mario Patrucco, Italy
Francesco Peduto, Italy
Giorgio Piaggio, Chile
Giovanni Plizzari, Italy
Sebastiano Rampello, Italy
Jan Rohed, Norway
Jamal Rostami, USA
Henry Russell, USA
Giampiero Russo, Italy
Gabriele Scarascia Mugnozza, Italy
Claudio Scavia, Italy
Ken Schotte, Belgium
Gerard Seingre, Switzerland
Alberto Selleri, Italy
Anna Siemińska Lewandowska, Poland
Achille Sorlini, Italy
Ray Sterling, USA
Markus Thewes, Germany
Jean-François Thimus, Belgium
Paolo Tommasi, Italy
Daniele Vanni, Italy
Francesco Venza, Italy
Luca Verrucci, Italy
Mario Virano, Italy
Harald Wagner, Thailand
Bai Yun, China
Jian Zhao, Australia
Raffaele Zurlo, Italy

Environment sustainability in underground construction

Tunnels and Underground Cities: Engineering and Innovation meet Archaeology, Architecture and Art, Volume 2: Environment sustainability in underground construction – Peila, Viggiani & Celestino (Eds)
© 2020 Taylor & Francis Group, London, ISBN 978-0-367-46579-7

Environmental sustainability in the construction of the Isarco river underpass for the Brenner Base Tunnel

U.B. Bacchiega, S.T. Torresani & R.Z. Zurlo
BBT SE Galleria di Base del Brennero, Bolzano, Italy

ABSTRACT: All possible planning and protective measures are being taken to safeguard the delicate environmental balance typical of a river, and to rationalize the use of natural resources. Spoil management, in terms of amounts and recovery/recycling methods of the materials, if well and carefully planned, can make a difference to the actual sustainability of the work. The most important issue as regards sustainability has to do with the interference of the works with the river and how the river is protected, since the tunnels will, in fact, pass under the Isarco river with low overburden. Planning optimizations were executed in agreement with the pertinent authorities as part of the complex authorization procedures carried out by the contracting authority, in constructive cooperation with all the actors involved.

1 INTRODUCTION

With the Brenner Base Tunnel, we see the birth of a railway that looks toward the future, crossing the Alps at the base of the mountains, with no more steep passes that are difficult to negotiate. The Brenner Base Tunnel is the key element of the new Brenner railway line from Munich to Verona. It will be a pioneering work of engineering for the twenty-first century and will significantly improve mobility in the heart of Europe, especially along the parts of the Scandinavian-Mediterranean Corridor that lie in Italy, Austria and Germany.

The environmental benefits of an underground railway infrastructure such as the Brenner Base Tunnel are not in question: shifting the volume of freight traffic that currently travels along the already overloaded Brenner highway, which will only increase in the future, onto the underground rail line will have a series of positive effects on the habitat of the narrow alpine valleys that the highway runs through, in terms of a drop in noise pollution and CO_2 pollution but also in terms of a reduced impact on the landscape and, finally, because of the improvement in the quality of life for the people living in the area.

The realization of a project of this size also requires a long and complex construction phase with all the associated impacts on the environment and on the local population. With the Isarco river underpass construction lot, the environmental management of the construction phase of the Brenner Base Tunnel will really get down to brass tacks. This is a very particular construction lot, also from an environmental point of view, and all possible planning and protective measures are being taken to safeguard the delicate environmental balance which typically characterizes a river, while rationalizing the use of natural resources as much as possible.

One of the main aspects to consider, from the very beginning of the project planning until the closure of the construction site, is spoil management, in terms of amounts and recovery/recycling methods of the materials themselves, that, if carefully and well planned, can make a difference to the actual sustainability of the work. In the Isarco river underpass construction lot the aim is to reduce excavation works to a minimum and maximize the re-use of spoil on the building site itself, considering as well the recent legal developments in this area. The treatment and reuse of spoil help to preserve natural resources, contributing to the sustainability of the activities during construction. The treated material is therefore recycled during construction work in the tunnel to produce the concrete needed for the first-phase and final linings.

A further factor determining the sustainability of the project is the protection of the Isarco river. The two main tunnels run, in places, no more than three metres below the river itself and its ecomorphological integrity must be safeguarded during the complicated construction phase, by virtue of protective measures agreed upon with the competent authorities and carried out through complex preparatory measures.

The effectiveness of the measures implemented is also verified by means of specific environmental monitoring activities. These are essential to regularly check the environmental impacts caused by the construction of the Isarco river underpass on the surrounding areas and all the environmental factors which might potentially be impacted by the construction lot.

The construction of this lot is therefore a strategic factor for a major project such as the Brenner Base Tunnel, particularly in terms of the sustainability of the project and of restoring the area concerned to its previous status. Thorough preventive analysis, correct and accurate planning, competent execution and the collaboration of all the involved parties, i.e. the client, the authorities, the contractors, have created a virtuous context that, through a conscious approach aimed at construction phase sustainability, pursues the realization of an important, not to say crucial piece of the Brenner Base Tunnel, which is a prerequisite for the sustainable development for the alpine environment being crossed.

2 PLANNING CHOICES

Under the river bed, four single-track underground tunnels will be built – with an overburden of 3 to 8 m and an excavation diameter of about 10 m – two for the line tracks and two for interconnections with the existing historical line.

The four tunnels, excavated starting from shafts located on both sides of the river bed, are 56 to 63 m long. The type of works and the construction methods have been planned in order to minimize the impact on the river and to avoid altering its normal flow; to protect the environment and the ecomorphological structure, only limited and partial damming of the river bed was done in the low water period to allow the construction of structures aimed to protect the river, such as the consolidation of the river banks and bed with boulders.

Figure 1.   Construction of the shaft for the north-bound rail track.

Considering the exiguous overburden under the river bed and the sensitive geotechnical and hydrogeological context, it was decided that underground excavation could only take place after carrying out soil consolidation measures, using several overlapping technologies, i.e. injections of cement and waterproofing mixtures in combination with ground freezing technology. The latter appears to be the safest technology to guarantee the hydraulic seal of the excavated cavity; at the same time, the mechanical characteristics of the frozen ground, in combination with the consolidation interventions and the first phase lining, are sufficient to ensure the tunnel statics in the short term.

Given that the river is located only a few meters from the ground to be frozen, the intermediate phase before freezing operations are started is the most critical because of the speed of the water flowing just below the river bed; therefore it is necessary to inject special mixtures into a strip of soil around the tunnels beforehand to reduce the permeability of the ground and, as a result, the flow speed below the riverbed and of the ground water.

According to the first phase of the final planning, which was completed in 2013, the underground works for the underpass were to be carried out by re-routing and partly damming, in several phases, an 800-metre stretch of the river in a bypass canal, then proceeding with cut and cover tunnelling with complex slab casting operations, which included works to be carried out underwater by divers, and the subsequent restoration of the original river bed. The final project, which foresaw the construction of tunnels by lowering the water level with a series of wells and re-routing the course of the river, was drawn up on the basis of the geological, geotechnical and hydrogeological model. The pumping of groundwater would have been enough to cause a significant sinking of groundwater levels, which could have caused an increase in the vertical filtration of the river water towards the groundwater, with a consequent possible increase in the transport and impoverishment of fine materials, thus generating impacts on the equilibrium and exchange dynamics between surface water and groundwater both in the short and long term.

The project approached was only reviewed during the executive planning phase in 2016, when shaft excavation and ground freezing were made part of the planning. The hydrogeological model was updated during the planning phase and a different planning solution was developed, aimed at simplifying construction methods and reducing environmental impacts. This solution essentially involves tunnelling under the valley floor after employing soil consolidation measures in the various areas such as jet-grouting, injections and soil freezing without lowering the groundwater level and temporarily rerouting the river, as originally planned. This choice has given fundamental advantages in terms of environmental sustainability in relation to spoil amounts and water management, reducing both impacts on the Isarco river and on the volumes of treated water.

The technical solutions adopted in the executive project made it possible to significantly reduce excavation volumes by around 450,000 cubic meters.

The executive project neither foresees the re-routing of the Isarco river, nor does it involve changes in the potamological conditions in the concerned area, thus increasing sustainability with regard to this aspect. As far as fish stocks are concerned, the sub-populations present in the affected area do not need to be captured and moved to another location as foreseen in the original project. The biotic interruption of the river can also be avoided, bringing advantages in terms of flora and fauna, in particular for fish, amphibians, birds and macroinvertebrate populations.

The executive project does not foresee any lowering of the ground water levels (no pumping is necessary) and therefore does not alter the existing underground water system.

3 SPOIL

3.1 *Spoil types*

In order to encourage the use of spoil for construction work, after treatment, it is classified into three classes of reuse, based on the qualitative and geomechanical characteristics of the spoil:

- Class A: high-quality material, suitable for aggregates for the concrete production;
- Class B: material that can be used for walls, filler and backfilling;
- Class C: material that cannot be used and is permanently disposed of.

This assessment is used to define the quantities of spoil that are temporarily stored for recycling and those that are disposed of permanently or will be used for landscape re-shaping of the construction site areas, as they are not otherwise recyclable.

Depending on the different timing of excavations compared to the construction of concrete works and the consequent relationship between the production of excavation materials and the demand for aggregates at the construction sites, there may be a temporary excess of available materials which are managed through temporary storage on the construction site or the sale of a part of it. The quantities of material which exceed the internal demand of the construction site can only be managed through sales programs.

Table 1. Amount of excavated material from the Isarco river underpass construction lot

Amount of spoil	Type A	Type B+C	Total
	mc	mc	mc
Excavated since works began*	412,439	54,244	466,683
Expected for this construction lot	656,003	582,379	1,238,381
Recycled within this lot for shotcrete production since works began*	139,006	-	139,006
To be recycled for shotcrete production	390,181	-	390,181
Recycled within this lot as filler material since works began*	5,613	8,215	13,828
To be recycled as filler material	-	285,684	285,684
Sold to third parties since works began*	97,785	-	97,785
To be sold to third parties within the construction lot	265,822	296,695	562,517
Temporarily stored on the construction sites*	168,873	53,957	222,831

* Amount stockpiled on June 30th 2018

The data on the production of excavated materials show an increase in materials suitable for the production of aggregates and less lower quality material due to the careful geological forecasts during the planning phase.

3.2 *Transport of spoil*

In order to minimize as much as possible the impacts deriving from the treatment and transport of spoil on the population living near the construction sites of the Brenner Base Tunnel, and on the surrounding environment, strategic decisions were made in the planning phase concerning the location and logistics of the individual construction sites, favouring electric traction, locating disposal sites and construction sites near the portals and using the exploratory tunnel as a transport route.

In particular, for the Isarco river underpass construction site, an access and an exit ramp was built to provide direct access from the construction site to the A22 Brenner motorway, to enable the supply of materials and the removal of spoil intended for sale, without interfering with traffic in residential areas, thus protecting the neighbouring population from further road traffic nuisance.

In addition, the almost complete reuse of spoil, not suitable for producing aggregates, for the landscape re-shaping of the construction site at the end of the works and the total satisfaction of the site's concrete demand by means of on-site crushing and mixing plants with good-quality aggregates from the site itself, further reduce transports outside the site.

The main results obtained with the logistics solutions adopted consist in reducing negative impacts due to construction site traffic (reduction of dust, noise, congestion of public roads) with advantages in terms of safety, quality of the surrounding environment, economy and overall performance, thus making the construction site logistically independent from other external factors (e.g. delays in transport due to traffic jams, frequent queues on the A22 motorway).

3.3 *Re-use of spoil as concrete aggregates*

BBT SE's award criteria for the tender procedure, which rewarded the best performing solutions in the field of post-processing, reuse and disposal logistics of spoil from the tunnel, allowed the best use of the know-how developed in these specialized fields by the building companies.

The treatment and reuse of excavated materials thus help to preserve natural resources, contributing to the sustainability of construction work. The treated material is therefore recycled during tunnelling work to produce the concrete needed for the first-phase and final linings.

As the geological composition of the spoil varies, the amount of reusable material varies greatly, depending on the location of the excavation. Where there is a spoil surplus compared to the demand at the BBT construction sites, it can be sold and used for other processes, such as concrete aggregates or other uses, thus avoiding quarrying operations.

The aggregates for the concrete for the construction of the tunnels are processed in crushing plants located directly on the construction sites, thus achieving further savings on transport of spoil and supply of aggregates.

The spoil can either be used within the production site, taken to other BBT construction sites or to other areas which are not on BBT construction sites. Depending on which of these three scenarios applies, specific procedures for transport, management and final reuse are complied with, in accordance with the most restrictive national and local regulations, as set down in the Spoil management and reuse plan for the construction of the Isarco river underpass construction lot. The solutions for spoil reuse, for example as aggregates for concrete production (shotcrete, etc.) and for backfilling as part of construction site operations, and the consequent reduced need to buy material from external stone quarries, significantly mitigate road transport and in the final analysis contribute to reducing overall ecological impact.

3.4 Re-use of spoil for site landscaping and reshaping

Among the works included in this construction lot, the preventive quarrying of the Rio Vallaga site was foreseen, which is close to the construction site, and the subsequent set up of a disposal site for the final disposal of spoil not suitable for producing aggregates. In order to improve environmental sustainability, the quarrying and the construction of the disposal site were cancelled during the executive planning phase, foreseeing that material not suitable for the production of concrete would be used to remodel the construction site area and cover the tunnels and shafts built with the cut and cover method, and for environmental restoration, while planning the site renaturation at the end of the works of all the areas to be restored to their original state. The forest margins will be restored with a rich and diversified vegetation in order to obtain, in a few years, a well-structured and improved forest, enhancing the ecotonal characteristics as compared to the adjacent rural areas, and benefiting the quality of ecosystems in this area. The forest restoration will be diversified according to the various areas, trying to achieve a phytocoenosis stratification as close as possible to the original natural structure and guaranteeing an environmental recovery that will ensure the continuity of the landscape physio type and the ecosystem integration of the native flora.

The spoil that is to be placed in the areas must comply with the threshold concentration values required by law and acceptable in terms of pollutants in the soil in relation to the specific use of the site; the percentages of anthropic materials must not be higher than the limits set by the Province of Bolzano (5% inert materials and minerals and 0.1% for incompatible materials such as PVC and fibreglass). In compliance with the provisions of the above prescriptions, the restoration of construction site areas into re-cultivated meadow areas where the root system only extends 30 cm downwards from the surface of the ground can be carried out using spoil containing elements of anthropic origin in the layers at a depth of more than 50 cm from the surface, while the restoration of areas to be re-forested can be carried out with spoil also containing anthropic elements to be placed at a depth of more than 2 meters from the surface.

4 WATER

4.1 Activities interfering with the Isarco river

Activities interfering with the Isarco river include the construction of a dam as a protection against flooding on both river banks in the area of the river underpass, waterproofing with jet-grouting columns carried out from the surface in the whole area, excavation of four ellip-tical shafts – accessed near the river banks and needed to excavate the underground tunnels to the north and to the south – and of the interconnections and the specific underpass under the river bed, preliminary treatment with injections with two-component chemical resins along the sidewalls of the shafts near the river, in order to contain the leakage of grout into the surface water and to allow an efficient execution of the columns even in the presence of running water.

The Isarco river underpass will be excavated as underground tunnels between the access shafts. The excavation involves the specific treatment of the material below the river bed to avoid potential infiltrations and create an adequate protective structure. The treatment includes low-pressure mortar injections and the ground freezing technique along the entire perimeter of the excavation. The ground water level will not be lowered, as the work will be carried out under hydrostatic conditions. In order to allow a dry excavation eliminating potential infiltrations and to create a suitable support structure, jet-grouting columns will be used along the perimeter and at the rock front.

Figure 2.   Works in the bed of the Isarco river.

With an eye to guaranteeing long-term tunnel safety and avoiding the possible exposure of the dome, proper erosion protection was laid in the river bed, using large boulders placed in such a way as to present an irregular surface that can effectively contrast the hydrodynamic action of the water flow and stabilize the bottom of the river, with slightly reinforced concrete slabs set below the boulders above the tunnel domes, to improve the binding effect of the boulder embedding material. A further measure was taken which is not directly linked to flood protection but was, rather, studied to improve the hydrogeo-logical conditions of the aquifer and prepare the ground for consolidation and freezing: jet-grouted column walls directly upstream and downstream of each tunnel and set into the concrete slabs.

The laying of the boulder coverage and the concrete slabs required working directly within the river bed, which was partly narrowed with a temporary coffer dam; the works were successfully completed over two consecutive years during the low-water period from December to March. In order to impact high-water run-off as little as possible, the coffer dams and small sluices were built on the right orographic side in the first phase and the following year, in a second phase, on the left orographic side, to constantly ensure regular drainage flow.

The subsequent preliminary treatment of the layer of material below the river for the exca-vation of the underground tunnels under the Isarco river will consist of a series of low-pres-sure injections of cement grouts with additives and of freezing the interstitial water along the entire perimeter of the excavation. This measure is meant to create a waterproof barrier and a solid support structure for the excavation.

4.2 Water treatment plant

During the complex execution of the underpass of the Isarco river, different types of waste water will be generated that need to be treated:

- water coming from the excavation of the tunnel carried out by drilling and blasting;
- water due to infiltrations and rain/snowfall coming from the cut and cover excavations of tunnels and shafts (infiltrations in shafts built during the excavation and sub-foundation phases reached peaks of up to 90 l/s);
- waste water from the vehicle and machine washing area;
- rinsing water from the washing of concrete mixers;
- rainwater from paved surfaces.

As already seen in similar circumstances, the following polluting parameters were considered at the intake of water treatment plant:

- turbidity due to suspended mineral solids (500 mg/l – 5,000 mg/l);
- pH from 9 to 12.5 due to the use of cementitious building materials;
- mineral oil due to accidental spillage from construction machinery;
- ammonium and nitrite due to the use of explosives to excavate the tunnels with drill and blast method (0 – 5 mg/l NO2-N).

On the basis of the hydrological assessments and the expected water inflows, an peak flow rate of 201.72 l/s was estimated. The plant has also been sized for a maximum flow rate of 250 l/s so as to be able to ensure the correct treatment of water, even when particularly high and unexpected water inflows occur at the plant.

Figure 3.   Water treatment plant at the Isarco river underpass construction site.

The physical chemical treatment plant has the following treatment sections:

- separation of coarse materials, floating materials and mineral oil;
- control of chemical and physical parameters at the intake;
- neutralisation with carbon dioxide to pH values between 5.5 and 9.5;
- flocculation and clarification of the water by sedimentation of the suspended solids by means of a dynamic decanter with a scraping bridge;
- residual removal of suspended solids by filtration on quartz sand;
- oxidation sections of nitrites with ozone;
- dehydration of thickened sludge using a chamber filter press.

In addition, the parameters of temperature, conductivity, pH and turbidity are measured and recorded automatically at the discharge and in the river both upstream and downstream of the discharge point into the Isarco river.

At the discharge point, the water must comply with the threshold value indicated in the authorization, which is 35 mg/l for total suspended solids and therefore lower than the national limit of 80 mg/l, whereas the maximum variation between the average temperatures in any section of the river both upstream and downstream of the discharge point may not be above 3°C.

5 ENVIRONMETAL MONITORING

5.1 *General approach*

It includes all the tests that are to be periodically or continuously carried out to analyse and describe the impacts on the environmental factors involved, during and following the construction and operation of the infrastructure on the territory in all the site areas involved. Monitoring activities are carried out on the basis of a specific environmental monitoring project, in which the sensitive environmental factor areas are laid out, based on the direct presence in the area of the cause of interference or a possible significant link that might cause a variation in the parameters connected to a certain environmental component.

5.2 *Characterization of spoil*

Analytical assessment of excavated soil and spoil are necessary to verify the absence of chemical and/or product contamination, to ascertain the regulatory regime to which the materials must be subjected (waste or non-waste), and to verify their suitability for use.

The spoil regularly undergoes shipping checks, chemical analyses and product sample analyses. The frequency of such sampling is determined on the basis of the quantities of spoil in each individual lot according to the most restrictive national and final planning requirements. Sampling shall be carried out on piles of spoil to obtain a representative sample, taking at least eight elementary samples, four of which shall be taken at depth and four at surface level, and resulting in a composite sample which shall, via quartering, provide the final sample for chemical analysis. The chemical parameters analysed include the set indicated in the Presidential Decree 120/2017 "Regulation laying down simplified rules for the management of excavated soil and spoil" plus some parameters related to specific tunnelling activities. The resulting data are compared with the acceptable concentration threshold values set forth in the provincial law on pollutants present in soil and the assessed product parameters are compared with the thresholds for anthropic material prescribed by the Province of Bolzano. Compliance with these limits is a discriminating factor when determining the possibilities for reuse and destination of the spoil. The limits for chemical parameters of the pollutants present in the spoil to be permanently stored or used for the restoration of certain areas are those set forth for sites for public, private and residential areas. If the material is to be used in areas for commercial and industrial use, or reused as aggregates, it must comply with the limits set forth for such use at sites for commercial and industrial use.

5.3 *Monitoring of the Isarco river*

Environmental hydro morphological and water quality monitoring is carried out at two measuring stations along the Isarco river, one upstream and one downstream of the underpass construction site. The baseline data are from the campaign that took place prior to the construction activities in 2014, after which biannual campaigns were carried out during the works.

In particular, the measured indices are hydro morphology I.F.F. (River Functionality Index) water quality, the I.B.E. indices. (Extended Biotic Index), STAR ICMi (MacrOper), ICMi (algae population monitoring), S.E.C.A. (ecological state of the watercourse) and chemical and bacteriological analyses of the water and, for the fish stocks, the calculation of the ISECI (index of the ecological state of the fish stocks). The monitoring carried out from October 2016 until today has not shown any worsening of the hydro morphological and water quality conditions of the Isarco river due to construction activities which included a vast consolidation campaign through jet grouting, the complete consolidation works of the Isarco river bed in the area of the future underpass and the continuous operations of the water treatment plant.

6 SUSTAINABILITY

As part of the executive planning, a qualitative analysis of sustainability was carried out, comparing the choices made in the executive planning with those of the final planning, according to the criteria of the NSW Sustainable Guidelines (Version 3.0), the guidelines of the Australian Protocol, created to promote the sustainable planning of rail transport infrastructure. The factors taken into account are energy and greenhouse gases, adaptation to climate change, materials and waste, biodiversity and environmental resources, pollution control, community benefits.

On the basis of the analysis carried out, the choices made in the executive planning showed greater environmental sustainability, respecting the objective set by the guidelines of the Australian Protocol, in terms of lower quantities of soil and spoil, less impacts on potamological, hydrological and hydro biological conditions of the area and reduced pumping of underground water, preservation of the construction site area from an ecological and hydro morphological point of view, less transport, lower greenhouse gas emissions, better logistical organisation of the construction site, less land occupation for disposal sites, energy savings, reduction of waste, shorter duration of activities and related impacts and improved acceptance by fishermen's associations.

7 CONCLUSIONS

The environmental policy implemented within the Brenner Base Tunnel project must be commensurate with the nature and scale of the project itself and, to this end, it has been contextualised through the identification of environmental objectives that the project aims to achieve in line with environmental policies at European level. These objectives were identified in the planning and further specified during the authorization process it has undergone.

The measures aimed at protecting the environment and the sustainability of the project were identified on the basis of the results of the environmental impact assessment. These measures were optimized during the process between the final planning and the executive planning of the specific construction lots and are implemented together with environmental monitoring to verify its effectiveness.

In fact, the comparative analysis of the two planning levels for the construction of the Isarco river underpass construction lot showed that the executive planning included technical proposals for improvement, which led to its enhancement in terms of sustainability.

The element that made the optimized executive planning most sustainable is the type of work planned for the Isarco river underpass, avoiding the temporary re-routing of a stretch of the river and a large amount of excavation. To date, monitoring has confirmed the positive ecological response of the Isarco river to the layout of the construction site and the choices made.

Carrying out the project in a sustainable way therefore means adopting initiatives and solutions that respect a very sensitive area with a high tourism appeal, protecting the landscape, the delicate existing Alpine ecosystem, and the health and well-being of the population living there.

REFERENCES

Italian National Agency for the Protection of the Environment A.N.P.A., 2000, River Functionality Index (I.F.F.), *Italian National Agency for the Protection of the Environment A.N.P.A. Manual/2000*: page 223, Rome.

Ministry of the Environment, for the Protection of the National Territory and the Sea, 2010, Decree no. 260 of 8 November 2010, Regulation containing the technical criteria for the classification of the state of surface water bodies, amending the technical standards of Legislative Decree no. 152 of 3 April 2006, containing environmental standards, drawn up pursuant to Article 75, paragraph 3, of the same Legislative Decree, Rome.

NSW Government, Transport for NSW, 2013, NSW Sustainable Design Guidelines Version 3.0, New South Wales Australia.

Tunnels and Underground Cities: Engineering and Innovation meet Archaeology,
Architecture and Art, Volume 2: Environment sustainability in
underground construction – Peila, Viggiani & Celestino (Eds)
© 2020 Taylor & Francis Group, London, ISBN 978-0-367-46579-7

Feasibility study for the thermal activation of Turin Metro Line 2

M. Barla, M. Baralis, A. Insana & F. Zacco
DISEG, Politecnico di Torino, Italy

S. Aiassa & F. Antolini
Geosolving srl, Torino, Italy

F. Azzarone & P. Marchetti
Systra SA succursale Italiana, Roma, Italy

ABSTRACT: An innovative energy segment for tunnel lining was patented by Politecnico di Torino in 2016 and tested in an experimental site installed in the tunnel of Turin Metro Line 1. Data were collected between 2017 and 2018 to form the basis of the performance assessment of the technology. Promising outcomes encouraged the authors to collaborate in this direction over the project of Turin Metro Line 2 to propose a methodology for assessing the thermal energy potentially exploitable by an underground infrastructure. This paper is intended to describe the methodology to be adopted for the feasibility study that will be conducted to assess the energy potential resulting from the thermal activation of the tunnel lining and how this can be distributed to surrounding users.

1 INTRODUCTION

Low enthalpy geothermal systems have always been largely used for space heating. Such applications offer one of the best ways for providing sustainable energy in urban environment where the ground immediately below a city can be used as heat source and/or energy storage reservoir. Shallow geostructures like foundations, diaphragm walls, tunnel linings and anchors, have recently been employed as heat exchangers (Laloui & Di Donna 2013). Their thermal activation is obtained by installing absorber pipes inside them, in which a heat carrier fluid circulates and transfers heat from the ground, for heating during winter, and to the ground, for cooling during summer.

A number of examples of foundation slabs, piles and diaphragm walls used for heating and cooling purposes of large buildings exist, particularly in Austria, Germany, the UK and Switzerland (Adam & Markiewicz 2009, Bourne-Webb et al. 2009, Brandl 2006, Pahud 2013, Xia et al. 2012). The application of this technology to tunnels is limited to a few case studies, i.e. the Lainzer tunnel of the U2 extension of the Vienna metro (Markiewicz & Adam 2003), the Austrian H8-Tunnel Jenbach (Frodl et al. 2010), the Germans Katzenberg railway tunnel (Franzius & Pralle 2011) and the Stuttgart-Fasanenhof Tunnel (Schneider & Moormann 2010). Nevertheless, an increasing number of scientific studies about feasibility and efficiency of geothermal tunnel activation reveals the growing interest in these applications (Adam & Markiewicz 2009, Barla et al. 2016, Barla & Perino 2014, Di Donna & Barla 2016, Franzius & Pralle 2011, Nicholson et al. 2014, Zhang et al. 2013). An innovative energy segment was patented by Politecnico di Torino in 2016 and tested in an experimental site installed in the tunnel of Turin Metro Line 1 (Barla et al. 2018). Data were collected between 2017 and 2018 to form a basis for the performance assessment of the technology in Turin subsoil. Promising outcomes have encouraged the authors to collaborate in this direction on the project of Turin Metro Line 2.

Line 2 will be a fully automated driverless light metro, similar to Line 1 and will extend, in a first stage, for about 15 km with 26 stations. It will connect the SW suburbs of the city to the NE ones. The excavation will be performed by means of TBM for roughly 11 km. This infrastructure project will represent an essential milestone for the city transportation system, because it will reach some decentralized areas of the city, making them livelier and connected, and revolutionize the surface space, particularly in the Northern zone.

This paper is intended to describe the procedure that will be adopted during the feasibility study to assess the energy potential of the thermal activation of the line. The objective is to quantify, by means of Finite Element Thermo-Hydro simulations, the heat that could be extracted from and injected into the ground by the geothermal activation of the structural elements supporting underground stations and tunnels. A study of the possible collectors for the heat produced will be also performed considering both existing buildings, future urban developments and the stations equipment.

2 LINE 2 OF TURIN METRO

The city of Turin is undergoing a renovation of its rail transportation system. Currently, the rail network is constituted by the underground metropolitan railway system and by the Line 1 of Turin Metro. The underground railway line connects the two main railway stations in the city (Porta Nuova and Porta Susa) with the suburban railway and the other 4 stations in the city area. Turin Metro Line 1 connects the North-West part of the city with the South-East one, from Fermi station to Lingotto multifunctional centre, for a total length of 13.4 km and 21 stations.

Turin Metro Line 2 will represent a fundamental line for metropolitan transport, connecting the southwest area of the city to the northeast zone. The total length of the new line will be about 14.5 km completely underground, crossing densely inhabited areas like Barriera di Milano, Aurora-Rossini, City Center, Crocetta, Santa Rita, Mirafiori Nord. Significant line extensions have been foreseen in basic design by the city of Turin towards San Mauro (North-East), Bengasi (South-East) and towards Orbassano (South-West) are currently in the feasibility study stage (Figure 1).

The main tunnel will be built by TBM from Anselmetti to Bologna station for a total of 12 km and Cut & Cover for a length of 3.0 km from Bologna to Rebaudengo (terminal station). The TBM tunnel depth ranges from 20 to 38 m below street level, while Cut & Cover one goes from 6.5 to 15 m. The stations track level ranges from 8.6 m (one underground level with at grade concourse) to 26.1 m (four level stations).

One of the most important design inputs for Line 2 is the adoption of state-of-the-art technology for the infrastructure, considering both line 1 experience and the eco-compatibility of the proposed geothermal system. The possible use of low-enthalpy energy resources with HVAC systems is strictly linked to the application of the "Smart station" concept in terms of environmental and passenger comfort related with reduced energy demand and the dynamic control of energy consumption. This target can be achieved by means of geothermal renewable resources and the efficiency maximization of the connected equipment. Conversely, the infrastructure can even be conceived not only as energy user, but as energy producer and conveyor.

Moreover, the connection between the tunneling technology and the thermal lining marks a new path that will necessarily tailor the metro system in the city grid providing a double effect of urban transport and energy resource.

Line 2 construction is also related to an extensive urban renewal project that will involve development of new parks, residential and commercial buildings over an area of 900000 m^2 (areas in red in Figure 1). Energy needs of the new settlements could be satisfied by the adoption of local district heating and cooling system grids based on the energy tunnel and integrated with other renewable energy sources.

Figure 1. Metropolitan transportation network and location of the development areas in Turin.

3 THERMAL ACTIVATION OF THE LINE

3.1 *Procedure for the study*

On the basis of existing information and additional investigation taking place during the design of the new Line 2, the following information will be collected:

- geometrical characteristics of the line. i.e tunnel shape and diameter, tunnel axis depth;
- construction technology and procedures;
- geological and geotechnical characteristics of the ground (Geotechnical Units - GU) along the line;
- hydrogeological characteristics of the different GU (permeability, trasmissivity, effective porosity, direction and velocity of groundwater flow, groundwater temperature);
- geothermal parameters of the different GU (conductivity, thermal capacity, diffusivity, ...).

The geometrical characteristics of the line, mainly the tunnel axis depth and the diameter of the excavation, will be of particular interest to assess where the excavation will be conducted below the groundwater level. The geological and geotechnical profiles along the line are further fundamental elements to be collected and reconstructed. The interpretation of these profiles will allow for the identification of the different GU in a significative area around the excavation and, consequently, to define a specific set of hydrogeological and geothermal parameters. Since the hydrogeological and geothermal parameters are typically measured or calculated pointwise (i.e. along boreholes) it is convenient to create interpolated maps with GIS tools to spatially extend the available information along the whole line. Figure 2 shows an example of map showing respectively the groundwater temperature and the direction of the groundwater flow for a specific area along the metro line alignment.

Once the global geometrical, geotechnical, hydrogeological and geothermal characteristics are gathered, following the procedure described in Baralis et al. (2018), the whole Metro Line

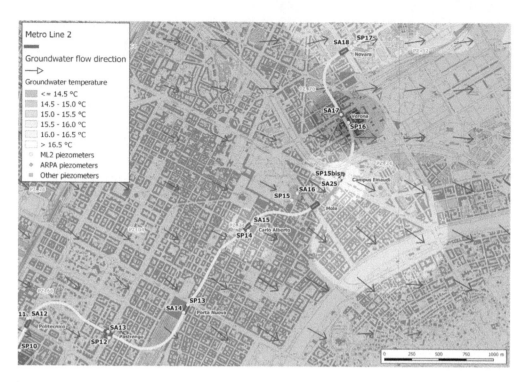

Figure 2. Sketch of groundwater temperature map and direction of groundwater flow along a section of the Turin Metro Line 2.

2 will be thus discretized into a finite number of homogeneous sections associated to a specific geometry of the tunnel to be excavated and to a specific set of geotechnical, hydrogeological and geothermal parameters (Figure 3).

Thermo-Hydro Finite Element numerical models will then be built for each representative section aimed at calculating the geothermal potential. Interpretation of the Thermo-Hydro numerical analyses will allow to determine the energy produced (kWh) and the energy efficiency for each homogeneous zone as well as the overall deliverable geothermal potential of Turin Metro Line 2.

As a final step, a spatial analysis on the area interested by the infrastructure realization will be performed in order to evaluate potential clients of the heat and cooling loads. This will require a GIS analysis to be carried out with the intention of identifying potential customers located in the surroundings of selected extraction points such as the stations and the ventilation shafts, where enough room for heat pump installation can be easily carved out. Based on GSHP plant technical feasibility issues, potential clients have to be located sufficiently close to the above-mentioned energy extraction points. Hence, a buffer zone of 25, 50 and 100 m around the stations and ventilation shafts will be evaluated. Cost effective analysis will allow to compare and select the best receivers among those identified.

The following sections will show an example of geothermal potential assessment for two selected construction processes, TBM tunnel excavation and cut and cover.

3.2 TBM tunnel excavation

3.2.1 Enertun segmental lining
Given the increasing number of tunnels excavated by means of TBMs, a novel energy tunnel precast segmental lining, named Enertun, has been designed and patented (Patent number: 102016000020821) at the Politecnico di Torino (Barla & Di Donna 2018, 2016). Compared to

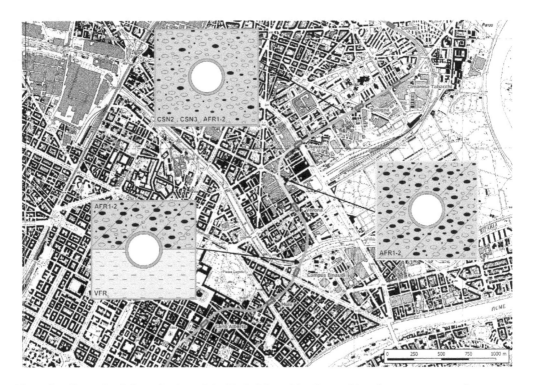

Figure 3. Example of discretization of the Turin Metro Line 2 tunnel into homogeneous sections.

previous configurations (Franzius & Pralle 2011), Enertun suggests a more efficient layout of the net of pipes. In fact, they are arranged so that their main direction results to be perpendicular to the tunnel axis, thus reducing bends and, subsequently, hydraulic head losses by about 20-30%. Moreover, thermo-hydraulic numerical analyses have shown that such a configuration results to be more efficient in terms of heat exchange when a ground water flow is present. As it is well known that underground water flow contributes significantly to the energy efficiency of geothermal systems (Di Donna & Barla 2016), the interest in installing energy tunnels increases in zones where a ground water flow exists. Improving the energy efficiency of this application in such conditions becomes consequently of primary importance.

Enertun segments were recently tested in an experimental site installed in the tunnel of Turin Metro Line 1 (Barla et al. 2018) allowing to evaluate the performance of the technology in Turin subsoil and anticipate technological details for the integration into the tunnel construction cycle. Promising outcomes from the experimental site encouraged to consider the Enertun technology for the thermal activation of the tunnels of Line 2.

3.2.2 *Numerical modelling of tunnel thermal activation*
With the purpose of quantifying the exploitable heat and studying the influence of the geothermal activation of the tunnel on the environment, a three-dimensional Thermo-Hydro finite element model can be used to reproduce a limited number of ENERTUN tunnel rings equipped with heat absorber pipes. The FEM software FEFLOW® can be used to this purpose, allowing to consider the advection and dispersion heat transfer mechanism. Neglecting these mechanisms would in fact lead to underestimation of the geothermal potential and overestimation of the thermal alteration in the surrounding soil (Alcaraz et al. 2016). The reader can refer to the software manual (Diersch 2009) for the mathematical formulation. For the simulation of the absorber pipes installed in the tunnel lining, one-dimensional highly conductive elements can be adopted. The use of these elements to simulate pipes in geothermal systems has been validated and good agreement was found compared to analytical solutions

(Diersch 2009). The conservation of mass and energy is also satisfied for these elements, while the fluid flow inside them is described by the Hagen–Poiseuille law. Accordingly, fluid particles are assumed to move in pure translation with constant velocity, similar to what occurs in circular tubes.

To this aim, the geometry of the problem needs to be reproduced accurately, together with the groundwater regime. Figure 4 shows the groundwater conditions in the South-West area of the city of Turin along the alignment of the Line 2. The 3D model shown in Figure 5 is representative of this specific cross section. Based on the assumption that the groundwater flow can be considered perpendicular to the tunnel axis, the 3D model reproduces only the geometry of one single lining ring. Adjacent rings should not influence the results from the thermal point of view. Thus, the adiabatic condition was set on lateral boundaries except for the groundwater inflow and outflow sides and the upper surface where representative air temperature was imposed. Other areas of the city will have different conditions with respect to the direction of the groundwater flow, therefore this assumption will not apply and larger models, including a sufficient number of tunnel rings, will be needed.

A difference in piezometric head is given between the two sides of the model causing a flow perpendicular to the tunnel axis as a function of ground hydraulic conductivity. Moreover, the water table is considered at 26 m from the surface, which corresponds to the level of the invert arch of the tunnel.

A temperature contour plot of the model at the end of a 30-days winter/summer simulation is shown in Figure 6. Due to the thermal activation, groundwater flow induces lower temperature variations in the ground below the tunnel than in the part above the water table.

Moreover, with additional Thermo-Hydro numerical simulations it was possible to verify that heat exchange conditions are more favorable if the whole tunnel is under the water table, because the flow induces a greater energy supply in the ground around the tunnel.

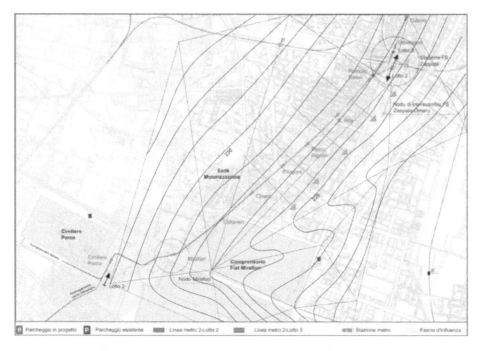

Figure 4.   Piezometric lines in the area of the South-West section of Line 2.

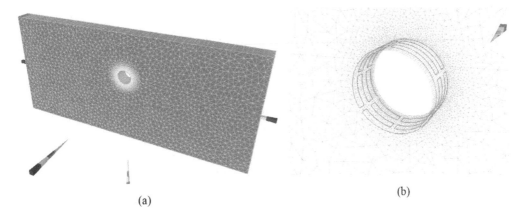

(a)   (b)

Figure 5. 3D finite element numerical model of the tunnel (a) and close view of the mesh adopted to simulate the pipe circuit in the Enertun segments (b).

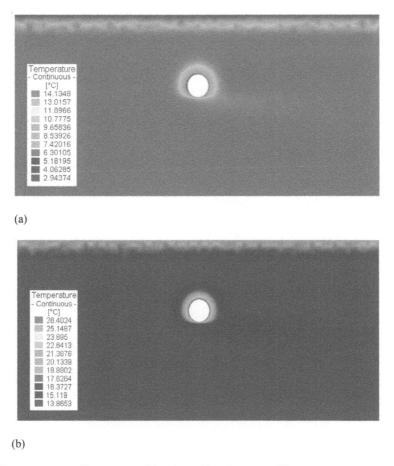

(a)

(b)

Figure 6. Temperature profiles computed in winter (a) and summer (b).

3.3 Cut & Cover excavation

Similarly to what described in 3.2.2, the thermal behavior of the concrete structures can be studied in the case of the Cut & Cover technique. This is planned to be used for the construction of the metro stations and for some shallow sections of the line.

Again, a 3D model can be constructed reproducing the geometry of one single diaphragm wall panel and slab when the groundwater flow is perpendicular to the tunnel axis and up to five blocks when the groundwater flow intercepts the tunnel with a different angle.

An example of a 3D FEM model is shown in Figure 7. It has a total volume of 6876.64 m^3 and dimensions of 70 m × 40 m × 2.5 m with 414900 prismatic elements and 20 layers of 12.5 cm. The model is checked for mesh sensitivity. Figure 5 represents the front view of the mesh with the boundary conditions imposed in order to reproduce the Turin underground flow and external temperature (on the left) and the detail of the schematic representation of the hydraulic circuit (on the right). Water table is assumed at approximately 11 m depth.

Figure 8 shows the results of the numerical analysis in terms of heat exchanged during one year of thermal activation. Because of a higher difference between input and output temperature, the thermal power produced by the system is higher in summer than in winter. In both cooling and heating modes, the slab is shown to have a higher contribution than a single diaphragm wall, probably because the average ground temperature below the slab is less influenced by external air temperature variations than that of the ground close to the diaphragm walls.

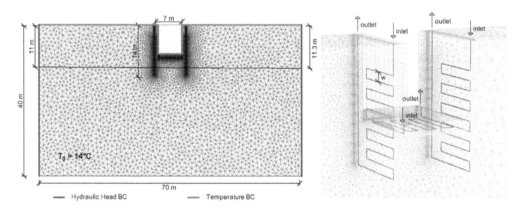

Figure 7.   Geometry and dimensions of the 3D model.

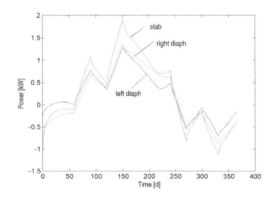

Figure 8.   Example of the result obtained in terms of heat exchanged during one year of operation.

4 CONCLUSIONS

This paper discussed the beneficial opportunity offered by the thermal activation of Turin Metro Line 2. Using 3D FEM models, the heat exchange between the ground and the structure can be simulated taking into account a considerable number of influencing factors. These include mainly the geometry of the excavation and the hydrogeological setting. Numerical analyses show that when groundwater flow exists, higher heat exchange rates and lower temperature variation in the surrounding soil are predictable. The preliminary results confirm those obtained by Di Donna and Barla (2016). Investigations performed in the past showed that 1 MW of thermal energy can be exploited per kilometer of tunnel (Barla et al. 2016). This would allow to satisfy the energy demand of a large number of new standard residential buildings along the new metro line. Further favorable perspective can be envisaged as it is generally known that the metro line realization can boost land value and new building construction.

An added value to the thermal activation of the metro structures is predictable in those areas of new urbanization or complete renovation. In these areas in fact connection to existing district heating is difficult due to technical reasons. Geothermal heat can thus provide an interesting option to reach the minimum law requirements in terms of energy from renewable sources. In this context, the Metro line itself can be seen as a local district heating and cooling network.

Further work will be performed to address systematically the different section of the new metro line and collect several datasets to be organized in a specific georeferred database in order to evaluate energy and urban planning policies for districts that will undergo transformation as a result of the metro line construction. These data will also be used to obtain an overall quantification of the exchangeable heat with regards to the existing and future potential clients.

REFERENCES

Adam, D. & Markiewicz, R. 2009. Energy from earth-coupled structures, foundations, tunnels and sewers. *Géotechnique* 59: 229–236. https://doi.org/10.1680/geot.2009.59.3.229

Alcaraz, M., García-Gil, A., Vázquez-Suñé, E. & Velasco, V. 2016. Advection and dispersion heat transport mechanisms in the quantification of shallow geothermal resources and associated environmental impacts. *Science of the Total Environment* 543: 536–546. https://doi.org/10.1016/j.scitotenv.2015.11.022

Baralis, M., Barla, M., Bogusz, W., Di Donna, A., Ryzynski, G. & Zerun, M. 2018. Geothermal Potential of the NE Extension Warsaw Metro Tunnels. *Environmental Geotechnics*. https://doi.org/10.1680/jenge.18.00042

Barla, M. & Di Donna, A. 2018. Energy tunnels: concept and design aspects. *Underground Space*. https://doi.org/10.1016/j.undsp.2018.03.003

Barla, M. & Di Donna, A. 2016. Conci energetici per il rivestimento delle gallerie. *STRADE & AUTO-STRADE* 5: 2–5

Barla, M., Di Donna, A. & Insana, A. 2018. A novel real-scale experimental prototype of energy tunnel. *Tunnelling and Underground Space Technology* (under review)

Barla, M., Di Donna, A. & Perino, A. 2016. Application of energy tunnels to an urban environment. *Geothermics* 61: 104–113. https://doi.org/10.1016/j.geothermics.2016.01.014

Barla, M. & Perino, A. 2014. Energy from geo-structures: a topic of growing interest. *Environmental Geotechnics* 2: 3–7. https://doi.org/10.1680/envgeo.13.00106

Bourne-Webb, P.J., Amatya, B., Soga, K., Amis, T., Davidson, C. & Payne, P. 2009. Energy pile test at Lambeth College, London: geotechnical and thermodynamic aspects of pile response to heat cycles. *Géotechnique* 59: 237–248. https://doi.org/10.1680/geot.2009.59.3.237

Brandl, H. 2006. Energy foundations and other thermo-active ground structures. *Géotechnique* 56: 81–122. https://doi.org/10.1680/geot.2006.56.2.81

Di Donna, A. & Barla, M. 2016. The role of ground conditions and properties on the efficiency of energy tunnels. *Environmental geotechnics* 1–11. https://doi.org/10.1680/jenge.15.00030

Diersch, H.J.G. 2009. DHI wasy software – Feflow 6.1 – Finite element subsurface flow & transport simulation system: Reference manual.

Franzius, J.N. & Pralle, N. 2011. Turning segmental tunnels into sources of renewable energy. Proceedings of the ICE - Civil Engineering 164: 35–40. https://doi.org/10.1680/cien.2011.164.1.35

Frodl, S., Franzius, J.N. & Bartl, T. 2010. Design and construction of the tunnel geothermal system in Jenbach/Planung und Bau der Tunnel-Geothermieanlage in Jenbach. Geomechanics and Tunnelling 3: 658–668. https://doi.org/10.1002/geot.201000037

Laloui, L. & Di Donna, A. 2013. Energy Geostructures: Innovation in Underground Engineering, Energy Geostructures: Innovation in Underground Engineering. https://doi.org/10.1002/9781118761809

Markiewicz, R. & Adam, D. 2003. Utilisation of Geothermal Energy using Earthcoupled Structures – Theoretical and Experimental Investigations, Case Histories. In: *XIIIth European Conference on Soil Mechanics and Geotechnical Engineering.* 25-28th August 2003, Prague.

Nicholson, D.P., Chen, Q., de Silva, M., Winter, A. & Winterling, R. 2014. The design of thermal tunnel energy segments for Crossrail, UK. *Engineering Sustainability* 167: 118–134. https://doi.org/10.1680/ensu.13.00014

Pahud, D. 2013. A Case Study: The Dock Midfield of Zurich Airport. In: Energy Geostructures: Innovation in Underground Engineering. 281–296

Schneider, M. & Moormann, C. 2010. GeoTU6 – a geothermal Research Project for Tunnels. *Tunnel Geothermics* 2: 14–21

Xia, C., Sun, M., Zhang, G., Xiao, S. & Zou, Y. 2012. Experimental study on geothermal heat exchangers buried in diaphragm walls. *Energy and Buildings* 52: 50–55. https://doi.org/10.1016/j.enbuild.2012.03.054

Zhang, G., Xia, C., Sun, M., Zou, Y. & Xiao, S. 2013. A new model and analytical solution for the heat conduction of tunnel lining ground heat exchangers. *Cold Regions Science and Technology* 88: 59–66. https://doi.org/10.1016/j.coldregions.2013.01.003

Tunnels and Underground Cities: Engineering and Innovation meet Archaeology,
Architecture and Art, Volume 2: Environment sustainability in
underground construction – Peila, Viggiani & Celestino (Eds)
© 2020 Taylor & Francis Group, London, ISBN 978-0-367-46579-7

Testing of an Enertun segment prototype in Turin metro line 1

M. Barla, A. Insana & F. Zacco
DISEG, Politecnico di Torino, Italy

ABSTRACT: Since the 1980s, a number of underground geotechnical structures have been turned into heat exchangers, the so-called energy geostructures, by embedding absorber pipes in their concrete mass. A carrier fluid circulating along the pipes extracts or injects heat from or into the ground with the aim of conditioning buildings. A novel energy tunnel precast segmental lining was patented at Politecnico di Torino in 2016. Compared to previous configurations, this patent suggests an innovative and more efficient layout of the net of pipes. The South Extension of Turin Metro Line 1 represented an extraordinary opportunity to test the novel energy segment. This paper is intended to describe the experimental site setup, by outlining the operation modes and the experimental campaign carried out. A preliminary coupled thermo-hydraulic, finite-element numerical analysis was also developed with the intention of performing a comparison with the experimental measurements.

1 INTRODUCTION

The global need to propose new solutions and to improve technologies able to explore and apply the benefits offered by renewable energy sources appears crucial, considering that these sources represent one of the most effective tools against climate change. Among the solutions available, shallow geothermal energy is getting more and more attractive, given that it is accessible everywhere in the world (below 10–15 m up to 400 m depth). Today, the use of shallow geothermal energy is common to many countries. Usually, heat is extracted from the ground for domestic heating or to produce hot water by means of open or closed loop systems. Therefore, it can potentially contribute to decrease the heating and cooling carbon footprint, leading to important environmental benefits (European Commission 2016, REN21 2017). A major advantage of shallow geothermal energy lies in the reliability of its supply as well as its nearly unlimited availability.

Since the 1980s, a number of underground geotechnical structures (piles, diaphragm walls, tunnel linings) have been turned into heat exchangers, the so-called energy geostructures, by embedding absorber pipes in their concrete mass (Barla & Di Donna 2016, Laloui & Di Donna 2013). A carrier fluid circulating along the pipes extracts or injects heat from or into the ground with the aim of heating/cooling buildings respectively. The integration of a structural element together with an energetic purpose requires little effort, though resulting in great environmental and economic benefits. Most practical applications are related to energy piles and diaphragm walls and are already operational especially in Austria, UK, Germany and Switzerland (Brandl 2006). Feasibility studies and an increasing number of new applications are under way.

Recently, an interest in applying this technology to tunnels has emerged, as witnessed by the number of studies that investigate thermal exploitation through tunnels (Adam & Markiewicz 2009, Barla et al. 2016, Barla & Perino 2014, Di Donna & Barla 2016, Franzius & Pralle 2011, Frodl et al. 2010, Lee et al. 2012, Moormann et al. 2016, Nicholson et al. 2014, Schneider & Moormann 2010, Unterberger et al. 2014, Wilhelm & Rybach 2003, Zhang et al. 2013). Compared to building foundations, an advantage of turning tunnels into energy tunnels is the larger volume of ground and surface for heat exchange. Furthermore, when mechanized tunnelling is used, the thermal activation process can be optimized in factory.

This paper will focus on the description of a prototype of energy tunnel installed in Turin Metro Line 1 South Extension, by outlining the operation modes and the experimental campaign carried out. A preliminary coupled thermo-hydraulic, finite-element numerical analysis was also developed with the intention of performing a comparison with the experimental measurements. The parameters of the model were calibrated to adequately reproduce the system performance obtained in situ.

2 DESCRIPTION OF THE PROTOTYPE

In order to test the thermal performance of a novel energy tunnel precast segmental lining (ENER-TUN) designed and patented (Barla & Di Donna 2016, 2018; Patent number: 102016000020821) at Politecnico di Torino, an experimental site of Enertun segmental lining was installed in the tunnel of Turin Metro Line 1 South Extension under construction, about 42 m northwards from Bengasi station, in the Lingotto-Bengasi section as shown in Figure 1. Turin metro tunnel can be classified as a cold tunnel, that is internal air temperature is similar to ground surface temperature and thermal influence due to fast-moving trains operation is negligible.

Figure 1.   Map of Turin Metro Line 1 along with a picture of the Enertun experimental site.

Turin Metro Line 1 connects the North-West part of the city with the South-East one, from Fermi station to Lingotto multifunctional centre, for a total length of 13.4 km and 21 stations. The construction started in 2000, in view of Turin 2006 Winter Olympic Games. In 2006 the first section from Fermi to XVIII Dicembre was opened, immediately followed by the second one up to Porta Nuova in 2007. The last part of the line towards Lingotto was completed in 2011. Two additional stations located at the Southern border of the city, Italia'61, planned to serve the future Piedmont Region Headquarters, and Bengasi, are currently under construction.

The Lingotto-Bengasi section is characterized by a 1.9 km length, 2 intermediate ventilation shafts (PB1 and PB2), a terminal shaft of end section (PBT) located approximately 200 m beyond Bengasi station and an intersection to allow for the future construction of the branch line to Lingotto railway station. The terminal tunnel will allow the inversion of trains behind Bengasi station (as it happens today at Fermi), in the section between the station itself and the terminal shaft. This solution will make it possible to fully exploit both the station (currently only one platform is used at Lingotto) and the potential of the VAL system (automatic light

242

vehicle): at peak times it will be possible to guarantee a train frequency every 69 seconds. Additionally, the length of the concerned section enables parking of four trains, ready to get in service depending on the users flows fluctuations.

The tunnel was excavated by an EPB TBM below the groundwater table at an average depth of the tunnel is in the range 16–20 m. The tunnel lining is 30 cm thick and each ring is made of 6 precast concrete segments (5 plus a key) mounted by the TBM itself while excavating and permanently sealed to protect tunnel tube against groundwater. Each ring is 1.4 m long and the internal diameter is 6.88 m.

In the following a detailed description of the prototype preparation and installation stages is given.

2.1 Energy segments manufacturing

Two rings of segmental lining were fully equipped with the ground&air net of pipes (Barla et al. 2018, Barla & Insana 2018) for a total of 12 Enertun segments. The two nets of pipes, one close to the extrados (tunnel surface in contact with the ground), the other close to the intrados (tunnel surface in contact with the air) will allow to test alternatively all above-mentioned three different configurations.

Segments manufacturing was the result of a 5-months phase (from December 2016 to April 2017) characterized by several meetings with the client, the contractors, the consultants and the staff in charge of precasting. Figure 2 shows an example of the design of a segment equipped with pipes in the Enertun ground&air configuration. While the air circuit already lays at the level of box-outs, the first and fifth rows of the ground circuit need to bend to reach the same level and exit at the intrados. As pipes configuration design took place after reinforced concrete design, their exact location and spacing was dictated by rebars position.

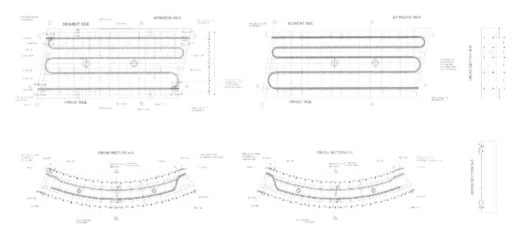

Figure 2. Example of the design of a segment equipped with pipes in the Enertun ground&air

The pipes used are fabricated in reticulated polyethylene (PE-Xa) with antioxygen barrier, have an external diameter of 20 mm and a thickness of 2 mm. They were simply tied to rebars through wire by hand in the precast concrete plant on a separate line so not to delay segment production schedule. About 116 m/ring of pipes were laid at the extrados and 110 m/ring at the intrados. The bends between each row of pipes took into account the minimum allowable bending radius based on the pipes' specifications (five times the external diameter).

Each segment represents an individual hydraulic circuit, in which the air net of pipes runs at the intrados and the ground one at the extrados. However, each ring must eventually form a single, continuous, completely connected circuit. The ground&air nets of pipes in each segment need to be connected to the adjacent segments' ones once the TBM has mounted them on site. To do so, four special coupling box-outs were envisaged, specifically designed and included at the intrados of each segment along longitudinal joints.

Once all the steel cages with embedded pipes were ready, they were moved inside moulds, where casting of the twelve Enertun segments took place in May 2017. It should be noted that great care was taken during this process and the flow rate of the concrete mix was a bit lower than usual, to avoid any damages such as pipes collapse. After casting, energy segments were demoulded and circulation tests were carried out to ensure pipes integrity, with a successful outcome. Continuity of the water flow through the segment, from one side to the other, indicated no collapse of pipes had occurred during casting. Then, at the end of the 28-days curing period, they were transported to the construction site.

2.2 *Site implementation*

Energy rings were placed on site by the TBM at the beginning of July 2017 (rings 179 and 180, as shown in Figure 1). Installation chainage was decided in accordance with the construction site managers with the intention to minimize impact on the construction operations. As of this date, fiber optics and electrical sensors connections together with hydraulic connections between adjacent segments were completed. This task was achieved by taking advantage of the TBM backup to reach the highest, hardly accessible points.

An 80 m cable was used to connect fiber optics to a control room out of the tunnel. Electrical sensors cables were linked through a fireproof, multipolar cable to a multiplexer fixed at the tunnel wall close to the energy rings, and then from here to the datalogger, located outside the tunnel.

As concerns pipes, connections between segments were done by placing a union tube connector in between the two opposite portions of pipes exiting each segment. The handling of pipes, albeit limited, compensated for the possible alignment imperfections. Afterwards, the pockets, hosting pipes and sensors connections, were protected by means of a layer of insulation to avoid heat losses along the circuit.

Two circuits per ring were obtained, air and ground, characterized by one inlet and one outlet each. After segments installation and hydraulic connections, the two rings were connected in parallel to the main conduit, made up of two flow and two return pipes (one for the ground circuit and one for the air circuit), for a total of four header pipes. These pipes are 32 mm in external diameter, 2.9 mm in thickness, are made of the same material of the 20 mm pipes and are located outside the segments, hanging on steel brackets fixed at the tunnel Eastern springline. They have the aim to collect heat carrier fluid coming from both rings and to lead it to the heat pump. Their length is about 70 m from the energy rings to the heat pump.

An 8.8 kW heat pump device was installed close to the previously mentioned datalogger. Two hydraulic pumps circulate the heat transfer fluid along the primary circuit. This fluid is a propylene glycol mixed with water allowing to work down to a temperature of -20°C.

The "control room" hosting both the heat pump and the datalogger is an open space located close to the future elevator shaft (at present stairs are located here to allow access to the tunnel entrance). In this space, the ground&air hydraulic circuit starting and ending points can be found and management of the heat pump and of the data acquisition system are possible without interfering with the muck train going back and forth from the TBM backup in the tunnel. A view of the control room is depicted in Figure 3.

Because of the experimental nature of the project, the secondary circuit of the heat pump is represented by a fan coil unit located close to the heat pump. Therefore, at present there are no real end users benefitting from the tests, but the heat is dissipated in or extracted from the air.

2.3 *Monitoring and acquisition system*

A comprehensive monitoring system was installed to monitor the energy tunnel performance both from a thermal and a structural point of view. The two energy rings were instrumented with a specifically designed monitoring system to observe stresses, strains and temperatures in the lining. Plan and cross-sections view of the sensors layout in the two energy rings are shown in Figure 4. It is possible to divide the description of the sensors installed based on their type, that is i) conventional sensors, ii) single-mode fiber optics and iii) multi-mode fiber optics.

Figure 3. The control room, hosting the heat pump, the datalogger and the secondary circuit, view from the mezzanine (left) and a close-up view (right).

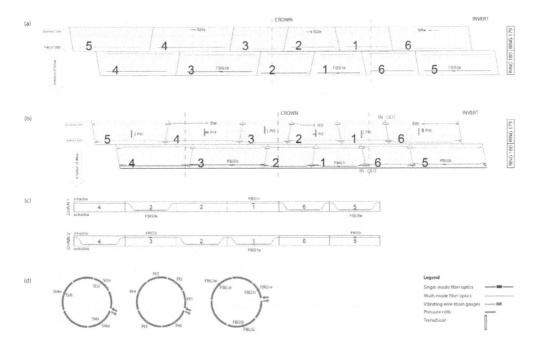

Figure 4. Design layout of sensors in the two energy rings: (a) extrados plan view, (b) intrados plan view, (c) single-mode fiber optics chains and (d) vertical cross sections showing the sensors location.

Type i) is represented by vibrating wire strain gauges and pressure cells. The energy ring 179 was equipped with these sensors. On one hand, vibrating wire strain gauges were welded to reinforcement bars in the precast concrete plant and placed longitudinally (segment 2) and tangentially (segments 4 and 6), both close to the extrados and to the intrados. The strain gauges have a built-in NTC thermistor to provide temperature data for thermal corrections. Strain and temperature correspondent to their location can be obtained. On the other hand, hydraulic pressure cells were tied by wire to the steel cage before casting and placed radially (segments 4 and 6), tangentially (segments 1, 3 and 5) and longitudinally (segment 2) in the central section of the concrete elements. Data are recorded by the data logger every 10 minutes and are transmitted every morning by FTP to a server located at Politecnico di Torino, so that they can be easily and remotely downloaded.

Type ii) is constituted by single-mode embedded Fiber Bragg Grating (FBG) sensors with temperature compensation, that were installed tangentially in ring 180 (segments 1, 3 and 5) both at the extrados and at the intrados. Two chains were designed to avoid losing all data in case of malfunctioning, by diversifying extrados and intrados sensors on each chain (Figure 4c). Embedment took place by means of zip ties around reinforcement bars, so that the sensors are able to extend or contract following concrete movements in a solidary way. Straps with no sensors run along the non-equipped segments, just for providing connection continuity with the following one. Measurements of temperature and strain take place locally thanks to an interrogation system connected to a computer. Wavelengths are measured with time and a post-processing phase is needed to convert them into quantities of physical significance.

Type iii) comprises multi-mode fiber optics. Both rings were provided with two chains of multimodal fibers, one at the extrados and one at the intrados. They are supposed to be able to return distributed measurements of strain and temperature, all along their length.

Not only what happens in the lining is observed, but also what occurs in the pipes and in the ground. Five temperature probes were devoted to the measurement of tunnel air temperature (T1), inlet/outlet temperature in the ground circuit (T2, T3) and inlet/outlet temperature in the air circuit (T4, T5). One more probe (T6) was located inside a piezometer well within the construction site to monitor the upstream groundwater temperature continuously, whose depth is in the range 11.70–12.40 m.

3 OPERATION OF THE SYSTEM

The aim of the experimental campaign is to evaluate efficiency and reliability of the prototype thermal activation together with its possible impacts on the lining and the surrounding ground.

Monitoring started in September 2017 with the assessment of undisturbed conditions at the site. Differential stresses, differential strains and temperatures in the lining were recorded under natural fluctuations of tunnel air temperature. The purpose of this one-and-a-half-month phase was also to verify the regular functioning of the system. A preliminary phase of flushing was necessary to eliminate most of air bubbles from the hydraulic circuit.

The reversible heat pump makes it possible to simulate summer and winter heating and cooling conditions. Depending on the fluid inlet temperature, this will be warmed or cooled by the surrounding ground. The possible types of tests that can be performed are:

a) Ground configuration, heating mode
b) Ground configuration, cooling mode
c) Air configuration, tunnel cooling mode

During winter 2017/2018 type a) tests were completed with both rings operating. At the end of each test the heat pump was turned off for long enough to ensure returning to the initial undisturbed thermal and mechanical conditions. Test types b) and c) were performed during summer 2018. For the three types of test the thermal power exchanged with the ground was computed and the mechanical thermally-induced effects in the lining investigated. The following quantities were recorded:

1. inlet and outlet primary circuit temperatures measured by the heat pump probes (*ENER-TUN in, ENERTUN out*)
2. inlet and outlet secondary circuit temperatures measured by the heat pump probes (*U1, U2*)
3. temperatures and stress-strain state in the lining (*Sxny, Pxn* for strains and stresses respectively, where *x* is the initial of the sensor orientation – longitudinal, tangential or radial -, *n* is the number of the sensor – 1 to 6, corresponding to each segment, and *y* indicates *i*ntrados or *e*xtrados)
4. temperature inside the tunnel (*T1*)

5. inlet/outlet ground temperature (*T2, T3*) and inlet/outlet air temperature (*T4, T5*) close to the energy rings
6. upstream groundwater temperature (*T6*)
7. external air temperature close to the heat pump (T_{air})and the following differences of temperature were calculated:
8. between primary circuit inlet and outlet close to the heat pump, considering also header pipes ($\Delta T_{ENERTUN}$)
9. between primary circuit inlet and outlet close to the energy rings, without considering the header pipes (ΔT_{ground})

The remote control of the heat pump assists the user that can set the secondary circuit return temperature that will govern the overall functioning. The heat carrier fluid inlet temperature to the Enertun circuit and its velocity are automatically adjusted by the device.

4 NUMERICAL MODELLING

4.1 *Geometry, boundary conditions and initial conditions*

A 3D numerical model was built with the FEM software Feflow (Diersch H.J.G. 2009) to reproduce the thermo-hydraulic behaviour of the two Enertun rings installed in the experimental site. The model, shown in Figure 5, is 75 m high and 150 m wide, with a thickness of 8.4 m. The pipes inside the two equipped rings were modelled reproducing the real geometry with mono-dimensional elements, the so called "discrete features" (shown in blue in Figure 6), with a cross section area of 201 mm². Inside the pipes a constant fluid flow velocity and a variable inlet temperature were imposed as boundary conditions. Fluid flow velocity inside the pipes depends on the test conditions, whereas the inlet temperature is given by the monitoring data.

Additional boundary conditions were also set, both thermal and hydraulic. As shown in Figure 5, the time series of the external air temperature was applied on the upper boundary of the model, which represents the free surface, whereas on the lower boundary the temperature was assigned a value of 14°C, constant throughout the year. On the tunnel internal boundary the temperature was fixed following the data coming from the monitoring system. The hydraulic boundary conditions consist of a constant hydraulic head on the left and right sides, with a different value from the two sides that allows a groundwater flow of 1.5 m/day from east to west.

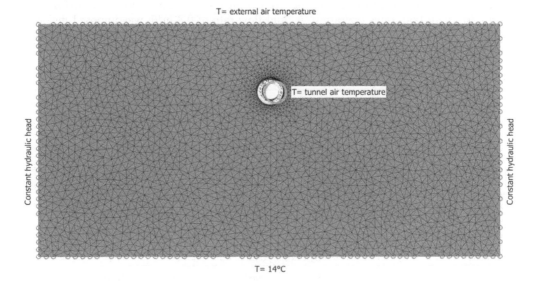

Figure 5.   Cross section of the FEM model with thermal and hydraulic boundary conditions.

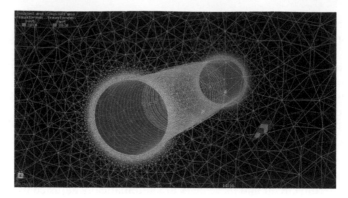

Figure 6.   Tri-dimensional view of the tunnel and, in blue, the Enertun pipes.

For the thermo-hydraulic initialization of the model, once a constant temperature of 14°C had been set along the whole model, a 30-days preliminary simulation was carried out without the thermal activation of the lining, in order to obtain the thermo-hydraulic initial condition before the simulation of the heating/cooling phases.

4.2   *Hydraulic and thermal properties*

Hydraulic and thermal parameters were obtained by previous studies, apart from the concrete thermal parameters, which were obtained by laboratory tests on the concrete used for the pre-cast Enertun segments. The parameters used in the numerical model are illustrated in Table 1.

4.3   *Back-analysis of an illustrative test*

The numerical model was calibrated by comparing the simulation output with monitoring data. The inlet temperature inside the pipes was imposed equal to the temperature measured by the monitoring sensors. A 15 cm thick layer of grout was also simulated between the ground and the segments, with thermal parameters which were varied to obtain the same monitored outlet temperature. The thermal parameters of the grout used at the end of the calibration process are equal to the 40% of the corresponding values of concrete.

The numerical simulation was carried out for a ground only heating mode test performed in March 2018 with both rings operating. A fluid flow rate of 0.8 m³/h was imposed inside the pipes. The duration of the energy rings thermal activation is 2 days. In Figure 7 it is possible

Table 1.   Parameters used in the numerical model

Parameter	Symbol	Unit	Ground	Concrete
Horizontal hydraulic conductivity	$K_{xx}=K_{zz}$	[m/s]	$4.15*10^{-3}$	10^{-16}
Vertical hydraulic conductivity	K_{yy}	[m/s]	$2.075*10^{-4}$	10^{-16}
Specific storage	S	$[m^{-1}]$	10^{-4}	10^{-4}
Porosity	n	[-]	0.25	0
Volumetric heat capacity of fluid phase	$\rho_w c_w$	$[MJ/(m^3 \cdot K)]$	4.2	-
Volumetric heat capacity of solid phase	$\rho_s c_s$	$[MJ/(m^3 \cdot K)]$	2	1.05
Thermal conductivity of fluid phase	λ_w	$[W/(m \cdot K)]$	0.65	-
Thermal conductivity of solid phase	λ_s	$[W/(m \cdot K)]$	2.8	1.12
Longitudinal dispersivity	α_L	[m]	3.1	-
Transverse dispersivity	α_T	[m]	0.3	-

Parameter	Symbol	Unit	Pipes
Fluid flow velocity	v	[m/s]	variable for different tests
Pipes internal diameter	φ	[mm]	16

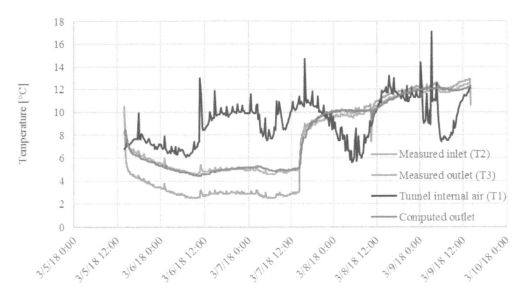

Figure 7.   Comparison between measured and computed data.

to notice that the outlet measured temperature and the computed one are highly comparable, testifying a good calibration of the numerical model.

5   CONCLUSIONS

The advantage of integrating thermally active systems into geotechnical structures such as tunnels is that the structures are being built anyway and the added cost is limited with respect to the overall cost. The thermal activation of tunnel linings is therefore an interesting opportunity that may allow to exploit the energy stored in the ground with economic and environmental benefit.

A real scale prototype of energy tunnel system constituted by two rings of Enertun segments, able to provide additional improvements with respect to previous applications of similar technologies, was realised and is undergoing a complex and detailed testing campaign. The results obtained so far allow one to draw the following conclusions:

– The prototype is capable of reproducing real scale behaviour of an energy tunnel with both the attention focused on the thermal performance and the structural behaviour of the lining. The amount of data collected will allow to deepen insight into the real behaviour of such systems.
– The construction procedures and details to implement energy segments into the tunnel construction scheme adopted are shown to be relatively simple and not matter of delays or drawbacks of the overall construction scheme. It is envisaged that improvements can be made for industrial application, e.g. to speed up segments preparation, pipes can be included in the steel cage before reaching the precast concrete plant or fiber reinforced concrete could be used; connection between segments can be done by heat sealing, etc.
– The FE 3D numerical model built to reproduce the thermo-hydraulic behaviour of the two Enertun rings and calibrated based on the measured results proved to be able to adequately simulate the conditions existing in situ. It could be used as a tool to forecast the system operational behaviour in conditions different from the tested ones, involving a different number of rings, different inlet temperatures or tunnel air temperatures, e.g. during the final metro tunnel operating phase, or the activation of the air circuit alone. It will also be of valuable help for Turin ML2 thermal activation feasibility studies.

REFERENCES

Adam, D. & Markiewicz, R. 2009. Energy from earth-coupled structures, foundations, tunnels and sewers. *Géotechnique* 59: 229–236. https://doi.org/10.1680/geot.2009.59.3.229

Barla, M. & Di Donna, A. 2018. Energy tunnels: concept and design aspects. *Underground Space*. https://doi.org/10.1016/j.undsp.2018.03.003

Barla, M. & Di Donna, A. 2016. Conci energetici per il rivestimento delle gallerie. STRADE & AUTO-STRADE 5: 2–5

Barla, M. Di Donna, A. & Insana, A. 2018. A novel real-scale experimental prototype of energy tunnel. *Tunnelling and Underground Space Technology* (under review)

Barla, M. Di Donna, A. & Perino, A. 2016. Application of energy tunnels to an urban environment. *Geothermics* 61: 104–113. https://doi.org/10.1016/j.geothermics.2016.01.014

Barla, M. & Insana, A. 2018. Energy Tunnel Segmental Lining: an Experimental Site in Turin Metro. In: World Tunnel Congress 2018. Dubai, UAE, 12

Barla, M. & Perino, A. 2014. Energy from geo-structures: a topic of growing interest. *Environmental Geotechnics* 2: 3–7. https://doi.org/10.1680/envgeo.13.00106

Brandl, H. 2006. Energy foundations and other thermo-active ground structures. *Géotechnique* 56: 81–122. https://doi.org/10.1680/geot.2006.56.2.81

Di Donna, A. & Barla, M. 2016. The role of ground conditions and properties on the efficiency of energy tunnels. *Environmental geotechnics* 1–11. https://doi.org/10.1680/jenge.15.00030

Diersch H.J.G. 2009. DHI Wasy Software – Feflow 6.1 – Finite Element Subsurface Flow & Transport Simulation System: Reference Manual

European Commission 2016. Overview of support activities and projects of the European Union on energy efficiency and renewable energy in the heating & cooling sector. Publications Office of the European Union, Luxembourg. https://doi.org/10.2826/607102

Franzius, J.N. & Pralle, N. 2011. Turning segmental tunnels into sources of renewable energy. *Proceedings of the ICE – Civil Engineering* 164: 35–40. https://doi.org/10.1680/cien.2011.164.1.35

Frodl, S. Franzius, J.N. & Bartl, T. 2010. Design and construction of the tunnel geothermal system in Jenbach/Planung und Bau der Tunnel-Geothermieanlage in Jenbach. *Geomechanics and Tunnelling* 3: 658–668. https://doi.org/10.1002/geot.201000037

Laloui, L. & Di Donna, A. 2013. Energy Geostructures: Innovation in Underground Engineering, Energy Geostructures: Innovation in Underground Engineering. https://doi.org/10.1002/9781118761809

Lee, C. Park, S. Won, J. Jeoung, J. Sohn, B. & Choi, H. 2012. Evaluation of thermal performance of energy textile installed in Tunnel. *Renewable Energy* 42: 11–22. https://doi.org/10.1016/j.renene.2011.09.031

Moormann, C. Buhmann, P. Friedemann, W. Homuth, S. & Pralle, N. 2016. Tunnel geothermics – International experience with renewable energy concepts in tunnelling/Tunnelgeothermie – Internationale Erfahrungen zu regenerativen Energiekonzepten im Tunnelbau. *Geomechanik und Tunnelbau*. https://doi.org/10.1002/geot.201600048

Nicholson, D.P. Chen, Q. de Silva, M. Winter, A. & Winterling, R. 2014. The design of thermal tunnel energy segments for Crossrail, UK. *Engineering Sustainability* 167: 118–134. https://doi.org/10.1680/ensu.13.00014

REN21 2017. Renewables Global Futures Report: Great debates towards 100% renewable energy. Paris

Schneider, M. & Moormann, C. 2010. GeoTU6 – a geothermal Research Project for Tunnels. *Tunnel Geothermics* 2: 14–21

Unterberger, W. Hofinger, H. Grünstäudl, T. Markiewicz, R. & Adam, D. 2014. Utilization of tunnels as sources of ground heat and cooling – Practical applications in Austria. *iC consulenten Ziviltechniker GesmbH Publications*, www.ic-group.org 1–6

Wilhelm, J. & Rybach, L. 2003. The geothermal potential of Swiss Alpine tunnels. *Geothermics* 32: 557–568. https://doi.org/10.1016/S0375-6505(03)00061-0

Zhang, G. Xia, C. Sun, M. Zou, Y. & Xiao, S. 2013. A new model and analytical solution for the heat conduction of tunnel lining ground heat exchangers. *Cold Regions Science and Technology* 88: 59–66. https://doi.org/10.1016/j.coldregions.2013.01.003

Tunnels and Underground Cities: Engineering and Innovation meet Archaeology,
Architecture and Art, Volume 2: Environment sustainability in
underground construction – Peila, Viggiani & Celestino (Eds)
© 2020 Taylor & Francis Group, London, ISBN 978-0-367-46579-7

Online identification of the excavation materials on the Saint Martin La Porte site

A. Barrel
Spie batignolles GC, Boulogne Billancourt, France

C. Salot
Tunnel Euralpin Lyon-Turin (TELT), Le Bourget du lac, France

ABSTRACT: As part of the excavation by TBM of a 8,7km long tunnel in the heterogeneous rock of the Briançonnaise carboniferous zone, an online identification installation allows a continual categorization of the materials for their future use. This innovating device is installed on the conveyor belt which transports 900 000m3 of evacuated material to disposal sites.

This system allows the classification of excavated materials into Cl1/Cl2 – into recoverable materials – and secondly, Cl3, – into unsuitable for utilization as road backfill – (according to the AFTES's recommendation GT35).

Two optical measurement devices have been installed, a color measurement device and a granulometric measurement device.

These devices allow a continuous view of the quality of the waste rock. It then controls the diverting of the materials on the conveyor belt to sort them at distinct drop areas to optimize the temporary stocks.

1 OVERVIEW OF LYON-TURIN PROJECT

1.1 *The Lyon-Turin project*

The cross-border section of the new railway link Lyon-Turin is 67 km long between Saint-Jean de Maurienne (France) and Bussoleno (Italy). Its main part is the 57,5 km Mont Cenis base tunnel.

1.2 *Our approach to the management of excavation materials*

The excavations will produce approximately 37 million tonnes of material. The prime objective is to optimise the utilisation of the excavated materials, primarily to use as aggregate for concrete within the tunnel and as backfill in external structures.

This process, implemented throughout the project, represents an environmental and territorial approach resulting in the re-use of the excavated materials within the structure. This allows us to limit the necessary impact of the final disposal of these materials thus preserving the resources and reducing green-house gas emissions through local use of the materials.

In addition, TELT joined both the United Nations Global Compact programme and Global Compact France in November 2015. The commitment was therefore made to respect 10 universally recognised principles in the aim to further corporate social responsibility. TELT thereby undertaking the promotion of greater environmental responsibility and precautionary principle in the development and diffusion of environmentally friendly technologies.

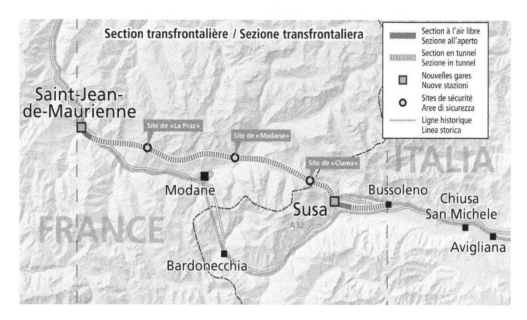

Figure 1.   Overall view of the cross-border section of the new Lyon-Turin rail link.

1.3   *Our specific approach applied to reconnaissance works*

The completion of the Saint Martin de la Porte access tunnel in 2010, had revealed a geological formation containing, in particular, significant convergence phenomena and carbonaceous areas. Further reconnaissance works were therefore launched in 2014 with the aim, in particular, of verifying the geological hypotheses and acquiring the necessary experience for the Base Tunnel using a tunnel boring machine.

This work is naturally in keeping with the approach relating to the management of excavation materials, the objective of which is to allow the storage of materials presenting potential use of aggregate for concrete or backfill. These materials will then be processed and used as part of the work's main structure.

2   PRESENTATION OF THE PROJECT AND THE GEOLOGY OF THE SITE

2.1   *The SMP4 project*

Work on the Saint-Martin-La-Porte reconnaissance gallery began in early 2015. The project owner is the Franco-Italian company TELT (Tunnel Euralpin Lyon Turin) which took over last February from LTF (Lyon Turin Ferroviaire, in charge of preliminary design studies and most of the reconnaissance work since 2001).

The construction of the reconnaissance works at Saint-Martin-La-Porte 4 was entrusted to a consortium of six companies: Spie batignolles GC (consortium leader), Eiffage Génie Civil (manager), Ghella SpA, CMC di Ravenna, Cogeis SpA and Sotrabas.

The work is divided into 4 main parts:

The work consists essentially of the construction of an 8.7 km reconnaissance tunnel, excavated with a tunnel boring machine, towards Italy (Part 2 of Figure 2). Carried out in the axis and diameter of the south tube of the future tunnel, this reconnaissance tunnel is excavated with a hard rock tunnel boring machine and will connect the foot of the Saint-Martin-La-Porte tunnel to that of La Praz.

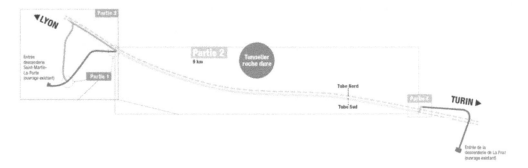

Figure 2.   Overall view of the SMP4 project.

2.2   *P2 – Tunnelling with a TBM*

The tunnel boring machine that will construct this tunnel was designed to cope with the particular geological constraints of the area. The 138m long tunnel boring machine is equipped with a cutter head of 11.26 metres in diameter and 76 wheels, with an operational power of 5 megawatts.

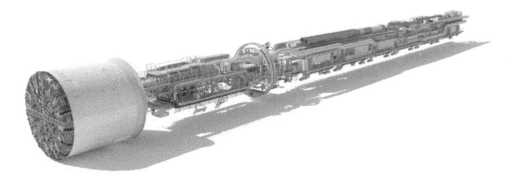

Figure 3.   TBM Federica.

The excavation by tunnel boring machine represents 870,000 m3 of materials present becoming 1,000,000 m3 once excavated.

2.3   *The geology*

Concerning the tunnel boring with a TBM, the lithostratigraphic sequence of the Briançonnaise coal zone to be crossed consists of an alternation of sandstone, black shale and carbonaceous layers.

Based on field studies and drilling analysis, a gradual increase in the sandstone fraction is emphasised within the Briançonnaise Coal Zone from West to East, i.e. from the Brequin-Orelle unit towards the La Praz unit. This is accompanied by a reduction, also gradual, in the quantity and thickness of the carbonaceous levels.

3   THE STAKES OF SORTING ON THE SITE SMP4

It is required in the SMP4 reconnaissance works contract that pre-grading of materials according to the classes prescribed by the AFTES GT35 is provided, i.e.:

– Recoverable materials: C11 and C12,
– Materials not suitable for road backfill: C13.

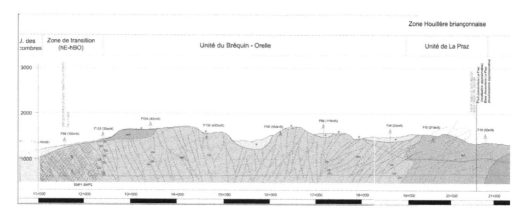

Figure 4.   Geological section of SMP4 project.

Given the geology of the area and the diameter of the cutting wheel, the excavated front is mostly mixed. It is generally composed of shale, sandstone and coal in different proportions depending on the advancement.

In principle this will be the actual configuration at least until contact with the La Praz unit. After this limit the sandstone will be in much higher concentration than the shale and the coal will be almost absent, until the end of part 2.

3.1   *Potentially reusable materials C11/2*

C11/C12 materials are materials of good or average quality, suitable to be transformed into aggregate for concrete, fill or for certain industrial uses.

Sandstone and shale are potentially usable materials. Depending on the degree of schistosity, they can have good geotechnical properties. Concerning the actual quantity of usable materials, this depends on the digging mode and the fracturing state of the terrain, in fact the percentage of fine particles can vary significantly.

3.2   *Non reusable materials Cl3*

Coal is an organic rock that evolves over time having poor to bad geotechnical characteristics. The coal or any other material containing it is permanently stockpiled

4   SOLUTION IMPLEMENTED

Two optical analysers have been positioned to the right of the belt conveyor at the tunnel entrance to define the physical properties of the materials:

– A colorimetric analyser,
– A particle size analyser.

4.1   *Colorimetric device*

4.1.1   *Principle*
The Colobserver® CL150 sorting system, by the company « Iteca Socadei », provides material identification based on a visual difference between useable and non-usable materials.

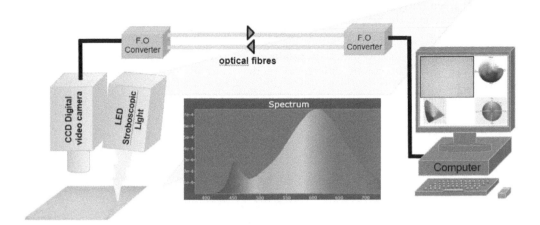

Figure 5. Principle of the colorimetric device.

Analysis consists in measuring the intensity of transmitted or reflected light and relating this measurement to the nature and composition of the materials. This analysis is performed in the visible range of the electromagnetic spectrum.

The product's colorimetric readings are measured using a strobe light source and a high-resolution digital colour sensor.

Figure 6. The device set up on the SMP4 site.

A self-calibration system compensates for drifts in the lighting and the measuring system. A supervision equipped with specific software, link-attached to the control room, calculates the colorimetric coordinates of each pixel and displays the colour trends of the image.

4.1.2 *Results*
The data collected are divided into 3 axes:

– The b* axis: blue tints (negative) to yellow tints (positive)

– Axis a*: green tints (in negative) with wheel (in positive)
– L* axis: represents the brightness from black (0) to white (100)

The thresholds were defined during the calibration of the device, first in the laboratory then on site with the materials produced by the excavation. An on-line moisture meter compensates for colour differences in the materials depending on the materials' water content.

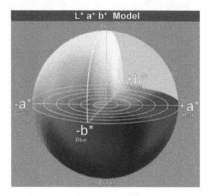

Figure 7.   Examples of colorimetric data.

The colorimetric analyser currently enables us to discriminate carbonaceous materials.

4.2   *Granulometry device*

4.2.1   *Principle*
The measurements are obtained by triangulating the light rays produced by a laser beam and captured by a camera, both directed onto the moving belt.
The measuring principle is based on optical triangulation:

– A laser illuminates the aggregate vein passing on the conveyor belt
– The vein's characteristics are calculated using the light signal recorded by the camera.
– The granulometric curve is obtained by statistical modelling of the size of the aggregates and their layout.

4.2.2   *Results*
The data makes it possible to obtain a granulometric curve from 5 to 300 mm of the materials passing on the belt, to the moment or an average over a defined period of time.

Figure 8.   Principle of the colorimetric device.

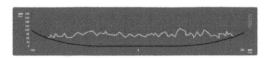

Figure 9.   Vein's view example.

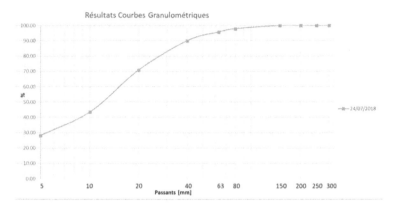

Figure 10.   Example of a granulometric curve.

The Tamisoft® particle size analyser by the company « Autom'elec » makes it possible to discriminate materials with a too great a proportion of particles. The thresholds can be adjusted on all measured portions according to the re-use of materials.

4.3   *Mechanical tests*

Along with the on-line devices, mechanical characterization tests are carried out on a weekly basis in order to check the choice of on-line analysers, to gain better understanding of the materials and potentially to adapt the thresholds.
   The characterization of Cl1/Cl2 pre-classified materials consists of:

–   Continuous analyser grain size data,
–   On-site tests: Point Load Test,
–   Laboratory tests for mechanical characterization
–   Laboratory tests to characterize sulphate content

5   LOGISTICS ASSOCIATED WITH SORTING

The excavated material is removed from the gallery via a conveyor belt. The nominal discharge rate from the conveyor is 800t/hour.
   Both measuring devices have been integrated into the automated conveyor system to:

–   recover information on the running condition of the conveyor belts,
–   provide information allowing activation of the sorting system.

   The materials are automatically pre-classified into 2 separate piles on a material transit platform. For safety reasons, as soon as one of the two devices detects some C13 the conveyor belt switches over and sends the materials to the corresponding pile. In order to limit too-frequent switching a time delay system has been set up for when the device returns to Cl1/Cl2 sorting.

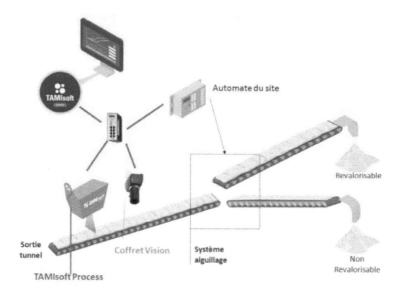

Figure 11.   The operating principle of sorting automation.

The materials are then transported by truck to specific depot sites.

The C11/C12 materials are partially re-used on the SMP4 site, after having been crushed and screened, as backfill for the lower arch segments of the tunnel or for draining layers of the external structure. The remaining material is stored for future use.

6   ONGOING RESULTS AND IMPROVEMENTS

The first results which make it possible to discriminate Cl3 non-usable materials have been conclusive in the heterogeneous geology of the project. It allows automatic pre-classification of the materials at the conveyor belt exit which will be systematically put in final storage owing to their evolutionary nature or their too great a proportion of particles.

The main problems encountered since the beginning of this work have been the calibration of the on-site devices and the drift of the results during times of high water levels.

Firstly, when the devices were set up, the question arose as to the sampling representability on the very large diversity of materials and the large range of particle sizes. A large quantity of materials and several tests on different lithologies were necessary.

Secondly, the presence of strong underground water inrushes or the requirement to add water for work purposes triggered reflectance which falsified the measurements of the colorimetric device. A digital treatment for the deletion of these zones was carried out by the devices' supplier.

Transport and depot logistics are also a sensitive issue in material sorting management. These must be anticipated as far ahead as possible to provide the necessary surfaces and facilities in order to maximize the reusability of materials.

Areas for improvement are of course still being studied on the SMP4 site. For example, we are currently looking at the possibility of adjusting our devices to differentiate C11 materials (for use in concrete aggregate), C12 materials (backfill) and ways to improve the general appearance of materials (less water and less mud) before they pass under the devices, thus obtaining more accurate and reliable measurements.

Improving logistics and making all personnel aware of the issues involved in the sorting and the usage of materials are also two essential elements for successful management and reuse of excavated materials.

REFERENCES

AFTES, 2016, Recommandation de l'AFTES GT35R1F2 La gestion et l'emploi des matériaux excavés
LTF, 2012, Mémoire de synthèse géologique er hydrogéologique
LTF, 2014, Lithographie, géologie structurale et tectonique

Tunnels and Underground Cities: Engineering and Innovation meet Archaeology,
Architecture and Art, Volume 2: Environment sustainability in
underground construction – Peila, Viggiani & Celestino (Eds)
© 2020 Taylor & Francis Group, London, ISBN 978-0-367-46579-7

Construction of portals in difficult conditions in the Andes, Peru

M. Boisán, M. de Cabo & C. Quiroga
Subterra Ingeniería, Spain

ABSTRACT: Yanango Tunnel is located in Junín, Peru; enhancing the connectivity between Lima and the Central jungle between Tarma and La Merced. The construction was awarded to a JV (MPM and Balzola) in a 25 MUSD contract, and took place from 2015 to 2016, and its inauguration was last September 2017. The tunnel has a length of 1,062m with three traffic lanes, in a section of the 96m². The geology is composed by fair to good geomechanical quality granodiorites. The main challenge was the construction of both portals, requiring changes in their initial designs, constructing two false tunnels structures before starting the excavation. The Western Portal was excavated in colluvial soils, with only 15° between the hill side orientation and the tunnel axis. To ensure the stability of the lateral slope, a soil nailing was necessary as well as a heavy micropile umbrella.

The Exit portal is located in a vertical relief and competent rock mass. The solution adopted has been a previous bolting reinforcement and a sequential excavation of the tunnel: false tunnel, a transition section, and afterwards starting the tunnel excavation.

1 EXIT PORTAL

The construction of the tunnel started by Exit Portal on February 2016. It was located in an almost vertical hill side, with good quality rock mass with RMR between 59 and 68 points.

Initial solution proposed the excavation of slopes 1H/10V reaching 80 meters height (Figure 1).

In order to avoid this complex excavation and to save time and costs, the solution adopted was to reinforce with bolts the natural slope without any excavation, and to construct sequentially the tunnel: false tunnel, a transition section, and afterwards starting the tunnel excavation (Figure 2).

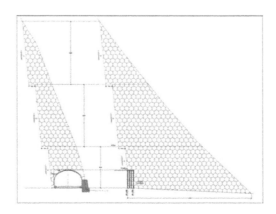

Figure 1. Inicial design of exit portal.

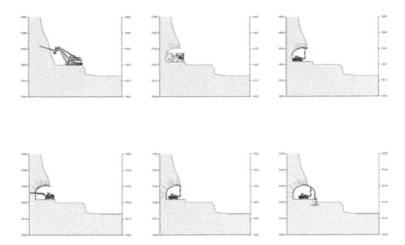

Figure 2. Excavation sequence in exit portal.

Based on geologic mapping, six families of joints were defined (Figure 3); and were used to analyse and define the proper reinforcement of the natural slope and the support of the part-tunnel.

As result, a pre- reinforcement, the execution of a lateral forepole executed perpendicular to the natural slope with φ32 mm bolts of 10 m length was proposed. This pre-reinforcement consisted of two rows of bolts spaced 90 cm horizontally and 1 m vertically. These bolts, in addition to providing a protection to the vault at the time of execution of the partial tunnel, served to stabilize possible wedges and planar breaks that might appear. (Figure 4)

In the part-tunnel, the support consisted on:

– Fiber reinformcement shotcrete: 5+5 cm
– Φ25 mm and 4.5 m length bolts spacing 2.3 m (T) x 1.23 m (L)

The construction of the 37 meters long of this part-tunnel was in steps of 1.23 m and last one month.

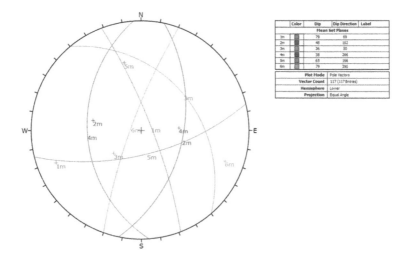

Color	Dip	Dip Direction	Label
Mean Set Planes			
1m	79	69	
2m	48	102	
3m	36	80	
4m	38	266	
5m	65	166	
6m	79	290	

Plot Mode	Pole Vectors
Vector Count	117 (117 Entries)
Hemisphere	Lower
Projection	Equal Angle

Figure 3. Joints in exit portal.

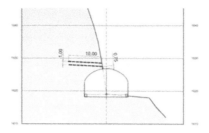

Figure 4.   Pre-reinformcement in natural slope of exit portal.

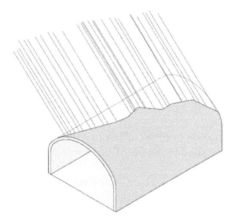

Figure 5.   Finished structure scheme.

Photography 1.   Slope pre-reinforcement.

As finished, a false tunnel structure was built. This structure, composed by steel arches and shotcrete, was calculated to support the fall of a block of 5 tons weigth. (Figure 5).

2   ENTRANCE PORTAL

The Entrance Portal (West portal) was located entirely in colluvial soils. The original design defined a lateral slope of 45m high in three banks with a slope 1H/3V. (Figure 6). In order to try to decrease the high of the slope, based on the low quality of the soil, a new geometric design was carried out.

Photography 2.   Part-tunnel and structure construction.

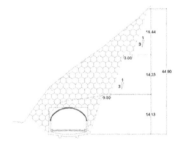

Figure 6.   Inicial design of entrance portal.

The new design consisted on a soil nailing for a lateral slope 35 m height, composed by four banks 4m high with a slope 1H/4V. In order to get the less height possible, frontal slope was define with only 15° between the strike of the hill side and the tunnel axis.

In order to define soil properties and the rock contact, three pits and a 40 m length borehole was carried out. In addition, and based on the complex of the portal, an inclinometer was installed in the borehole in order to control the stability during the portal excavation.

Based on the stability analysis carried out, the soil-nailing was composed by (Figure 8):

– Self-drilling bolts of 40ton, spacing 2.0m (H) x 1.5m (V) with lengths between 6 and 9 meters
– 20 cm of shotcrete
– Double steel mesh
– Drains

After this soil nailing, the false tunnel structure was built and the landfill above completed, before starting the construction of a double canopy tube forepole and the excavation of the tunnel. Figure 9.

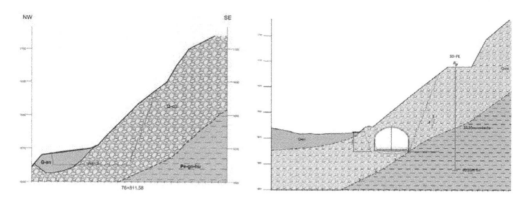

Figure 7. Geological sections in entrance portal.

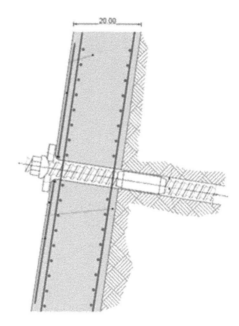

Figure 8. Slope reinforcemente detalis in entrance portal.

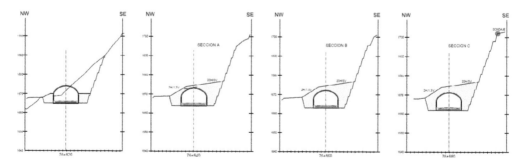

Figure 9. Land fill of false structure.

During the construction, the inclinometer was measured frequently in order to control deep movements. The results showed some movements that was possible controlled modifying slightly the construction process (Figure 10). In addition, topographical landmarks was controlled in the slope to detect superficial movements.

The main excavation and reinforcement of entrance portal started on February 2016 and ended on June 2016, having placed 808 self-drilling bolts and 135 drains.

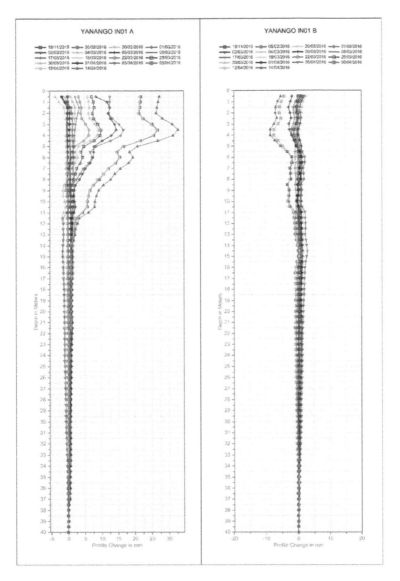

Figure 10.   Inclinometer measurement.

Photography 3.　View of the excavations.

3　CONCLUSIONS

The complex Andean orography sometimes requires unconventional solutions to achieve an adequate design of the portals of the tunnels in order to get viable designs.

Tunnels and Underground Cities: Engineering and Innovation meet Archaeology, Architecture and Art, Volume 2: Environment sustainability in underground construction – Peila, Viggiani & Celestino (Eds)

Tunnel design approach during Saudi Landbridge Railway Project

F. Caranti, M.E. D'Effremo, L. Utzeri, F. Sacchi, S. Ciufegni & E. Bianchini
ITALFERR S.p.A., Rome, Italy

ABSTRACT: The aim of this article is to describe the engineering approach followed in the design phase of the Saudi Landbridge Railway Project, applying the best international practice for tunnel design in a non-conventional environment. The alignment has been designed in order to fully comply with the specific functional and technical requirement requested by the Client. To this purpose, a systematic socio economical and cost-benefit analysis has been studied to identify the most suitable civil work, including tunnel, for the optimization of the railway line. Furthermore, due to the presence of extreme environmental condition characterized by desert and dune area, specific sand mitigation solutions have been studied. Specifically for the design of cut & cover tunnels in dune area, non-conventional topics have been addressed, such as the variation of the asymmetrical load distribution due to the dunes displacement or the need to prevent the entrance of the sand into the tunnel.

1 INTRODUCTION

The Kingdom of Saudi Arabia (KSA) has planned the upgrade and expansion of its railway system through several projects collectively known as the "Saudi Railway Expansion Project". The Saudi Railway Expansion Project includes a new rail link across the country known as the 'Saudi Landbridge Project which includes:

- Construction of approximately 1300 Km new line between Jeddah and Jubail;
- Integration of the new lines with Jeddah Islamic Port, King Abdul Aziz Port (Dammam), Riyadh Dry Port and King Fahad Industrial Port and Commercial Port (Jubail).

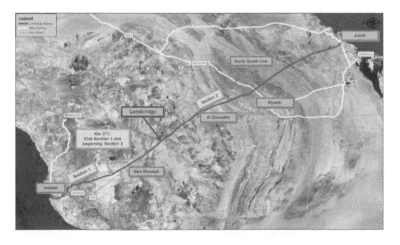

Figure 1. General map of Saudi Landbridge Railway Project.

The Saudi Landbridge Project has been designed to expand the existing rail network and transform the existing Railway into a world-class freight and passenger line linking the east and west coasts of the Kingdom of Saudi Arabia (KSA). It will provide a capability to move large quantities of cargo over long distances at competitive rates and offer safe and comfortable overland passenger transport.

The Saudi Landbridge Project is divided in two different sections:

- Section 1 – from Jeddah to km 271, including JIP terminals, Jeddah Dry Port, Marshalling Yard and Wayside maintenance.
- Section 2 – from km 271 to Jubail, including Al Duwadimi station, Riyadh Dry Port, Marshalling Yard, freight and passenger maintenance, Jubail Dry Port and Marshalling yard.

The Public Investment Fund (PIF) has commissioned Italferr S.p.A. as Consultant to develop Preliminary Design and Detailed Engineering phases.

2 ENGINEERING APPROACH

The design phase has been developed on the basis of the tender documents and project requirements and it has included three separate phases (Concept, Preliminary and Detail design phase).

As depicted in the following flow chart (Figure 2), the definition of the optimal alignment and of the most suitable civil works has constituted a critical step of the entire design process since it was the result of a multitasking analysis that has involved all the experts of the design team.

In particular, after the definition of the 1300km of optimal alignment in the extremely heterogeneous Saudi environment, several analysis have been conducted in order to select the civil works that can guarantee the best mix of quality, durability, and functionality in compliance of the project requirements. For tunnels, the design has been developed on the basis of the three main requirements:

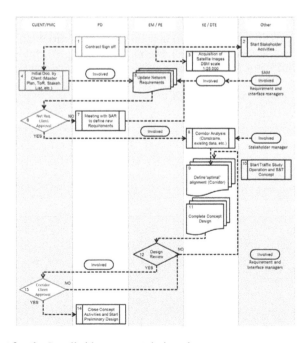

Figure 2.   Flow chart for the Landbridge concept design phase.

- *SYS.CWT.01. Tunnels longer than 2 Km will be double bore single track tunnel*
- *SYS.CWT.02. Tunnels shorter than 2 km will be single bore double track tunnel*
- *SYS.CWT.03. The inner cross section for double track tunnel and for single track tunnel will be designed around AREMA Clearance Envelope and taking into account the clearance envelope specified by TSI standards for High Speed trains.*

The social and economic comparison has constituted the more reliable tool to select the optimal alignment and the most appropriate civil works.

Due to the heterogeneity of the lands crossed by the Landbridge Project, several civil works are present along the railway route in order to assure the optimal resolution of topographical and hydrological interferences but also to minimize the fragmentation of human and ecological communities affected by the Project.

The main inputs of these analysis are the following:

- stakeholder issues (social and cultural impact);
- environmental constraints;
- construction cost;
- maintenance cost at different years;
- safety and security issues, included the technical and economic aspects related to the equipment provision in remote area (i.e. power and water supply);
- the result of these analysis the number of civil works along the route are summarized in the table below.

Table 1. Number of civil works along Landbridge alignment.

Viaducts	Flyovers	Tunnel/Flyovers	Underpasses	Camel Crossings	Culverts	Tunnels
176	29	10	42	10	1006	6

3 WINDBLOWN SAND AND SAND MITIGATION MEASURES

The Saudi Landbridge Railway Project crosses a huge variety of landforms along the route, with different wind, sediment, and geomorphologic conditions.

A detailed description of the surface local conditions was obtained by systematic field observations and measurements. Many dune area and sand sheets were identified by satellite images and detected along the railway alignment during site visits (Figure 3).

The sand area were classified as *Dunes area* only where the wind generates the displacement of the dunes, typically moving along a preferred direction. In fact, through the analysis of satellite images available at different time instants, it was possible to identify the directions and the average speeds of dunes displacement, so as to grossly estimate the net dune drift ($m^3/m/yr$).

In such environmental conditions, windblown sand can have a wide range of undesired effects on the infrastructure, concerning Safety issues, that require operation interruption, and Serviceability/Maintenance issues, that reduce the performance of the system requiring intensive and expensive maintenance plans (Figure 4).

Figure 3. Desert area and Dunes area detected along the railway alignment.

Figure 4.   Undesired effects of the windblown sand on railways.

Combining sand covering and grain size obtained in geological and geomorphological surveys, with wind data collected from anemometric stations in the Kingdom of Saudi Arabia, a mapping of the Actual Sand Drift along the alignment was obtained, showing values higher than 100 m^3/m/yr in desert and dunes area (Figure 5).

A peculiar Sand Mitigation Measure was conceived and designed for Sand Drift lower than an acceptable value, consisting of a ditch trapping the windblown sand at the upwind side of the railway (Figure 6).

The ditch walls are retained by means of a Reinforced Earth System; the top layer of the track side R.E.S. wall is a gabion laying over the original ground level and bearing a precasted reinforced concrete cantilever, aerodynamically shaped to increase sand trapping (Figure 7).

This type of Sand Mitigation Measure is mostly efficient for Actual Sand Drift up to 25 m^3/m/yr for each side of the railway, requiring a single cycle of sand disposal per year.

The analyses cost-benefit lead to adopt this type of Sand Mitigation Measures where Total Actual Sand Drift was lower than 80 m^3/m/yr, since the sand disposal from the ditch provided in maintenance plan was still economically convenient comparing to other drastic solution such as viaducts and tunnel.

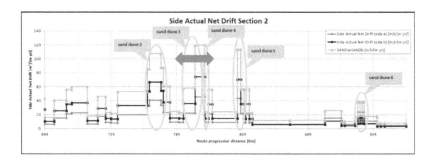

Figure 5.   Stretch of the map of Actual Sand Drift along the alignment.

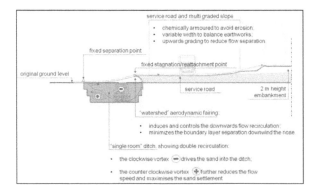

Figure 6.   Conceptual design of Sand Mitigation Measure for Saudi Landbridge Project.

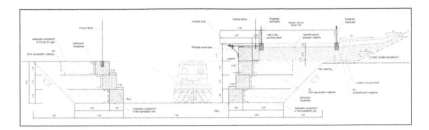

Figure 7.   Detail design of Sand Mitigation Measure for Saudi Landbridge Project.

In flat desert area, for Actual Sand Drift higher than 80 m³/m/yr the only alternative solution to the Standard Sand Mitigation Measure is the viaduct.

For each Sand Dunes area, where the Total Actual Sand Drift is higher than 80 m³/m/yr, different solutions, depending on the geomorphological features of the area, were studied by varying the vertical alignment.

For each area were analyzed:

a) different possible technical solutions:
   • viaduct;
   • large cutting and a low viaduct at the bottom;
   • cut and cover tunnel;
   • traditional tunnel.

b) specific economic assessments. The estimated cost for solution was defined on a stretch wider of the extension of dunes area, to consider the variability of the works before and after the dunes.

4   TECHNICAL SOLUTIONS TO CROSS DUNES AREA

As described in the paragraph 3, the engineering solutions to cross dune systems are the following: viaduct, large cutting with low viaduct, deep tunnel, cut and cover tunnel.

a) Large cutting with low viaduct (Figure 8)

The first solutions characterized by slight slopes, has been conceived to allow the passage of the windblown sand from upwind side to downwind side and to provide an available space for residual sand accumulation below the viaduct: cut slope 4:1, base of the cutting 50 m, high pier minimum 4 m. The available volume of sand accumulation is about 350 m³/m. According to the approximate values of the net dunes drift defined during the design, the maintenance will be necessary from 4 to 15 years.

b) Deep tunnel

In this peculiar geological environment with sand, the only chance to realize a tunnel was to carry out an intensive consolidation work around the tunnel and on the tunnel face. The aim was

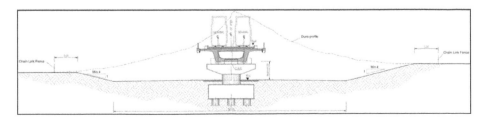

Figure 8.   Viaduct in large cutting.

to realize a consolidated arch around the tunnel to achieve the stability condition during the excavation of tunnel itself. The typical cross sections shown in the following figures have been thought with the aim not only to achieve the stability condition during the excavation operations, but even to protect the tunnel in case of a very high variation of orography due to the sand drift. In fact, if the high variability of sandy drift gives a serious reduction of the overburden on final concrete lining of the tunnel, the consolidated arch around the tunnel will protect the lining itself.

In particular, the way of proceeding could be to realize consolidation works constituted by jet grouting, i.e. injections of grout with high values of pressure (500–600 bar). In case of overburden on tunnel's crown minor than 25 m, the jet grouting should be executed from ground level. In next figures some schemes of the proposed solution are represented. That scheme is suitable if the overburden on the tunnel's crown is less than 25 m, cause to the length of boring for jet grouting (the suggested maximum length for jet grouting is 30/35 m).

In case of overburden greater than 25 m, the consolidation works with jet grouting should be executed in progress, during the excavation (consolidation work with jet grouting at the tunnel face and around the cavity).

In case of overburden on the tunnel's crown minor than 20 m, it would be suitable to foresee a cut and cover solution.

However, many issues have been raised against the realization of deep tunnels in dunes area and in particular:

- in this context, the realization of the jet grouting solution has a big difficulty, that is water supply (a big amount of water should be transported in desert area);
- the realization time is much longer than the one for an usual tunnel. In particular, in case of consolidation works during the excavation at the face and around the cavity, on the basis of some past experiences, the construction rate advancement, with full face conventional excavation, could be equal to 0.5 m/day for single bore double track tunnel and equal to 0.8 m/day for double bore single track tunnel. In case of consolidation works from ground level, the excavation time is faster, but the total construction time depends on the available number of machines that will be utilized on the ground level to realize the consolidated arch around the bore;
- the overburden could change because of the sand drift. For this reason the variation of the entity of loads acting on linings and the possible asymmetry of loads itself are other aspects to care, especially in case of mobile dunes with a considerable drift;
- it's necessary to guarantee, with maintenance operations on the portals, the total length of tunnel, that could change because of the sand drift;
- in dune area, the orography vary during the lifetime of the structure. This variation has to be considered in the definition of the safety configuration of tunnel. In particular, because of the possible modification of the geometric configuration of exits by the sand drift, the aim is to avoid emergency transversal and/or vertical exits towards ground level.

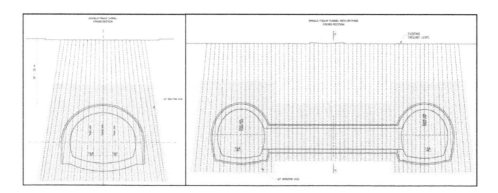

Figure 9.   Consolidation works from ground level.

In the end, another issue to take into account is the extension of the dunes transversely to the tunnel axis. The realization of a tunnel can't be recommended if this extension is minor than 200–300 m (i.e. 100–150 m on every side respect to tunnel axis).

c) Cut and cover tunnel

This solution has the advantage of having no need of the consolidation works. The structural section must take into account the presence of asymmetric loads due to the presence of dunes. Therefore, we consider a section of the polycentric characterized by a structural thickness greater than the usual tunnel and a cement bound crown for the variation of the entity of loads acting on linings and the possible asymmetry of loads.

In the Figure 10 is depicted the description of the main phases of the construction for the calculation assumptions. The phase A is the existing morphology of the ground. Typically there is a first layer of moving dunes and a second layer of sand material having better characteristics. In this second layer the Cut and Cover Tunnel will be founded. The temporary excavation of phase B allows the built the Cut and Cover Tunnel and the Cement Bound coating. The coating is not considered in the calculations, it only provide a better distribution of the stress between the vault concrete and the ground. In the phase C the backfill of sand material will be done. The characteristics of this material will be at list as the Sand layer. Phase D describes the moving, and the maximum expected thickness, of the dunes during the life of the structure, up to totally cover it.

The maximum expected backfill has been hypothesized by analyzing the maximum height of the dunes ridges in this area. Earth pressure on the side walls and on the top of the tunnel has been calculated studying two different scenarios:

- A: maximum expected backfill (40 m – 4 m are the backfill design)
- B: asymmetric distribution of the ground (40m – 4 m are the backfill design)

In both cases the minimum backfill design is 4 m over the vault, at least 1 m over the cement bound, which have the maximum thickness of 3 m. The backfill design shall have at least the characteristics of the material and shall be well compacted. The favorable effect of the Cement Bound coasting is not considered in the calculation.

For the scenario A Terzaghi's ground arch concept has been adopted. The formulation for the calculation of the vertical and lateral pressure can be seen in the Figure 11.

For the scenario B it has been considered that the moving dune partially cover the designed backfill. This provide an asymmetric pressure on the Cut and Cover Tunnel.

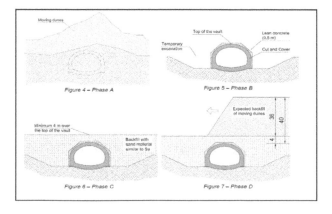

Figure 10.   Cut & cover tunnel – Section.

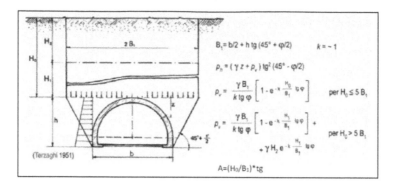

Figure 11. Terzaghi ground arch concept.

5 SAND DUNE AREA N.1. COMPARISON OF TECHNICAL SOLUTIONS

As an example, in the sand Dunes area identified between km 553+600 and km 554+900, the Concept Design solution foresees a low alignment with standard trench. This solution has been discarded due to the level and the variable direction of sand drift expected. These reasons don't allow the use of standard mitigation measure because of the extreme high cost for maintenance.

5.1 Comparison of technical solutions

Three hypothesis of alternative alignment have been studied where change only the top of rail in the profile:

- Solution 1. Vertical Alignment medium solution (Concept solution). Keeping the vertical alignment at the Concept design solution:
  - with a cut and cover tunnel (solution 1a). The solution 1a foresees a cut and cover tunnel to pass under the sand dunes to avoid any type of windblow sand problem, Figure 12. For this solution, the detail of the different civil works from km 548+560 to km 558+650 are shown in the following Table 2.
  - with a large cutting with a small viaduct (solution 1b). In the solution 1b the alignment and the civil works are the same of the solution 1a, but the cut and cover tunnel is replaced by a small viaduct, Figure 13.

Figure 12. Civil works for sand dunes area solution 1a.

Table 2. Number and features of civil works for solution 1a.

Description	From chainage	To chainage	Length (m)
Embankment	548+560	554+037	5.477
Cut and Cover	554+037	555+197	1.160
Small Viaduct	555+197	556+000	803
Embankment	556+000	558+650	2.650
Trench	558+650	561+450	2.800

Figure 13. Civil works for sand dune area solution 1b.

- Solution 2. Vertical Alignment high solution. Moving higher the vertical alignment, the viaduct and small viaduct solution has been considered, Figure 14.

For this solution, the detail of the different civil works from km 548+560 to km 558+650 are shown in the following Table 3.

5.2 Economic evaluation

Based on the previous description, it has been evaluated the construction and maintenance cost for the different solutions. The construction and the total cost (construction + maintenance + sand removal) has been calculated in three different scenario: 25, 40 and 100 years.

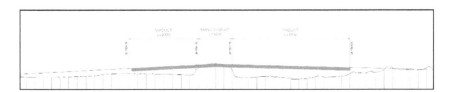

Figure 14. Civil works for solution 2.

Table 3. Number and features of civil works for solution 2.

Description	From chainage	To chainage	Length (m)
Embankment	548+560	552+150	3.590
Viaduct	552+150	554+150	2.000
Small Viaduct	554+150	555+150	1.000
Viaduct	555+150	558+650	3.500
Embankment	558+650	559+650	1.000
Trench	559+650	561+450	1.800

Table 4. Economic comparison of the alternative solutions.

Section: Dunes Area 1. km 548 +560–561 +450	Construction costs SAR	Total cost (Construction + maintenance + sand mitigation for 25 years) SAR	Total cost (Construction + maintenance + sand mitigation for 40 years) SAR	Total cost (Construction + maintenance + sand mitigation for 100 years) SAR
Solution 1a	561.843.089	1.186.508.284	1.558.148.399	3.068.401.375
Solution 1b	474.287.862	1.147.663.226	1.543.966.003	3.187.095.425
Solution 2	925.988.700	1.4530.553.77	1.765.361.383	3.044.090.408

275

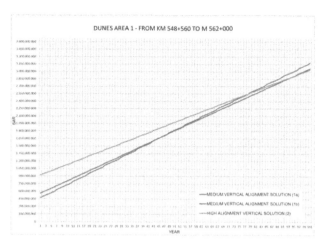

Figure 15. Cost curves for the different analyzed alternatives.

Based on the analysis developed, it is possible to define the solution that combines the technical issue and the most economical one. It is evident that the solution on viaduct (solution 2) is more expensive than the others. The solutions 1a and 1b are comparable in term of cost taking in consideration the total cost comparison from 25 years to 75 years. Based on these considerations, the suggested solution has been the solution 1a with a cut and cover tunnel.

The length of the tunnel is below 2 km and the cost is comparable with the other solution, but the protection of the tunnel against the moving sand dunes is greater than the small viaduct in trench.

6 CONCLUSION

The engineering approach followed in the design phase of the Saudi Landbridge Railway Project has been developed applying the best international practice, even in a non-conventional environment, and with customized socioeconomical assessment for the selection of the most suitable civil work, including tunnel, for the optimization of the railway line.

Specifically in the several dune area interfered along the route, cut & cover tunnels has been identified as the most suitable and reliable solution in order to avoid the risks generated by the sand dunes during the railway operation phase. After the technical and economical assessment, the cut & cover solution has been utilized to cross the 6 dunes area.

Furthermore, the design of a tunnel in this specific environment has included lots of non-conventional topics, such as the variation of the asymmetrical load distribution due to the dunes displacement or the need to prevent the entrance of the sand into the tunnel.

REFERENCES

Final Report – Rev 2, Etihad Rail Stages 2 & 3. RWDI report, (2011).
Edgell S., Arabian Deserts, Springer, (2006).
Fryberger S.G., Dean G. *Dune forms and wind regime*. In: McKee, E. (Ed.), A Study of Global sand Seas. University Press of the Pacific (1979) 141–150.
Fryberger S.G. & Al-Sari A.M. & Clisham T.J., Rizvi S.A.R., Al-Hinai K.G. *Wind sedimentation in the Jafurah sand sea, Saudi Arabia*. Sedimentology 31–3 (1984) 413–431.
Kawamura R. Study on Sand Movement by Wind. Reports of Physical Sciences Research Institute of Tokyo University, Vol. 5, No. 3–4, (1951).
Pye K. & Tsoar H. *Aeolian Sand and Sand Dunes*. Springer, 2008.

*Tunnels and Underground Cities: Engineering and Innovation meet Archaeology,
Architecture and Art, Volume 2: Environment sustainability in
underground construction – Peila, Viggiani & Celestino (Eds)
© 2020 Taylor & Francis Group, London, ISBN 978-0-367-46579-7*

Introduction of biopolymer-based materials for ground hydraulic conductivity control

I. Chang
School of Engineering and Information Technology, University of New South Wales (UNSW), Canberra, ACT, Australia

A.T.P. Tran & G.-C. Cho
Department of Civil and Environmental Engineering, Korea Advanced Institute of Science and Technology (KAIST), Daejeon, Republic of Korea

ABSTRACT: The use of a hydraulic barrier to prevent or constrain the water flow or/and residual contaminant-containing water from leaking, flowing into underground constructions such as tunnels has been addressed in the past decades. There are different types of barrier materials used to improve hydraulic properties such as soil – bentonite, cement-bentonite, soil admixes using bentonite, cement, and asphalt, chemical and other additives mixed with the natural soil. In fact, the hydraulic barrier materials used for tunnels need to work well under the earth pressure and hydrostatic water pressure acting on the tunnel lining. In this study, a linear polysaccharide gellan gum, which has been investigated in the fields of pharmaceutical technology, biomedical applications, and food products, will be used to improve the hydraulic behavior of sand. The advantage of gellan gum biopolymer is its capable of forming hydrocolloid gels when mixed with heated water and limiting water flow through the gel performance in soil hydraulic conductivity control at various depth and pore pressure conditions. A pressurized hydraulic system allows performing various pore water pressure and confinement condition to observe the pore clogging behavior of gellan gum biopolymer treated sands. Furthermore, soil hydraulic conductivity variations due to changes in confinement pressure and pore water pressure will be observed.

1 INTRODUCTION

Recently, the use and application of biopolymer-soil treatment in geotechnical engineering practices is actively investigated and attempted by numbers of research. A considerable amount of literature has been impressed the promising future of the use of biopolymers in practical geotechnical engineering. Biopolymer can enhance the inter-particle cohesion of soil (Lee et al., 2017, Im et al., 2017, Chang and Cho, 2018), therefore, it shows a good performance in soil strengthening (Chang and Cho, 2012, Chang et al., 2016b, Chang et al., 2015b, Khatami and O'Kelly, 2012, Chang et al., 2015a), in dust controlling (Chen et al., 2015, Miękoś et al., 2017, Larson et al., 2010), and anti-desertification (Chang et al., 2015c). Another advantage of the biopolymer is its water holding capacity, therefore, it can lend positive performance in hydraulic reduction (Chang et al., 2016b, Bouazza et al., 2009).

The use of a hydraulic barrier to obstruct the water flow or/and residual contaminant-containing water leaking and flowing into underground constructions such as tunnels has become common method in the past decades. There are different types of hydraulic barrier materials used to improve the hydraulic properties of the ground, such as soil-bentonite, cement-bentonite, soil admixes using bentonite, cement, and asphalt, chemical and other additives mixed with the natural soil (Karol, 2003, Pusch, 2015, Warner, 2004). In fact, the hydraulic barrier materials used for tunnels need to work well under the earth pressure and hydrostatic water pressure acting on the tunnel lining. In this study, the applicability of a linear polysaccharide (gellan gum) as a new admixture to soil for ground hydraulic conductivity control is investigated. The advantage of gellan gum biopolymer is its capable of forming hydrocolloid gels when mixed with heated water and limiting water flow through the gel performance in soil hydraulic conductivity control at various depth and pore pressure conditions.

As the hydraulic barrier is constructed at a certain depth with an appearance of groundwater, hydraulic barrier materials will be subjected to quick and high effective stress and water pressure. A pressurized hydraulic conductivity device, which allows performing various pore water pressure and confinement condition, has been suggested (Chang et al., 2016b). For the advantage, the device was used to perform hydraulic tests on sand and sand/clay mixture to see the effectiveness of gellan gum on hydraulic reduction of different type of soils.

2 MATERIALS AND EXPERIMENTAL PROCESS

2.1 *Biopolymer and soil*

2.1.1 *Soil types*
Jumunjin sand is a standard sand in Korea, which is classified as poorly graded sand (*SP*) according to USCS classification. It has an average particle size of 0.46 mm, specific gravity (G_s) of 2.65, and the coefficient of uniformity (C_u), and the coefficient of gradation (C_c) are found to be 1.39 and 0.76 respectively (Chang et al., 2018, Chang et al., 2017, Chang and Cho, 2018).

A commercial kaolinite – Bintang kaolin, which is classified as *CH* according to the USCS classification is used as an additive to enhance the rheology of gellan gum. It has the specific gravity of 2.7 and average particle size of 44 μm. The clay powder was mixed with jumunjin sand with a ratio of clay to sand at 1:9 to obtain sand-clay mixture.

2.1.2 *Gellan gum*
Gellan gum is a linear polysaccharide produced by the bacterium Pseudomonas elodea, which has been investigated in the fields of pharmaceutical technology, biomedical applications (Osmałek et al., 2014), food industry (Morris et al., 2012, Saha and Bhattacharya, 2010, Imeson, 1992). Moreover, gellan gum applicability to geotechnical engineering such as soil strengthening (Chang and Cho, 2018, Chang et al., 2017, Chang et al., 2015b), hydraulic conductivity control (Chang et al., 2016a). In this study, low acyl gellan gum biopolymer supplied by Sigma Aldrich (CAS No.71010-52-1) has been used.

2.2 *Sample preparation and experiment procedure*

2.2.1 *Sample preparation*
Biopolymer hydrogels were mixed with soils at target biopolymer to soil contents in mass as 0.5% and 1.0%. To allow thorough mixing the initial water content has been set at 33%. Gellan gum powder was first dissolved and hydrated into deionized water heated at 100°C to obtain uniform gellan gum solution. Thereafter, dry soil and heated gellan gum solution were uniformly mixed.

2.2.2 *Experimental procedure*
The hydraulic conductivity of soil was determined by using a pressurized hydraulic conductivity test apparatus (Figure 1). The gellan gum hydrogel – soil mixtures were set into a cylindrical cell which is 9.3 cm in height and 8.0 cm in diameter. At the top and

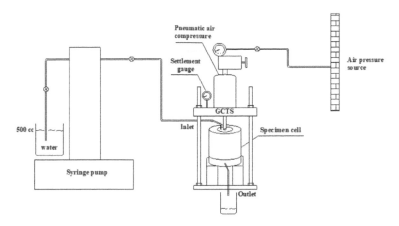

Figure 1. Schematic diagram of hydraulic conductivity setup.

bottom of the specimen, filter papers were placed so that water can evenly distribute within specimen during the experimental process. After the specimen was fully set up and cooled down, confining pressure was then applied to the soil under drained condition using a pneumatic air compressor so that the effective stress acting on the soil should be 100, 200 and 400 kPa. The consolidation process lasted for 24 hours at where vertical strain of soil reached constant. Wet curing was carried out for de-airing and saturation purpose under undrained condition. A constant water pressure of 70 kPa into specimen using a high-pressure precision syringe pump. As the flow rate of water reached zero, the authors assumed that the specimen was fully saturated, and the permeability test was conducted. After drained equilibrium was completed, the permeability of soil was then observed at varying water pressure, which started by 70 kPa. At each confining stress (i.e effective stress), the constant water pressure was increased until the water pressure was significantly higher than effective stress.

The saturated permeability is calculated based on Darcy's law:

$$k = \frac{V \cdot L}{A \cdot h \cdot t} \qquad (1)$$

where V is the collected volume of water, L is the height of soil specimen, A is the area of soil specimen, h is the head difference, and t is the time required to the V volume

3 RESULTS AND ANALYSIS

3.1 *Effect of gellan gum on hydraulic reduction for soils*

The hydraulic conductivities of gellan gum-treated sand and sand/clay mixture are shown in Figure 2 and Figure 3. For untreated soils, average permeability of pure sand and sand/clay mixture are 7.15 x 10-7 m/s and 6.18 x 10-7 m/s, respectively. The presence of kaolinite did not improve the hydraulic reduction of sand. It is due to the low water adsorb-ability of kaolinite particles, which is classified kaolinite as a non-swelling clay mineral (Osacky et al., 2015). Furthermore, water pressure was high enough to even flush clay particle out of the soils.

As the soils were treated by gellan gum, hydraulic conductivity of the soils decreased significantly by at least 10 times. The hydraulic reduction of soils was due to the water absorption ability of gellan gum which shows different mechanism regarding soil types. For pure sand, the gellan gum film coating sand surface (Chang et al., 2016a) adsorbed, held water, which controlled the flow rate of water passing through the sand specimen. For sand/clay mixture,

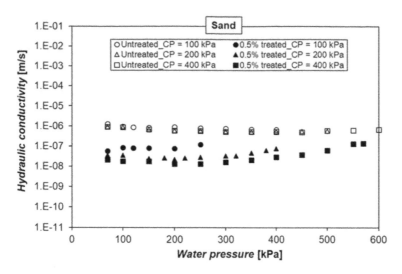

Figure 2. Hydraulic conductivity of gellan-treated sand.

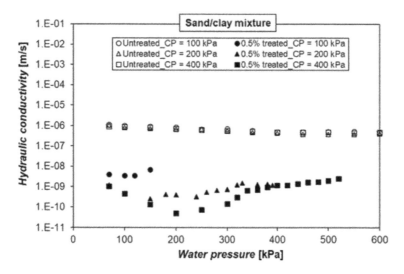

Figure 3. Hydraulic conductivity of gellan-treated sand/clay mixture.

the gellan – kaolinite matrix forming via hydrogen bonding between gellan gum biopolymer and kaolinite (Chang and Cho, 2018) performed a different contact with water molecules. The inter-particle interaction via hydrogen bonding with clay particles of gellan gum could reduce the loss of kaolinite during permeability test (Figure 4). Gellan- kaolinite matrix obstructed water flow via water adsorption of gellan gum and kaolinite, reduced flow speed.

The confinement pressure does not show any effect on the hydraulic conductivity of untreated soils, however, a slight decrease in hydraulic conductivity with confinement pressure can be seen in the case of treated soils (Figure 2 and 3). The confinement pressure arranged soil particles and reduced soil pores during the consolidation process. In other words, a slight difference in dry density (Figure 5) led to the difference in hydraulic reduction.

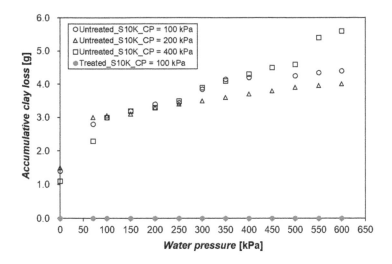

Figure 4. Effect of gellan gum on kaolinite controlling.

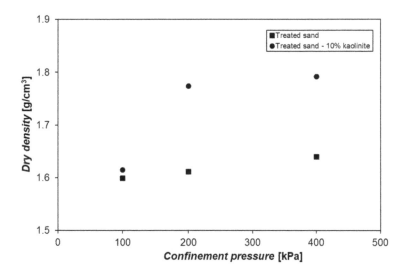

Figure 5. Dry density of soil before hydraulic test.

3.2 *Role of clay in hydraulic reduction effectiveness of gellan gum*

Figures 6 and 7 show the hydraulic conductivity of sand and sand/clay mixture as they were treated at the same condition of gellan gum concentration and confinement pressure. When the soils
were treated with 0.5% gellan gum, sand/clay soil showed lower conductivity compared to sand.
However, 1% gellan gum performed a slight difference in the conductivity of the soils. It is believed
that the ratio of gellan gum and kaolinite exhibited different performance in the hydraulic conductivity reduction. At 0.5% gellan gum concentration, a number of kaolinite and gellan gum produced a strong ion bonding, which is along with water adsorption of gellan gum and kaolinite
particles triggered higher reduction of the flow rate within the soil. The hydraulic conductivity of
sand treated 0.5% gellan reduced by 100 times, which could drop by 1000 times with the presence
of kaolinite. However, when gellan gum concentration of 1.0% was used, the presence of kaolinite
showed inconsiderable effect on the water adsorption of gellan gum, and gellan gum seemed to
play a dominant role in the hydraulic reduction of soil.

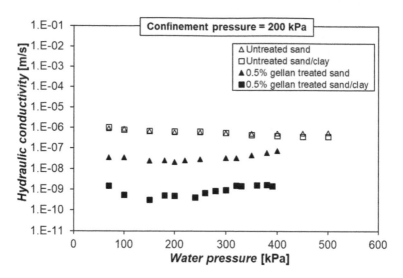

Figure 6.   Role of clay in hydraulic reduction effectiveness of 0.5% gellan gum.

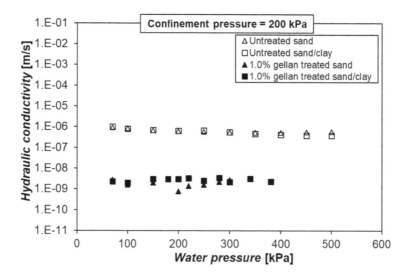

Figure 7.   Role of clay in hydraulic reduction effectiveness of 1.0% gellan gum.

4   CONCLUSION

In conclusion a decrease in the permeability was observed with the addition of gellan gum into the soil regardless of the type of soil used. However, the presence of kaolinite can improve the effectiveness of gellan gum on hydraulic reduction, which depends largely on the number of ion bondings within the gellan-kaolinite matrix. The finding from this study can be suggested as a new hydraulic materials not only for the tunnels but underground constructions in general. For further study, higher gellan concentration should be tested to observe a general trend of hydraulic conductivity of gellan gum treated soils as gellan gum concentration increases.

ACKNOWLEDGEMENTS

This research was supported by a grant (19AWMP-B114119-04) from the Water Management Research Program funded by the Ministry of Land, Infrastructure, and Transport (MOLIT) of the Korean Government; a National Research Foundation of Korea (NRF) grant funded by the Korean Government (MSIP) (No. 2017R1A2B4008635); a grant (19SCIP-B105148-05) from the Construction Technology Research Program funded by the MOLIT of the Korean Government.

REFERENCES

Bouazza, A., Gates, W. & Ranjith, P. 2009. Hydraulic conductivity of biopolymer-treated silty sand. *Géotechnique*, 59, 71–72.

Chang, I. & Cho, G.-C. 2012. Strengthening of Korean residual soil with β-1,3/1,6-glucan biopolymer. *Construction and Building Materials*, 30, 30–35.

Chang, I. & Cho, G.-C. 2018. Shear strength behavior and parameters of microbial gellan gum-treated soils: from sand to clay. *Acta Geotechnica*, 1–15.

Chang, I., Im, J. & Cho, G.-C. 2016a. Geotechnical engineering behaviors of gellan gum biopolymer treated sand. *Canadian Geotechnical Journal*, 53, 1658–1670.

Chang, I., Im, J. & Cho, G.-C. 2016b. Soil-hydraulic conductivity control via a biopolymer treatment-induced bio-clogging effect. *Geotechnical and Structural Engineering Congress 2016*. ASCE.

Chang, I., Im, J., Chung, M.-K. & Cho, G.-C. 2018. Bovine casein as a new soil strengthening binder from diary wastes. *Construction and Building Materials*, 160, 1–9.

Chang, I., Im, J., Lee, S.-W. & Cho, G.-C. 2017. Strength durability of gellan gum biopolymer-treated Korean sand with cyclic wetting and drying. *Construction and Building Materials*, 143, 210–221.

Chang, I., Im, J., Prasidhi, A.K. & Cho, G.-C. 2015a. Effects of Xanthan gum biopolymer on soil strengthening. *Construction and Building Materials*, 74, 65–72.

Chang, I., Prasidhi, A.K., Im, J. & Cho, G.-C. 2015b. Soil strengthening using thermo-gelation biopolymers. *Construction and Building Materials*, 77, 430–438.

Chang, I., Prasidhi, A.K., Im, J., Shin, H.-D. & Cho, G.-C. 2015c. Soil treatment using microbial biopolymers for anti-desertification purposes. *Geoderma*, 253–254, 39–47.

Chen, R., Lee, I. & Zhang, L. 2015. Biopolymer stabilization of mine tailings for dust control. *Journal of Geotechnical and Geoenvironmental Engineering*, 141, 04014100.

Im, J., Tran, A.T.P., Chang, I. & Cho, G.-C. 2017. Dynamic properties of gel-type biopolymer-treated sands evaluated by Resonant Column (RC) tests. *Geomechanics and Engineering*, 12, 815–830.

Imeson, A. 1992. *Thickening and gelling agents for food*, London; New York, Blackie.

Karol, R.H. 2003. *Chemical Grouting and Soil Stabilization*, New York, Marcel Dekker, Inc.

Khatami, H.R. & O'Kelly, B.C. 2012. Improving mechanical properties of sand using biopolymers. *Journal of Geotechnical and Geoenvironmental Engineering*, 139, 1402–1406.

Larson, S., Ballard, J., Griggs, C., Newman, J.K. & Nestler, C. An innovative non-ptroleum Rhizobium Tropici biopolymer salt for soil stabilization. ASME 2010 International Mechanical Engineering Congress and Exposition, 2010 Vancouver, Canada. Volume 5: Energy Systems Analysis, Thermodynamics and Sustainability; NanoEngineering for Energy; Engineering to Address Climate Change, Parts A and B: American Society of Mechanical Engineers, 1279–1284.

Lee, S., Chang, I., Chung, M.-K., Kim, Y. & Kee, J. 2017. Geotechnical shear behavior of xanthan gum biopolymer treated sand from direct shear testing. *Geomechanics and Engineering*, 12, 831–847.

Miękoś, E., Zieliński, M., Kołodziejczyk, K. & Jaksender, M. 2017. Application of industrial and biopolymers waste to stabilise the subsoil of road surfaces. *Road Materials and Pavement Design*, 1–14.

Morris, E.R., Nishinari, K. & Rinaudo, M. 2012. Gelation of gellan – A review. *Food Hydrocolloids*, 28, 373–411.

Osacky, M., Geramian, M., Ivey, D.G., Liu, Q. & Etsell, T.H. 2015. Influence of Nonswelling Clay Minerals (Illite, Kaolinite, and Chlorite) on Nonaqueous Solvent Extraction of Bitumen. *Energy & Fuels*, 29, 4150–4159.

Osmałek, T., Froelich, A. & Tasarek, S. 2014. Application of gellan gum in pharmacy and medicine. *International Journal of Pharmaceutics*, 466, 328–340.

Pusch, R. 2015. *Bentonite clay: environmental properties and applications*, Boca Raton, CRC Press, Taylor & Francis Group, CRC Press is an imprint of the Taylor & Francis Group, an Informa business.

Saha, D. & Bhattacharya, S. 2010. Hydrocolloids as thickening and gelling agents in food: a critical review. *Journal of Food Science Technology*, 47, 587–597.

Warner, J. 2004. *Practical handbook of grouting: soil, rock, and structures*, Hoboken, N.J., John Wiley & Sons.

*Tunnels and Underground Cities: Engineering and Innovation meet Archaeology,
Architecture and Art, Volume 2: Environment sustainability in
underground construction – Peila, Viggiani & Celestino (Eds)*
© 2020 Taylor & Francis Group, London, ISBN 978-0-367-46579-7

Management and use of materials excavated during underground works

L. D'Aloia Schwartzentruber & F. Robert
CETU (Tunnel Study Centre), Bron, France

ABSTRACT: Future transportation infrastructure projects will include significant underground
sections. These projects will result in huge quantities of excavated materials that could not be
entirely deposited on landfill. The use of excavated materials during underground work will contribute to preserve natural resources and to reduce the impacts related to the transport of materials over long distances. Thus, alternative solutions can be found by producing for example new
eco-designed building materials. The owner is responsible for the management of excavated
materials. It is in charge of defining a specific policy and must specify the requirements for reuse
and recycling. To help it, three general scenarios are described and a management method is proposed, from the project definition study stage to the construction phase. The role of each project
stakeholder is detailed. Finally, some of the main aspects are illustrated by case studies.

1 INTRODUCTION

The management of the materials excavated during underground works must be integrated
into the concerns of the owner at the earliest stage of the studies. Nowadays, excavated materials are too often disposed without any real reflection on their potential reuse and recovery.
These practices lead to a loss of valuable natural resources and have strong socio-economic
and environmental impacts. The management of excavated materials must henceforth partici-
pate to the circular economy by proposing the optimization of the use of these natural resources.
Eco-design, industrial and territorial ecology, reuse and recycling are pillars addressed by the
management of excavated materials (Figure 1).

The reuse and recovery of these materials must be properly studied and sufficiently early to
facilitate the access to the resource and hence, to meet societal expectations and regulatory
requirements.

Therefore, in view of the great underground projects for the coming decades in France,
CETU (Tunnel Study Centre) has published an information document in collaboration with
Cerema (Centre d'études et d'expertise sur les risques, l'environnement, la mobilité et l'aménagement). This document addresses first and foremost the project owner to make him aware
of the necessity to enhance the management of natural geological excavated materials during
underground works (CETU 2016) Considerations on the specificities of excavated materials
and on the influence of the excavation method (conventional or mechanized tunneling) are
also dealt with.

Three scenarios are presented depending on whether or not the owner uses the excavated
materials. The main uses are stated. Finally, the information document focusses on the role of
actors throughout the project course, from the definition studies to the construction stage.

This information document and the recommendation of the AFTES working group n°35
(the French tunneling and underground space association) are both complementary documents. The latter provides a more detailed description on the technical and economic aspects
of the management and use of excavated materials (AFTES 2016).

CIRCULAR ECONOMY
3 areas
7 pillars
ADEME

Recycling
(materials and
organic matter)

Extraction/Manufacturing and
Sustainable Supply Chain

Eco-design (Products and
procedures)

WASTE
MANAGEMENT

SUPPLY FROM
ECONOMIC
STAKEHOLDERS

Industrial and
Territorial Ecology

Functional Economy

CONSUMER
DEMAND AND
BEHAVIOUR

Responsible Consumption
- Purchasing
- Collective Consumption
- Use

Extension of Product
Lefespan
- Reuse
- Repair
- Recycle

The circular economy is defined as an "economic system based around of exchange and production methods that, at every stage of the product life cycle (goods and services), aim to increase the efficiency of resource usage and diminish environmental impact, while also improving the wellbeing of individual citizens." The circular economy is based around three areas of action and seven pillars.

Figure 1. The circular economy according to the French Environment and Energy Management Agency (Ademe).

2 PRACTICES AND KEY CHALLENGES

2.1 *Current French context and abroad*

In France, as it was already stated, very few documents and recommendations specifically target natural geological materials excavated during the construction of tunnels. Besides, practices have been so far largely focused on reuse and recovery for pavement structure and railway platform. In addition, excess materials have often found a way out in the balance between excavating including earthwork) and refilling. However, for some urban projects this balance cannot be reached anymore. More over as regards the large volumes nowadays targeted, the identification of other sectors of reuse and recovery such as the production of concrete aggregates became necessary.

In the case of on-site aggregates production by reusing excavated materials, the impacts due to the transportation of brand new aggregates (Green House Gas emissions, crossing of urbanized areas, etc.) are avoided. Natural resources are saved and landfill disposal is minimized. This option is rarely chosen in France, but has been widely used for twenty years in Switzerland (Thalmann et al. 2005, Lafranchi et al. 2016) where changes in regulations accompanied the development of practices and knowledge. However, extensive studies are carried out for major projects as for example the base tunnel of the Lyon-Turin rail link (Colas 2012).

In the case of a tunnel, a quick assessment of the material quantities shows that, in the best case, only 10 to 20% of the excavated materials can be reused for concrete production. Thus, other solutions for reuse and recovery must be found through an economic and territorial analysis. It is indeed possible that some specific materials that we intend to deposit, could bring added-value by supplying new economic sectors. Appropriate characterization and potential pretreatment may be required to comply with industrial specifications. This will led us beyond the simple current characterization that is usually tunneling-oriented but not reuse- and recovery-oriented. Finally, the progress of studies, together with the implementation of dedicated on-site facilities will enable to secure supplies through a relevant formalization of commitment.

For European countries, this approach must comply with the Directive "waste" 2008/98/EC of the European Parliament and Council (19th November 2008), which repeals other Directives.

2.2 Some key figures concerning the main French underground projects

In France, most of the current and future linear infrastructure projects will be built underground. Huge quantities of excavated materials are expected. An average length of about 20km of new tunnels will be built each year, in the next decade.

The future French major transport infrastructure projects, which are partly or entirely underground, are mainly railway, or subway projects. They exhibit significant underground structures and facilities such as tunnels, cut and covers and underground stations. These projects are sometimes urban ones. The dense and restricted environment raises the question of technical feasibility and spatial occupation. The non-urban projects can be also technically constrained by the relief and the crossing of areas difficult to access. Then, underground route makes it possible to overcome the topography, to preserve the living environment on the surface, to ensure the continuity of the green and blue belt network. Hence, the surface can be dedicated to other uses.

These projects exhibit strong territory and mobility issues. They generate large volumes of excavated materials. Estimated volumes from a few major projects currently under study are as follows (cf. Table 1).

2.3 The legal framework

To cope with these large volumes of excavated materials, project owners must define a management approach from the earliest stages till the end of works. This approach must take into account the regulations in force.

In 2008, the European Union parliament introduced a five-step waste hierarchy to its waste legislation Directive 2008/98/EC, which member states must introduce into national waste management laws. Priority order: waste prevention, as the preferred option, is followed by reuse, recycling, recovery including energy recovery and as a last option, safe disposal.

Then, disposal should be used as the last resort whereas it is often the first studied option.

In France, the owner is responsible for the management of excavated materials until their recovery or final disposal (Article L.541-2 of the Environment Code). The owner has to provide administrative files dealing with environmental protection: transit facilities and for instance facilities to produce aggregates (and every other mineral-based product) on site. As a priority, the owner has to plan how to reuse and recover excavated materials. (Article L.541-2-1 of the Environment Code).

The French regulation and law promote circular economy and have set a number of quantitative targets in the fight against waste. For instance, Article L541-1-1 of the Environment Code now includes the following obligations:

Table 1. Main features of major underground projects in France.

Project	Grand Paris Express	The Lyon-Turin link (TELT) Base tunnel (Ambin)	French access tunnels	New railway « Provence Côte d'Azur (LNPCA) »	Lyon Part Dieu Station	CIGEO (ANDRA)
Means of transport	Subway	Rail network		Rail network (priority 1→first step)	Rail network	Nuclear waste storage
Cumulative Length (CL)	180km (90% of CL) + 68 stations	57km	86km (61% of CL)	≈60km (33% of CL)	Underground station	≈270km of galleries (including storage cells)
Excavated Materials	45 Mt (≈17M m3)	16M m3	19M m3	≈15M m3	600 000t	≈10M m3

- To recover 70% of the waste stemming from the building and public works sectors by 2020;
- To reduce the amount of non-hazardous non-inert waste admitted to a storage facility by 30% in 2020 compared to 2010, and by 50% in 2025;
- To reduce by 50% the quantities of non-recyclable manufactured products placed on the market before 2020.

These measurable goals demonstrate the government willingness to put into practice the circular economy principle. As a production model, the circular economy is an alternative to the linear one: "take, make, dispose". A circular economy is a regenerative system in which resource input, waste, emissions, and energy are minimized by slowing, closing, and narrowing energy and material loops. This can be achieved through long-lasting design, maintenance, repair, reuse, recovery and recycling. Regulatory changes are still required in France to develop the concept of by-product introduced by the "Waste" European Directive. Indeed, excavated materials should be regarded as a potential resource without referring anymore to the concept of waste.

3 SPECIFICITIES OF EXCAVATED MATERIALS

3.1 Impact of the construction methods

The nature of the materials excavated during underground works varies according to the geological layers and the construction method.

Rock materials from limestone massif for example, and unconsolidated ones made of sand and clay or any other natural material can be found. The way to manage excavated materials depends, among others, on the mechanical properties of rock and soil, as well as their physical and chemical characteristics. Some materials may lead to practical difficulties in finding specific uses: traces of asbestos, carboniferous shale, gypsum or anhydrite. Polluted soil due to anthropogenic activities can also be encountered within the different surface layers. The soil identified as polluted comes under specific regulation.

The construction method can affect the quality, shape and dimension of excavated materials, and thus, their reuse and recovery potential. Pre-processing operations such as crushing, screening, washing, liming can be required and would modify the cost/benefit ratio from both the economic and environmental points of view. Construction methods belong to either "mechanized" or "conventional" tunneling.

The choice of the construction method depends on the one hand on the geological nature of the ground, and on the other hand on the tunnel length. Table 2 describes possible choices in relation to material nature.

Table 2. Construction methods: application areas according to the material nature (CETU 2016).

	Unconsolidated soil				Rock ground	
	Gravels	Sands	limes	Clays	Tender rock	Hard rock
Conventional tunneling (drill and blast)			Roadheader Machine			
	Roadheader machine + specific treatments					Explosive
Mechanized tunneling (Use of a Tunnel Boring Machine (TBM))		TBM (earth pressure balanced shield)				
					TBM (open shield) + segments	
	TBM (Slurry shield)					TBM (Hard rock + grippers)

A TBM is economically justified for long tunnels (length greater than 3,000 m) whereas conventional tunneling remains more competitive for short ones (length lower than 1,000 m). Both the conventional and mechanized tunneling will compete with each other in between. In practice, a combination of different construction methods will be usually implemented on the same tunnel.

3.2 *Material characterization*

Currently, geotechnical investigations are performed according to several phases to improve the geological model and to choose and better define the construction method. Even if this target remains of the utmost importance, one should take the opportunity to perform additional tests more oriented towards reuse and recovery of excavated materials. This raises a number of important questions. For instance:

- What additional tests should be performed? For what potential use?
- Are they relevant? Are they performed on a representative sample?
- What are the ranges of variation of the material main features?
- ...

4 CURRENT FRENCH RECOMMENDATIONS AND DOCUMENTS

4.1 *The CETU's information document*

4.1.1 *General description*
In collaboration with Cerema, CETU has published an information document aimed at the owner to help him manage the natural geological materials excavated during underground works (construction of a tunnel, a technical gallery, a cut and cover, or even an underground station). The management process has to be implemented from the project definition to the end of works. The information document describes the three possible scenarios. They take into account current regulations, various planning documents, as well as the professional voluntary commitments.

In this information document the following topics are addressed:

- Specificities of excavated materials (depending on the geological nature of the soil and on the construction method);
- Description and analysis of the three possible scenarios (responsibilities of stakeholders involved and administrative procedures to be followed);
- Possible general uses depending on the material nature;
- And finally, the role of the different stakeholders at each stage of the project (from the definition studies to the works) whether it is for a road or a railway project.

4.1.2 *The 3 different scenarios*
Following the quantitative and qualitative analyses of the natural geological materials performed during the definition studies of the project, three management scenarios are possible:

- Scenario 1: the construction site exhibits a deficit material balance. The excavated materials will be reused on the same site;
- Scenario 2: the excavated materials will be sent to another construction site of the same owner;
- Scenario 3: the owner doesn't use the materials excavated on the construction site.

Figure 2 shows the three possible scenarios.

The owner may implement a combination of these three scenarios depending on the properties and use of materials.

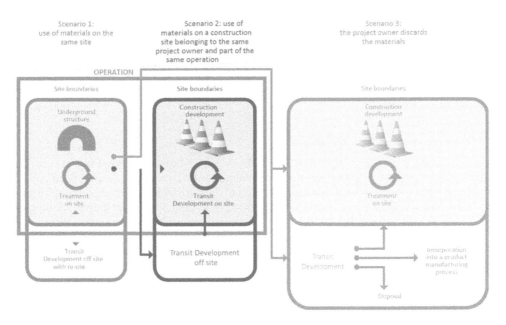

Figure 2.   The three possible scenarios for the management of excavated materials (CETU 2016).

In the case of scenario 1 or 2, natural geological materials excavated during underground works are not referred to as "waste" within the meaning of Article L.541-1-1 of the Environment Code. The owner does not discard them. However, he is responsible for their management.

For both these scenarios, excavated materials are used on a site of the owner after a pre-treatment on the site itself or on a dedicated one. In the latter case, a contractual document will be signed by the owner and the service provider to ensure that:

– The materials entering and leaving the facilities comply with the specifications set by decree;
– Outgoing materials fulfill all the qualities required by the owner for the identified use;
– Material traceability procedures are implemented. In particular, excavated materials are not mixed with other materials and all the excavated materials return to the owner after pretreatment.

In the case of scenario 3, the owner discards excavated materials that are referred to as "waste". Thus, he becomes responsible for their management until final recovery or disposal, even if they are transferred for treatment to a third party (Article L.541-2 of the Environment Code). He must check that the body to which he entrusts the excavated materials is an approved organization to take over waste.

The excavated materials management will depend on their characteristics (mechanical, physicochemical and environmental properties) and on on-site or off-site pretreatment possibilities (available space, compliance with the urban planning code, administrative records, etc.). The project owner will rely on recommendations for resource preservation and waste management and will have to define his own strategy to comply with the waste hierarchy imposed by the European regulation.

The choice of specific uses of excavated materials will be also guided by a technico-economic analysis. The main aspects of this approach are presented in the recommendations of the French working group GT35 (AFTES 2016).

4.2   The French Recommendations of the GT35

In France, the working group n°35 (GT35) of the French tunneling and underground space association (AFTES) was the first body to work on the management and use of excavated materials.

A recommendation dedicated to underground works and more especially in rocky terrain, was published in 2007. This recommendation was revised in 2016 to take into account works in soft materials, to reflect current thinking and to address economic and regulatory aspects.

The CETU information document is complementary to this recommendation. It raises the awareness of the actors of their responsibilities and the role they have to play at each stage of the study and construction process.

4.3 Other recommendations and their limits

In France, a new working group led by UMTM (Union of Earth and Sea Trades) is currently preparing recommendations dealing with the management of excavated soils in road construction and in linear transport-infrastructure projects. This guide is expected to be published in early 2019. The working group is composed of more than 20 members including building companies, design and supervision offices, owners, professional associations and unions, public bodies . . .

Four Subgroups are working on both geological and geotechnical aspects, environmental ones, legal framework and project management. CETU and AFTES have recently joined the working group as far as excavated materials during underground works (construction of tunnel, cut and cover, gallery, underground station . . .) will be included in the scope of this guide. However, from a quality and variability point of view, considering in the same manner excavated soils from earthworks and materials excavated during the construction of tunnels can be put into question.

Other French recommendations and technical documents have been published earlier. Their field of application may differ from case to case: road construction or development plan, as well as covered topics: environment, sanitary risk, geology, geotechnics, management... (SETRA 2011, Blanc et al. 2012). The main objective of the most recent guides is to tend to harmonize specifications and practices, to account for the recent evolution of the regulation framework and to foster the reuse and valorization of excavated materials. However, none of them except the GT35 recommendations and the CETU information document actually addresses the specificities of the management of excavated materials during underground works.

5 SKETCHING THE BROAD OUTLINES OF A SUSTAINABLE MANAGEMENT OF EXCAVATED MATERIALS

In France, there are very few past experiences of ambitious and successful management of excavated materials on underground construction sites. There is also a lack of methodology and consequently a reluctance to induce changes in practices. To comply with the waste management hierarchy established by the European Directive 2008/98/EC, reuse and recovery should be preferred over the simple disposal of waste materials.

European countries like Switzerland and Italy for example, have acquired valuable knowledge and have gained experience in the field of reuse and recovery of excavated materials. Besides, one can notice that most of the time, the regulatory framework of these countries significantly changed to favor reuse and recovery.

CETU initiates a research program on the management of excavated materials during underground works. The first step consisted in publishing an information document to increase awareness about the management of excavated materials and to make sure that they will be considered as a natural resource (CETU 2016). The second one will consist in developing a methodology to support the owner in applying a streamlined approach in the management of excavated materials. CETU also contributed to the French recommendations of the working group n°35 of AFTES, especially on the regulation theme (AFTES 2016).

Both these documents provide information on the regulation framework, on technical and economic aspects. However, even if international experiences have been also widely shared

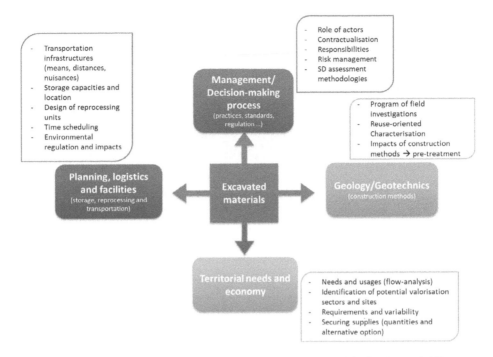

Figure 3.   Schematic description of the management of excavated materials (by CETU 2018).

within the profession of underground works, this is not currently sufficient. Additional elements are required to put into practice the recommendations and to make sure that all the actors feel concerned by this issue and develop sufficient skills.

Sketching the broad outlines of a sustainable management of excavated materials requires tackling the following four main topics:

– The knowledge of excavated materials (geological context and geotechnical characterization). The impacts of the construction methods should be accounted for,
– The analysis of territorial needs and economic considerations,
– Planning, technical facilities and logistics (including transportation),
– Management and decision making process.

Each of these topics is described in Figure 3 and is illustrated on the case study (cf. section 6.).

6   ILLUSTRATION

The management of excavated materials is illustrated bellow on the basis of the French railway project referred to as LNPCA (Ligne Nouvelle Provence Côte d'Azur) between Marseille and Nice in the south of the country. This project aims to improve commuting and in the long run to de-bottleneck the link between the two large metropolises, to encourage intermodality and finally to adapt high speed to the territorial constraints. The future LNPCA will be realized in three stages or "priorities" (Figure 4):

– Priority 1 concerns the node of Marseille and the Azurean one (before 2030),
– Priority 2 is dedicated to the improvement of the sections Aubagne/Toulon and Est Var Signe (between 2030 and 2050),
– Priority 3 will treat the sections Toulon/East Var and Nice/Italy (after 2050).

Only the study of Priority 1 will be addressed by this illustration.

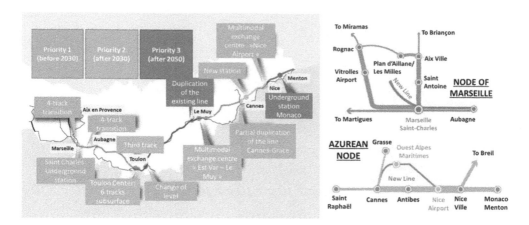

Figure 4. The 3 priorities of LNPCA (on the left) - The node of Marseille and the Azuean one (on the right) (Priority 1).

6.1 The expected materials

The knowledge of expected materials is essentially based on two test campaigns (geological reconnaissance), that have been performed during the preliminary stage, prior to the enquiry preceding the Declaration of Public Utility (DUP in French). Field visits (quarries) and rock outcrops analysis enabled to collect more details. Expected materials volumes have been then estimated for each variant, depending on the length of tunnels, their depth and the dimensions of the cross section determined by the speed of trains. Several data are summarized in Table 3. The maximal dimensions of blocks will depend on the construction method (600mm for conventional tunneling and 40 to 150mm for mechanized tunneling depending on the type of TBM).

6.2 Potential reuse and recovery

Potential reuse and recovery options have been proposed on the basis of past experiences (in France and abroad). This has to be enriched by a territorial analysis: quarries, future construction sites, industrial activities, infrastructure network... A material sheet has been established for each material category. Details about location, geological investigation, drilling campaign and main features have been gathered and summarized. Finally, possible and basic reuse and recovery options are proposed according to the GTR classification (pavement structure) (GTR 2000) and the recent one suggested by the working group GT35. In this case, additional tests have to be performed. Furthermore, reuse as aggregates for concrete is also considered. In the case of LNPCA, materials are as varied as possible in the following categories: clays, limestone, sand and gravel, magmatic rocks. For these categories, suitable recovery and/or reuse options are mentioned in the table of Figure 5.

Table 3. Geological reconnaissance, tests performed and expected volumes of materials.

	Phase 1		Phase 2		Excavated Mat.
	Nb of boreholes	Nb of tests	Nb of boreholes	Nb of tests	Volume (m3) x1000
Marseille node: North area East area	11	47	11	47	66 to 884 416 to 864
Azurean node West area East area	25	115	30	173	654 to 1141 1078 to 1528
Total	36	162	41	220	

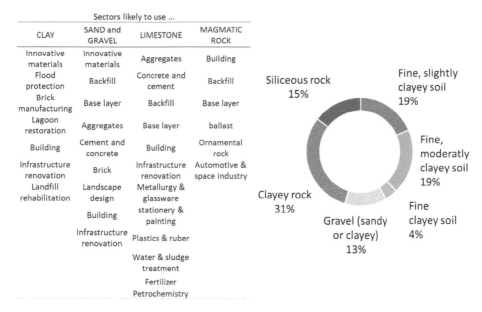

| | Sectors likely to use ... | | |
CLAY	SAND and GRAVEL	LIMESTONE	MAGMATIC ROCK
Innovative materials	Innovative materials	Aggregates	Building
Flood protection	Backfill	Concrete and cement	Backfill
Brick manufacturing	Base layer	Backfill	Base layer
Lagoon restoration	Aggregates	Base layer	ballast
Building	Cement and concrete	Building	Ornamental rock
Infrastructure renovation	Brick	Infrastructure renovation	Automotive & space industry
Landfill rehabilitation	Landscape design	Metallurgy & glassware	
	Building	stationery & painting	
	Infrastructure renovation	Plastics & ruber	
		Water & sludge treatment	
		Fertilizer	
		Petrochemistry	

Figure 5. Possible reuse and/or recovery according to the material category (on the left) - Material distribution for a given variant at the depth of tunnels (on the right) (Thuet 2016).

7 CONCLUSION AND PERSPECTIVES

The knowledge of the materials will be refined with the progress of the project. This will enable to quantify uncertainties and define variability in the characteristics. In addition, the study of the territorial economy will enable to better identify valorization sectors. Taking into account existing infrastructures and the need for new ones, is also necessary. Lastly, reuse and recovery-oriented materiel characterization should complete common geotechnical tests usually performed on boreholes.

REFERENCES

AFTES 2016. Recommandation de l'AFTES. Recommandation sur la Gestion et l'emploi des matériaux excavés. *"GT35R1F2" (2nd Ed). 2016: 80.*
Blanc, C., Lefevre, F., Boissard, G., Scamps, M., Hazebrouck, B. 2012 - Guide de réutilisation hors site des terres excavées en technique routière et dans des projets d'aménagement. *Guide BRGM, February 2012: 53.*
CETU 2016. Matériaux géologiques naturels excavés en travaux souterrains – Spécificités, scénarios de gestion et rôle des acteurs. *Document d'information CETU, March 2016: 31.*
Colas J. 2012. Étude de la valorisation des déblais de chantiers de tunnels riches en sulfates en granulats à béton. *PhD thesis, University Paris Est, December 2012: 282.*
GTR 2000. Réalisation des remblais et des couches de forme. Fascicule 1: Principes généraux. *Guide Technique SETRA/LCPC. July 2000 (2nd ed.): 211.*
Lafranchi, P., Catelli, E., Petitat, M., Vicentini, P. 2016. Spoil Management at AlpTransit Ceneri Base Tunnel - Key elements for a successful natural resource management. *Proceedings of the World Tunnel Congress: "WTC2016", San Francisco, 22–28 April 2016: 10.*
SETRA 2011. Acceptabilité environnementale de matériaux alternatifs en technique routière. Evaluation environnementale. *Guide SETRA, March 2011: 32.*
Thalmann, C., Caron, C., Brino, L., Burdin, J. 2005. Gestion et Valorisation des Matériaux d'Excavation de Tunnels - Analyse comparative de 3 grands projets: St. Gothard, Loetschberg, Maurienne-Ambin. *Proceedings of the International Symposium AFTES, Chambéry, 10–12 October 2005: 12.*
Thuet, L. 2016. Ligne Nouvelle Provence Côte d'Azur: Quel avenir pour les matériaux d'excavation ? Analyse stratégique et technique. *Mémoire de MASTER 1 Géologie appliquée Université de Franche Comté – UFR Sciences et Techniques de Besançon. September 2016: 41.*

Tunnels and Underground Cities: Engineering and Innovation meet Archaeology,
Architecture and Art, Volume 2: Environment sustainability in
underground construction – Peila, Viggiani & Celestino (Eds)
© 2020 Taylor & Francis Group, London, ISBN 978-0-367-46579-7

Tunneling projects: A focus on renaturalization

G. Dajelli
Italferr S.p.A, Rome, Italy

ABSTRACT: Huge soil's quantity are available in the underground infrastructures construction. A lot of soil producted can be used, as morphological modeling of quarries and requalification of disused sites are, through appropriate renaturalization works. Italferr Spa projects are focused on sustainability. The works related to the reuse and remodeling of exceeding soil materials coming from tunnels digging and the environmental restore projects take care about the need to guarantee the involved areas best insertion in the landscape, also properly considering the involved area environmental and landscape features in order to not perceived them as landscape waste. The deployment phase involves laying soil through the progressive levelling of the different site's layers, until the project goal and the expected morphology to. The main aim of this operation is to let start the slow evolution towards the climax without the need for subsequent actions.

1 INTRODUCTION

Italferr's Arcisate-Stabio Railway Line Upgrade Project is located in Italy, near Como Lake (Lombardy Region). The Project involves two renaturalization procedures, as described here within. The left over soil from the excavation foreseen in the Project has been used to:

1. rehabilitate the Femar quarry in Viggiù Municipality
2. morphologically remodel the deposit site CSFB02, in Arcisate Municipality.

The environmental re-naturalization design activities required a team of Key Experts with specific geotechnical and environmental skills, in order to be able to classify the project as a "methodological" project for the re-naturalization operations using material derived from underground railway works.

The assessment and identification of the optimal areas for the re-naturalization project was led by two main criteria:

– appropriate size and location (closeness to the location of the Project);
– geological/environmental compatibility with the excavated material The analysis that have steered the design activities have been based on a series of steps aimed at finding the correct site re-naturalization process. The first step has been the assessment of the geographical, climatic and phyto-vegetational profile of the prospect locations taken into account.

Figure 1.   Location of identified areas: deposit site CSFB02 in the municipality of Arcisate (blue boundary); Cava Femar area in the municipality of Viggiù (red boundary).

2   FEATURES OF THE AREA

The Project is located in the Northern part of Italy, in the Lombardy Region close to the Como Lake, in the Verbano, Ceresio and Lario Lakes whereabouts. The area is protected by the Alps on its North side (region of the great pre-Alpine lakes on the southern Alpine slopes), and it is subject to the direct influence of the winds stemming from the Po River Valley, which in turn get asaffected by the presence of the lakes. Said set up gives rise to the climate condition defined as 'Insubric temperate, sub-continental'.

The rain gauge regime may be defined as pre-Alpine,with heaviest precipitation in Spring (May), in Autumn (October and November). Extreme peaks of summer heat and winter cold which might be expected get instead mitigated by the presence of the lakes. Winters are usually dry and sunny. The lowest number of rainy days is, in fact, normally recorded in the months of January and February, when low air pressure lingers over the Mediterranean, and the region is protected by the Alps from the Northern cold winds.

The climatic trend, as assessed by Varese's measuring station shows a "sub-litoraneous" climate, characterised by a total absence of arid or sub-arid summer periods, and by relatively high rainfall (over 1,000 mm per year), with two equinoctial precipitation Maxima (in Spring and in Autumn) and two minima (in Summer and in Winter).

The average annual temperature at low altitudes is close to 12°C. The average temperature variation every 100 m of increase in altitude shows a decrease around 0.5°C in Winter, and 0.7°C in Summer.

The number of sunshine hours is proportional to cloud coverage, and it is closely linked to the evolution of the horizon. The average annual cloud coverage is about 50–60%, showing the highest values that can be assessed within the Alpine Region. Sunshine hours show an average total 55% of the highest possible Amount.

The area is mainly located within the Alps' foothills . It includes moraine hills and low marly-arenaceous mountains.

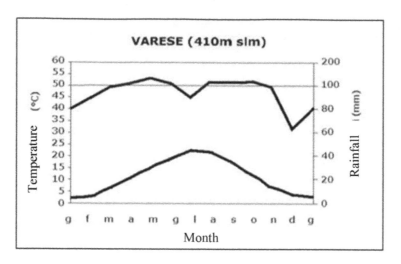

Figure 2.   Climatic trend – Varese.

This pre-Alpine area features the presence of broadleaf forests, that could potentially cover the entire mountain system. Truly, this pre-Alpine forest formations appears to be highly fragmented, consistently with having been often replaced by farmland or by communities of false acacia trees. . The hornbeam has found its optimal habitat here. It is found mixed with durmast and common oak, and together they establish hillside oak-hornbeam communities that often overlap the chestnut communities and the false acacia communities. in this area, the pre-Alpine region is limited to a narrow and discontinuous strip North of the Po river plain.

The Forestry Plan for the Mountain Community of Piambello and the actual Forest Types of the Lombardy Region that assess the vegetation in the area. In compliance with said rules and guidelines, the design has thus assessed the arboreal typological units that can be planted. that is, Field surveys conducted in the adjoining areas have complemented the data set on which the design has been based.

The possible arboreal typological units thus defined are:

– oak communities containing common oak and/or durmast in the higher lands
– pure False acacia communities
– mixed False acacia communities.

The field surveys show that *Castanea sativa, Fraxinus excelsior* can be found next to the two types of oak one and that *Fraxinus excelsior, Robinia pseudacacia* (an exotic species the distribution of which is due to significant anthropic action) are always found within the vegetation-dominated surfaces, along with *Carpinus betulus* (in an accessory function, located especially in areas next to micro-basins or at the bottom of small crevices cutting into the upland surface), *Betula pendula, Populus tremula, Acer pseudoplatanus, Tilia cordata* and *Prunus avium.*, even though in an ancillary function.

The undergrowth consists primarily of *Sambucus nigra, Corylus avellana*, and secondarily of *Euonymus europaeus, Viburnum opulus, Crataegus monogyna, Cornus sanguinea, Ligustrum vulgare* and *Cytisus scoparius*. The grass surface consists of *Vinca minor*, the creeping epiphyte *Hedera helix* and *Convallaria majalis* along with minority species such as *Vaccinium myrtillus, Pteridium aquilinum, Agrostis tenuis, Potentilla erecta, Humulus lupulus, Rubus sp.* and *Artemisia vulgaris*.

The typology is anyhow typical of the *Quercion robori-petraeae* habitat.

3   ENVIRONMENTAL REHABILITATION OF THE FEMAR QUARRY (MUNICIPALITY OF VIGGIÙ)

One ot the two areas chosen as subject to environmental rehabilitation activities, within Arcisate-Stabio Railway Line Upgrade Project, is the Femar Quarry area. The area is located between Verbano Lake and Ceresio Lake, in the Lombard Pre-Alps, about 2.2 km south-east of the town of Arcisate, and about 2.7 km south-west of the town of Viggiù.. The area is located in the strip of hills within Varese upper-Province, at altitudes ranging between 341 and 353 m a.s.l., with a prevailingly south-eastern exposure.

The surrounding area appears to be fairly urbanised (urban and industrial areas alike): the pressure on the forest is evident.

The privately-owned 600,000 sqm Femar Quarry had once been used as a quarry, and then as a backfill deposit. Its geometric conformation is a direct consequence of the extraction methodonce used. . Indeed, the area was used for the extraction of sand and gravel. At a later time, it had beenbackfilled with inert material from earth-moving and excavation works, thus leading to the partial reconstruction of the original slope.

The quarry bed is located at an altitude of about 340 m a.s.l., while the top of the rock wall, once the quarry face, is located at about 400 m a.s.l..

The site northern face was surrounded on all sides by a fairly steep slope, about 60 m tall w. r.t. ground level.

Figure 3.   Project plan – Femar Quarry.

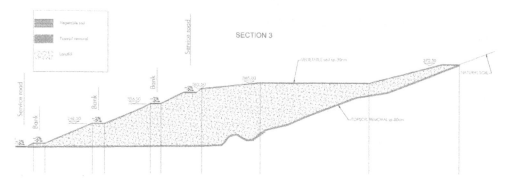

SECTION 3

Figure 4. Section – Femar Quarry.

Figure 5. Ante operam landscape.

The project foresees to use the soil made available by the excavations performed within the "Arcisate-Stabio" railway works to refill the quarry, thus morphologically remodelling it.

The environmental restoration design envisages filling the quarry by means of subsequent layers of soil, up to the ground level, located as said at about 400 m a.s.l.

Once the filling is completed, the design prescribes to profile the lain soil, in order to create a slope with a 24–26° inclination, till rejoining the ground level. To this extent, the design foresees a sequence of about 10 m high tall steps.. The total amount of soil that can thus be disposed of is about 400,000 m^3.

Finally, a 30 cm thick topsoil layer is disposed, to improve the landscape and to restore the vegetation cover.

Figure 6.   Post operam landscape.

4   MORPHOLOGICAL REMODELLING OF THE CSFB02 DEPOSIT SITE (MUNICIPALITY OF ARCISATE)

The second area chosen as subject to environmental rehabilitation activities, within Arcisate-Stabio Railway Line Upgrade Project, is the CSFB02 deposit site.

The area is located in the Lombard Pre-Alps, in the strip of hills within Varese upper province, in the portion of land entering the Ceresio valley.

It is located about 1.5 km south-east of the town of Arcisate,, at an altitude of about 355 m a.s.l.. In the 430,000 sm area, south-eastern exposure prevails, with an inclination of about 20 degrees. The surrounding is fairly urbanised (urban and industrial areas) overall, and the pressure on the forest is evident.

The area is at an average altitude of about 340 m a.s.l., and the surface concerned by the requalification design amounts to out 40,000 m^2.

The requalification design involves the reclamation of the relatively flat worksite area, by filling it with inert material made available by the excavations performed within the "Arcisate-Stabio" railway works, in order to create a natural profile guaranteeing adequate inclination is a safe and stable.

The soil to be used for the refilling, amounts to about 350,000 m^3, including a 30,000 m^3 topsoil layer to be used to landscape the area.

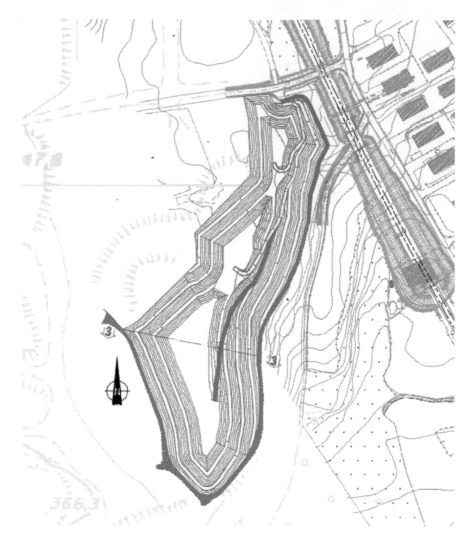

Figure 7.   Project plan – CSFB02.

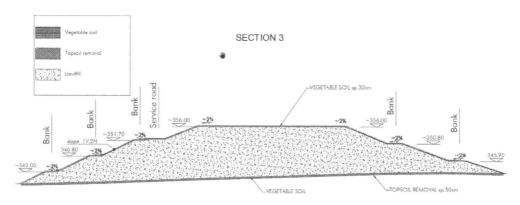

Figure 8.   CSFB02, Section.

Figure 9.   CSFB02 deposit site, ante operam

Figure 10.   CSFB02 deposit site, post operam.

5 GOALS AND AIMS OF THE RE-NATURALIZATION PROJECT

The main goal of the re-naturalization operation is to trigger a slow and gradual natural evolution process, without the need of subsequent further intervention. The final aim of the operation is to attain in a reasonably short time the re-naturalization of the concerned areas, by means of the reconstruction of a natural environment, a habitat that can host the largest amount of plant species. The environmental rehabilitation project takes into account the need to assure the harmonious blending in of the restored areas and the surrounding environmental and landscape features. Low maintenance costs, leading to the optimization of Project life-cycle costs, also led the design choices.

Also, the re-naturalization process has been designed in order to restore the original plant population, peculiar to the involved sites, as well as to reconstruct spatial continuity with adjacent habitats: the final aim of the project has been, in short, to restore the ecological continuity of the involved sites. Indeed and to this regard, the tendency is to reconstruct nuclei of natural vegetation in anthropized areas. The valence of said micro-environments is twofold: landscape-, as well as ecological-wise, as they favour the evolution of local biodiversity.

Figure 11.   State of landscape after the realization of the renaturalization projects.

6 CONCLUSIONS

The first step in a re-naturalization project is the assessment of the he species that spontaneously grow in the area, their organization into communities, and their evolution.

The assessment of the events that, in subsequent steps of time, have affected the vegetation of a specific territory is a *conditio sine qua non* for the correct design.

The approach adopted in assessing and defining the optimal association of vegetal species enables to reconstruct the vegetation communities that, within a given homogeneous area, are conducive to a specific mature stage.

This is a main phase, in that in environmental rehabilitation project plant species that are similar, in floristic composition and structure, to the successional pioneer stages must be chosen and put in place. Thus, the dynamic process of recovery towards the potential natural vegetation typical of the area is guaranteed.

Floristic, vegetational and syndynamical coherence enables to restore the landscape-related values of the area.

In both sites chosen for the morphological reconfiguration by means of the conferment of the available soil, the project envisaged the planting of a new wooded ecosystem .A thick strip of covering shrubs is foreseen on the edges of the newly planted mixed forest, in order to form a border capable of creating a protective sector for the forest as well as an area that enviromental restore projects enables the recreation and the stable presence biodiversity, in compliance with the rules of ecosystem transition areas. The higher production of edible fruits attractive to the wild fauna living in the shrub area boosts the function of the forest as support towards the increase of fauna biodiversity

A quincunx pattern of plantation reduces the artificial appearance of the newly planted patches of wood, especially in the first few years, along with the insertion of small clearings, and of diverse plantation patterns in the distribution of tree and shrub species. A naturally shaped structure over time is thus guaranteed. High density planting is a topsoil restoration strategy that falls within the scope of a method commonly applied to natural areas, allowing to achieve immediate colonization of the land as well as the natural growth of the forest according to natural selection processes. Many of the individual plants have been selected within the local environment, in order to oblige to the natural behaviour that favours plants that have succeeded in adapting quickly to the local edaphic conditions. In this sense, the death of some individuals is not to be read a sign of failure of the plantation, but rather as the expected response that will diminish the density of the plantation and will ensure that the plants that have rooted properly live on. Plants with greater vegetative vigour will be favoured.

Tunnels and Underground Cities: Engineering and Innovation meet Archaeology,
Architecture and Art, Volume 2: Environment sustainability in
underground construction – Peila, Viggiani & Celestino (Eds)
© 2020 Taylor & Francis Group, London, ISBN 978-0-367-46579-7

Management of the muck produced by EPB TBM tunnel excavation: Environmental aspects and consistency control for efficient handling

E. Dal Negro, A. Boscaro, A. Picchio & E. Barbero
Underground Technology Team Mapei Spa, Italy

ABSTRACT: A relevant aspect of TBM excavation with EPB technology, is the management and final disposal of the muck produced. The presence of standard foaming agents may cause the need to dispose the muck as a waste instead of by-product. MAPEI has developed a new line of products (POLYFOAMER ECO line) fully biodegradable, with the main goal to reduce the environmental impact of the muck. Another tough aspect is the muck handling when the excavation procedures or the presence of groundwater cause a too fluid muck. Even though a common solution is the use of lime, this practice may bring to some direct environmental impacts, especially in urban areas. Specific products such as MAPEDRILL SV have been developed to achieve the proper consistency of the excavated soil, without affecting the chemical characteristics of the muck. The paper describes the differences between the new line of products compared with the standard ones.

1 INTRODUCTION

During tunnelling with EPB-TBMs huge amount of underground soil is removed for the tunnel construction and the addition of special additives as foam and/or polymers is required to enhance the soil properties modifying its characteristics with the final aim to guarantee the TBM advance and the final realization of the civil work. This operation is called "soil conditioning" and the main chemical products used are the foaming agents, which are made by several chemical ingredients: surfactants, polymers, solvents, etc.

According to the latest environmental approach, a deep analysis of the biodegradability and toxicity of the foaming agents available on the market is strictly necessary to manage a TBM project selecting the optimal soil conditioning solution. Within the TBM excavation process, a huge amount of in-situ ground is excavated and mixed with chemical products obtaining a uniform material usually called "muck". As result of the excavation and the conditioning process, the final "muck" must have specific characteristics to meet both the technical and environmental requirements of the project and suitable to be re-use as by-product to avoid very high extra costs.

In some cases, huge amount of groundwater or wrong soil conditioning process can occur and consequently the muck consistency cannot be proper for a correct disposal: too liquid muck for an easy handling and transport. An additional treatment is usually made adding lime to the muck to modify its viscosity, but this affects the characteristics of the muck increasing its environmental impact and changing the persistency of the foaming agent used for the soil conditioning process. Furthermore, this operation has some risks for the workers and it has several other impacts (i.e. dust) especially in urban areas. A solution with new polymers has been achieved to keep optimal the environmental characteristics of the muck modifying its consistency.

The correct soil conditioning and muck handling are of paramount importance for the success of a TBM project and nowadays the tunnelling industry demands chemical products able to guarantee high standard performance with as low as possible environmental impact.

In accordance with this approach to a TBM project, the R&D Department of the MAPEI group has developed new products (POLYFOMER ECO line products for soil conditioning and MAPEDRILL SV for spoil treatment) and technologies for the mechanized tunnelling market with the goal to keep high technical standards and to provide innovative "environmental-friendly" solutions.

2 SOIL CONDITIONING

In the last decades, the EPB-TBM has become more and more common as excavation technique for a tunnel construction due to the several advantages it has compared to the traditional tunneling techniques (Drill&Blast, Road headers, etc.) and to the big improvements of the chemical products used for soil conditioning. The chemical products for soil conditioning developed by MAPEI and other chemical companies allowed the use of EPB-TBMs in much more kind of ground achieving better TBM advance, easier managing of difficult geological conditions and an overall cost reduction.

Among soil conditioning products the most common and used are the foaming agents. They are used to generate foam that is injected continuously towards the tunnel face, into the mixing chamber and/or along the screw conveyor during TBM excavation. The properties of the foam depend on the chemical composition of the foaming agents and their main components:

- Surfactants: principal ingredients, with different chemical nature (cationic, anionic, etc.)
- Polymers: enhance the technical characteristics of the foam. Both synthetic and vegetal types.
- Solvents: available in different compositions (glycols, alcohol, dichloromethane, etc.).

Each ingredient of the foaming agent has its own chemical composition and affect the final performances of the product both technically and environmentally.

The improving of the chemical products used in the TBM industry made a step ahead in the last years: the technical performance of the conditioning products which allowed a worldwide development of EPB-TBM have been reached by MAPEI new generation of foaming agent (POLYFOAMER ECO line) which have also high biodegradation rate and very low toxicity towards watery and land organisms.

For a well-practice design of a TBM project, the selection of the most suitable soil conditioning products should be carried out in accordance to the technical requests demanded from the geotechnical and hydrogeological conditions and to comply the environmental regulations of the area where the project is located, the muck consistency requests for a correct disposal and all parameters required for the spoil re-use (Figure 1).

The result of soil conditioning with POLYFOAMER ECO line products is a muck with the required technical characteristics for the successful TBM excavation and in the meanwhile, the environmental parameters for a correct disposal and/or to re-use the muck as by-product can be complied.

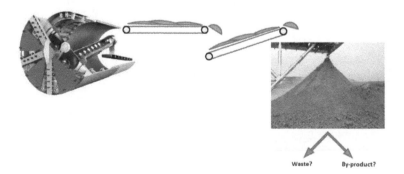

Figure 1.   Scheme for the TBM excavation process and muck disposal.

2.1 Technical characteristics of soil conditioning

The technical aspects related to soil conditioning are evaluated since the analysis of the projects documents as Geotechnical Report, Hydrogeological information about project area, geological profile, etc. Utilizing this information, combined with job-site experiences, a starting point regarding values for each conditioning parameter is reached. When representative soil samples are available, laboratory tests are very useful to have a preliminary idea of most suitable soil conditioning products and their parameters (Figure 2 and Figure 3) for the further application on TBM.

Technical cooperation with chemical supplier and Contractor is finally the most profitable manner to optimize the soil conditioning process from a technical and economical point of view directly during the TBM excavation.

2.2 Environmental aspect of soil conditioning

Foaming agents available in the TBM industry differ significantly not only from a technical point of view, but also from their environmental characteristics which depend on the product biodegradability (important to describe the persistence of a foaming agent along time) and its toxicity to terrestrial and watery organisms. Both biodegradability and eco-toxicity of the soil conditioning products ought to be evaluated and analysed to have a complete figure of chemical products environmental impact. All these environmental properties can be quantified with tests following international guidelines and standards (OECD, 2000 and OECD, 2005) and it is recommended and valuable for a TBM project to quantify the products impact from independent and accredited institutes or laboratories.

For more and more TBM projects the environmental impact of the "muck" is becoming a target for the selection of the conditioning agents and so the biodegradability and eco-toxicity of the chemical products must be well known before the TBM start.

The most complete analysis to investigate the environmental features of a chemical product include toxicity tests for several organisms (mammals, fishes, algaes and daphnias) and evaluation of biodegradability curve:

- Oral and cutaneous toxicity towards mammals.
- Toxicity towards fishes algaes and watery organisms.
- Biodegradability curve: Measured in the first 28 days, according to OECD 301 method (OECD, 2000).

The eco-toxicological data represent the amount of product required to make "effect" to the 50% of organisms' population used for the tests (mammals, fishes, daphnias, algaes, etc.). According on the different regulations "to make effect" means to block the growth of the organisms, stop their reproduction, etc.

Figure 2–3. Samples of the same soil conditioned with standard foaming agent and with new generation MAPEI foaming agent (POLYFOAMER ECO line products).

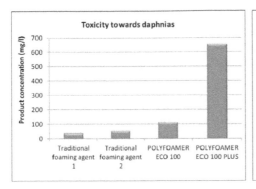
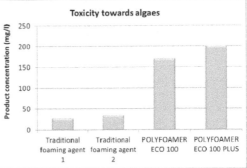

Figure 4–5. Comparison between standard foaming agents and new MAPEI POLYFOAMER ECO line products: toxicity towards watery organism and algaes.

If compared with traditional foaming agents, the eco-toxicity of the new POLYFOAMER ECO line products is extremely lower (Figure 4 and Figure 5).

The biodegradability of the foaming agents is evaluated considering two indexes:

• COD (Chemical Oxygen Demand): the amount of oxygen which can be consumed by chemical reactions in a defined system (chemical product, solution, conditioned soil, etc.).
• BOD (Biochemical Oxygen Demand): the amount of dissolved oxygen demanded by aerobic biological organisms to break down organic compound in a given sample at certain temperature over a specific time-period.

The biodegradability of a product at a defined time is indicated as a percentage of the BOD at that stage over the COD of the system: chemically it means how much oxygen is consumed compared to the total available for chemical reactions.

Even if the biodegradability value is of mandatory importance, also the BOD at a specific time (i.e. 28 days according to OECD 301 C) must be considered to define the environmental properties of a chemical product (Figure 6). Lowest is BOD and lowest is the "effort" for chemical reactions to consume the oxygen in the system.

The biodegradability and eco-toxicity of the foaming agents are very important for the selection of the products with the lowest environmental impact even though they are not enough for a "well-done" decision about the conditioning products for a TBM project. Even more attention should be paid to the environmental impact of the conditioned soil planning

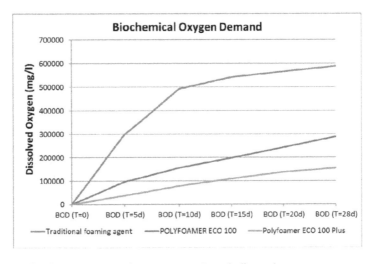

Figure 6. Scheme for the TBM excavation process and muck disposal.

its re-use as by-product according to the project area regulations and restrictions reducing the environmental impact of the civil work and avoiding extra-costs for the Contractor.

3   ENVIRONMENTAL IMPACT OF THE MUCK

The conditioned soil environmental impact is becoming a key factor for the success of a TBM project, especially from the Contractor's economical point of view, and for the project itself sustainability. A new and reasonable approach to the soil conditioning process should be taken to reduce the environmental impact and overall costs of the project.

In this way the following points should be considered:

- The soil conditioning process does not end when the muck is on the TBM conveyor belt, indeed it is a chain process from the tunnel face to the final use of the spoil.
- Planning the disposal and the re-use of the muck is nowadays part of the project design as tunnel face support pressure calculation, studies for the mix-design of concrete works, static calculation of tunnel, etc.
- For each specific TBM project the environmental parameters to be complied for a spoil re-use should be defined at the design stage. The foaming agents and all chemical products used and required for the correct soil conditioning should be selected in accordance with the technical and environmental requests.
- The muck consistency should be appropriate for its transport: too liquid muck would lead to higher transport cost. The common products used to reduce the spoil viscosity (i.e. lime) affect the environmental characteristics of the soil and so new type of products should be used for the further necessary treatment of the soil in the job-site area.

The possibility to re-use the muck allows to reduce the impact of project and the overall costs: to transport out from the job-site the muck as waste, a huge economic effort is demanded to the Contractor that is not sustainable considering the higher and higher amount of material extracted during a TBM project.

Starting from the preliminary design stage it is possible to design the whole process of soil conditioning in a sustainable way, from the TBM excavation to the final-destination of the spoil (Figure 7).

As part of the project design, the soil conditioning process has specific stages:

1. Analysis of conditioning products.
2. Preliminary evaluation of soil environmental characteristics (soil it-self and when chemical products are added).

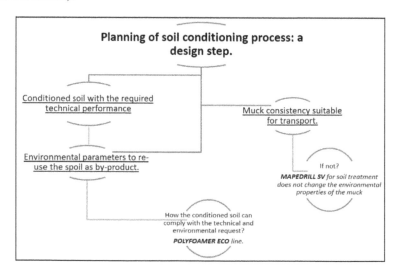

Figure 7.   Design approach for the soil conditioning process.

3. Design the environmental parameters for the specific project in that peculiar area that should be complied: the muck will be re-used only if the "site-specific" requirements are satisfied.
4. Planning the re-use of the muck as by-product for complementary civil works relate to the project as road substrate, re-filling material, etc.
5. Verify that also the technical characteristics of spoil (i.e. its consistency) are satisfied for its re-use and if not use the suitable products for further treatment.
6. Supervising during the tunnel construction phase that all parameters are satisfied to manage the soil as by-product.

The foaming agents and all chemical products used and required for the correct soil conditioning should be selected in accordance with the environmental and technical parameters set up at the design stage for a profitable TBM excavation.

The chemical products in combination with the site soil (i.e. the final muck) should comply the technical and environmental requests for the final re-use of the muck.

4   MUCK TREATMENT

When difficult hydrogeological conditions (i.e. big amount of underground water) or wrong soil conditioning parameters are used, the muck cannot have the right consistency to be disposed and later transported out from the job-site.

The most common and used product to face this situation within the job-site is the lime, which can reduce the soil viscosity if added in quite big quantities. Even though it has been largely used in the last decades, it has several disadvantages that make it no more the best solution in accordance with the most recent approach for soil conditioning and treatment:

- For the addition of lime to the muck bulk, several safety aspects should be taken from the workers and so avoiding its use it is possible to reduce the risk activities at the job-site.
- The lime amount to reach the right spoil viscosity is quite high and it should be mixed well with the muck: more operations are required and they are often highly time-consuming.
- The use lime in urban areas is limited due to safety and health reasons (i.e. dust).
- Adding lime into the muck, it is created an environment that is no more suitable for biodegradation of chemical products: the environmental parameters imposed to manage the soil as by-product cannot be complied.

The natural polymer MAPEDRILL SV has been formulated by MAPEI R&D laboratories which can successfully replace the lime. Adding to the soil treatment (Figures 8 and 9) the following advantages:

- The natural composition of the product does not affect the environmental characteristics of the spoil.

Figure 8.   Slump test with soil sample before the addition of MAPEDRILL SV.

Figure 9.   Slump test with soil sample after the addition of MAPEDRILL SV.

- The average dosage of the polymer is much lower than the lime.
- The polymer it-self has all required features to be approved for feasibility design of the TBM project: high biodegradation rate, lox toxicity.

5   CONCLUSIONS

All progresses made by EPB technology during the last decades made it the most common mechanized tunnelling technique and a great part of this success is due to the innovation of the soil conditioning products. In the more recent years, the approach to an EPB project is changed: no more only the technical features of a project are the main concern for its success.

A "well-practiced" and modern approach to an EPB project consists in a precise quantification of the overall environmental impact of the project and its feasibility with costs optimization. Within this innovative approach, the soil conditioning process became part of the design phase of a TBM project and so several site-specific technical and environmental parameters are set up and must be complied by the Contractor. The soil conditioning is now seen as a process from the TBM excavation to the destination of spoil (waste or by-product) that should be planned since the preliminary design stage.

Due to big investments and efforts in the Research & Development, MAPEI has formulated new foaming agents (POLYFOMER ECO line products) for soil-conditioning and new products for muck treatment (MAPEDRILL SV) to provide to the Contractor the solutions to fully comply both with the technical and environmental requirements of a modern EPB project.

REFERENCES

EFNARC (2005). Specification and guidelines for the use of specialist products for Mechanized Tunnelling (TBM) in Soft Ground and Hard Rock www.efnarc.org.
Guglielmetti, V., Mahtab A. & Xu, S. (2007) Mechanized tunnelling in urban area, Taylor & Francis, London.
Peila, D., Oggeri, C., Borio, L., (2009), Using the slump test to assess the behavior of conditioned soil for EPB tunnelling, Environmental & Engineering Geoscience, pp 167–174, 2009, Vol XV.
Pelizza, S., et al. (2011). Lab test for EPB ground conditioning, pp 48–50, Tunnels & Tunnelling, September 2011.
OECD 203, Acute toxicity testing, Organisation foreconomic co-operation and development, Paris 2000, ENV/JMMONO(2000)6.
OECD 301, Biodegradability testing, Organisation foreconomic co-operation and development, Paris 1995, OECD/GD(95)43.
OECD Guideline for testing of chemicals, ENV/JM/TG(2005)5/REV1, April 2005.

Tunnels and Underground Cities: Engineering and Innovation meet Archaeology,
Architecture and Art, Volume 2: Environment sustainability in
underground construction – Peila, Viggiani & Celestino (Eds)
© 2020 Taylor & Francis Group, London, ISBN 978-0-367-46579-7

Tunnel entrances integrated into a landscape: The example of the Gronda motorway bypass

E. Francesconi
Spea Engineering, Milan, Italy

R. Degni
Autostrade per l'italia, Rome, Italy

ABSTRACT: The Gronda motorway bypass crosses different valleys, where the landscapes are either protected or in any case are of high quality. The tunnels mouths portals are designed in a circular shape, inclined in relation to the slope of the earth cover to reduce the exposed concrete. The technical services building are integrated into the remodelling of the land, leaving only the front entrance visible. The earth covers use gently inclined artificial embankments capable of being planted with shrub and tree species, that will encourage the vegetation to return to its former state. The way in which the project is set into the landscape, and the evolution of the works over time, are fully documented by aerial and ground-level CGIs.

1 FROM DEFINITIVE DESIGN TO THE IMPLEMENTATION PROJECT

1.1 *Introduction*

The Implementation Project (IP) for the tunnel mouths of the Gronda motorway bypass was designed taking particular care with the landscape and architectural aspects, which are based on consultations, modifications, and refinements made at the Definitive Design (DD) stage in consultation with the competent public authorities. Whilst the works all fall within the municipal territory of Genoa, taken together they form a complex project, as is shown by the numbers in Table 1.

1.2 *The impact on the landscape at the approval stages*

Critical issues relating to the impact of the project on the landscape emerged at the very beginning of the EIA procedure; the EIA Technical Committee of MATTM (*Ministero dell'Ambiente e della Tutela del Territorio e del Mare* - the Ministry for the Environment and the Protection of Land and Sea) made a recommendation to "*carry out more in-depth studies in relation to the excavations, the subsequent reorganisation of the ground, and the placement of the spoil, particularly in areas where the landscape is still untouched, avoiding major morphological alterations and minimising backfills*" and to "*generally minimise the morphological transformation in the places where the works are to be carried out, by restricting the excessive height of retaining walls, harmonising the design of the new tunnel mouths with those of the tunnels already existing in the vicinity, and harmonising the structures of the new viaducts with the structures and forms of those already existing whilst keeping their decks at as low a level as possible compatibly with safety requirements*".

On the same matters, the General Directorate for Protection and Landscape of MIBACT (the Ministry of Cultural Heritage and Activities and Tourism) required the project documentation to include more detailed landscape design and analysis.

Table 1.   The numbers of the Gronda motorway bypass.

Conventional roads	28.2 km	
Main motorways	45.0 km	
Interconnecting ramps	21.3 km	
Total length, all stretches	66.3 km	
Main motorway tunnels	39.2 km	12 tunnels
Tunnels on interconnecting ramps	10.5 km	13 tunnels
Bridges/viaducts	5.0 km	21 viaducts (11 new, 10 existing)
Valleys crossed (principal)	5	Polcevera, Varenna, Leiro, Cerusa, Torbella
Tunnel mouth sites	23	
Tunnel portals	41	
Major tunnel mouth site	4.2 ha	Amandola Est, Monterosso Est
Minor tunnel mouth site	0.1 ha	Campursone sud/nord
Technological services cabins	17	
Ventilation plant rooms	6	
Types of technical plant rooms	5	

In response to these requirements an initial revision was made to the sites for the tunnel mouths, particularly those in areas where the landscapes are either protected or in any case are of high quality either because of the high value of their man-made or natural components, or because they are parts of panoramic settings.

In parallel with these revisions, a detailed photographic helicopter survey was carried out followed by a GIS study of the areas of intervisibility of the proposed works, which were then set out on a "Visual Interference Map". A ground-level photographic census was also carried out, descending from the large to the small scale, to survey the landscape characteristics of the valleys to be crossed.

The most significant aerial and ground viewpoints were then selected, enabling the project and its visual interferences with the landscape to be simulated in CGIs that were collected together in two A3 size albums. Because these images enable the works to be understood immediately they are an important supporting element for evaluating and forecasting the impact of the project, and are a valid tool for assessing their landscape compatibility within the meaning of Prime Ministerial Decree 12/12/2005.

All of these documents, prepared for the DD stage, were updated and implemented by the IP.

1.3   Compliance with Provisions B.1 and B.2 of the EIA

At a later stage the DD was again updated to meet new requirements introduced by Provisions B1 and B2 of EIA Decree 28/2014, which consolidated the guidelines for the architectural and landscaping design.

Provision B1 focused on the mouth of the Delle Grazie Tunnel:

– B1: "Architectural and landscape protection: a solution must be envisaged for agreement with the Superintendency [omissis]- in fact despite the variation that was incorporated into the original project, this design, as proposed, remains one of the most significant critical points [omissis]. The retaining walls will have a significant impact, and the substantial alteration to the vegetational structure of the Park, as proposed, still do not make the works compatible with the need to protect the historic Park and the landscape".

Provision B2 focuses on a number of topics. To ensure that it was exactly complied with it was split into three sub-provisions:

– B2: "For the whole stretch, the solutions adopted must be compatible with the context, avoiding in particular the visual impact of retaining walls, and must propose an approach to new planting that takes the existing vegetation into account. Therefore:

- *the impact on the landscape of embankments, vehicle aprons, technical buildings, and retaining walls [omissis] must be minimised, and new planting must have the ability to develop and harmonise with the vegetation already existing in that context;*
- *service areas adjacent to tunnels must be as small as possible;*
- *the size of works areas at the sites must be reduced to a minimum, and the areas used for works roads and work sites must be made good after the works have been completed, ensuring that the vegetation and the morphology of the terrain are effectively reinstated";*
- *B2 bis:* [omissis];
- *B2 ter:* "*solutions with lower impacts must be identified in the area of the Voltri Cemetery; the visual impact of the tunnel mouths at the ends of the viaduct connecting the Amandola and Voltri Tunnels, which must be minimised, are in fact not only within the cemetery area but also affect two areas that are still intact".*

Sub-provision B2 bis had no bearing on the landscape aspects. In order to analyse sub-provisions B2 e B2-ter, and to identify a solution that would meet with the approval of MIBACT, various meetings were held over a year to consider and define the technical solutions and design criteria that would guide the revisions at the DD stage, leading up to the design criteria for the IP.

2 THE IMPLEMENTATION PROJECT FOR THE TUNNEL MOUTH SITES

2.1 *Interpreting the landscape and the anthropic semiology*

For the IP, each tunnel mouth site was re-analysed in greater detail to catalogue its natural and man-made constituent elements by means of:

- desktop photographic interpretation to define the most important elements of each site;
- a site survey to verify the photographic interpretation and refine its perimeter;
- a large-scale photographic survey to identify the most important landscape elements at each site;
- a photographic investigation of the recurring architectural elements (low walls, enclosures, finishes, etc.), as a design reference for the completion and making-good works;
- graphical representation of the above as drawings (Figure 1), using mainly monochrome symbols.

Figure 1.   Interpretation of the landscape, anthropic semiology, representational drawing, plan.

2.2 *Design guidelines and solutions*

In compliance with the prescriptions issued by the authorities, and in continuity with the solutions that were agreed with the competent public bodies for EIA Provisions B.1 and B.2, the following design guidelines were adopted in the IP, adapting them case by case to the particular characteristics of each site.

Finishes for r.c.walls, variable depending on the context:

– a ribbed effect, for the vertical walls of the vehicle apron at the tunnel mouth;
– a sprayed plaster effect, for the walls supporting the apron at the tunnel mouth;
– natural stone.

Reinforced earth construction to:

– reduce the amount of surface removed for constructing the containing embankments;
– use the morphological design of some altered landscapes to create simulated terracing;
– possibly replace reinforced earth with earth strengthening and a grassed frontal surface.

Surface finishes and geometries of tunnel mouth areas and vehicle aprons:

– asphalt top layer finish to have a beaten earth colour;
– shapes should be optimised to avoid hard edges;
– verify: entrances to vehicle aprons from the motorway and/or other public roads;
– verify: manoeuvring spaces in the aprons, for maintenance and emergency vehicles;
– verify: overall dimensions of safety and acoustic barriers and variable message signs.

The design of the technical services building should:

– contain all the plant within a single building, leaving only the front entrance visible;
– bury the pump room and firefighting tank below ground, and reduce the number of above-ground spaces;
– preferably mask the technical services building with backfill or integrate it into the remodelling of the land.

Tunnel mouth portals to be constructed:

– using an inclined terminal r.c. segment at the sloping portals;
– make the inclination of the terminal r.c. segment variable in relation to the slope of the earth cover, to reduce the area of exposed concrete;
– if possible, make the terminal r.c. segment asymmetrical on plan, so that it follows the shape of the slope;
– homogenise the portals with the geometries of the existing tunnel mouths.

Reduce excessive changes of level by:

– using multiple small stepped and staggered layers of reinforced earth, where possible eliminating some stretches of the upper layers and introducing gentle artificial slopes or strengthened earth walls.

Cover artificial tunnels to:

– restrict the use of vertical structures. Instead, use gently inclined artificial embankments capable of being planted with shrub and tree species or, as necessary, strengthened earth walls with a grass finish;
– as much as possible, reinstate slopes to their original inclinations;
– provide dedicated routes to ensure that the tunnel cover is accessible for maintenance;
– where possible, reinstate the continuity of walking routes that were interrupted by the project.

Landscaping works should:

– minimise the asphalted surfaces of service aprons.
– provide for reforestation that will encourage the vegetation to return to its former state.

2.3 Definitive landscaping: the example of the west mouth of the Amandola Tunnel

The guidelines were interpreted differently for the different landscapes and in relation to the design of the road and the technical services, and resulted in a net improvement (deriving in part from applying the legislation) in terms of spaces, volumes, and geometries at the existing tunnel mouths. The west mouth of the Amandola Tunnel is one good example of how this was designed for the sensitive landscape at Voltri.

The tunnel mouth is laid against the face of a mountain slope that has a constant inclination and where trees are the predominant vegetation (Figure 2). Urban residential settlement along the road at the valley floor, and agricultural activities on the slopes, have already strongly transformed the landscape. A hard-paved service apron alongside the west carriageway gives access to the technical plant rooms - the electrical intake cabin and pump room, and the fire-fighting water tank. The generator unit is constructed above the apron.

Adhering to the EIA provisions, and beginning from the earliest studies for the excavation and site preparation stages, all the retaining walls for the site aprons, which are to remain as part of the final landscaping arrangements, were modelled as curvilinear terracing. To reduce the visual impact of these solutions and improve their insertion into the landscape, almost all the technical plant rooms are buried below ground with only their front entrances visible; the only elements emerging out of the terrain are their ventilation outlets.

Between the DD and IP stages, superimposed existing/proposed plans and sections (excavations/backfills) were prepared for each tunnel mouth so that the changes to the morphology could be evaluated.

2.4 The design of the technical plant rooms serving the tunnels

The technical plant rooms are of different types (Table 2) depending on the specific plant requirements of each tunnel and the available space. They are all constructed from fairfaced r. c., taking account of the position, in plan and section, of the anchored retaining walls behind them. The fire-fighting water collection tanks are placed forward of the cabin itself, to avoid the need to construct the footings for the temporary retaining walls at a greater depth below them (Figure 3).

The elevation of the technical cabins shows the facade partly set back from the external line of the eaves with an r.c. retaining wall to either side, aligning with the external eaves. The eaves and the lower part of the external walls have a fairfaced concrete finish; the mid-part has a vertically ribbed off-the-board concrete finish. The surface of the vehicle apron is

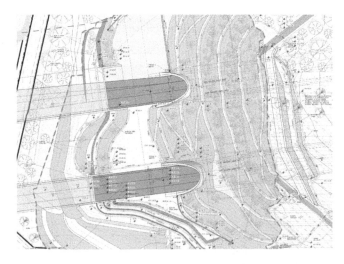

Figure 2.   Amandola Tunnel, west tunnel mouth, definitive arrangements, detail of plan (EIC Type 1).

Table 2.   The five types of technical plant rooms.

Type	Composition	Examples
1	Electrical Intake Cabin (EIC) only	Amandola Ovest, Borgonuovo Est, Monterosso Est
2	EIC + underground tank	Borgonuovo Ovest, Monterosso Ovest
3	EIC + above-ground tank	Genova Ovest, Amandola Est
4	EIC + underground tank + vent. plant + safety tunnel exit/entrance	Bric del Carmo South, Voltri Est
5	vent. plant + safety tunnel exit/entrance	Genova Ovest, Montesperone Est

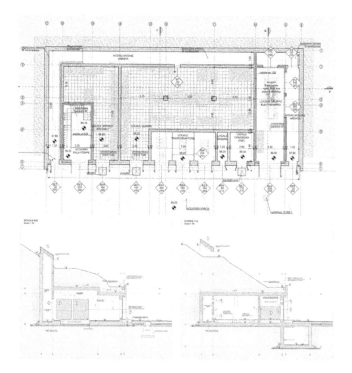

Figure 3.   Borgonuovo Tunnel, west tunnel mouth, Electrical Intake Cabin 1.1 (Type 2): above, detail of ground floor plan; .below, cross-sections.

bituminous conglomerate in a beaten earth colour. The fuel tanks for the pumps are buried below the ground.

The architectural and structural design of these technical rooms includes the openings in the walls and between the different rooms, the applied and floating floor finishes, the internal and external doors and windows, and the systems to prevent maintenance workers from falling from the roof.

2.5   *The landscaping design criteria*

The landscaping is designed to integrate the new works with their environmental context considering not only the landscape but also the ecological and environmental reinstatement aspects. The pedological requirements of the tree and shrub species for each site were analysed in order to define a planting scheme that would be capable, over time, of attaining a certain ecological balance whilst also increasing the stability of the artificial slopes. This guided the choice of the soil substrate to be used for the landscape reinstatement.

The design methodology, derived from the Environmental Impact Study, required:

- analysis of the potential vegetation and plant structure as they existed before the works;
- interpretation of the modifications in areas of the territory where earthmoving works, new surface water regimes, and new structures are to be implemented;
- botanical project: identification of the plant types.

2.6 *Planted areas: the planned solutions*

Polyphyte meadow

In addition to ecological criteria, the selection of the type of herbaceous grass seed to be used takes account of its ability to colonise and rapidly form a continuous surface, whilst improving the soil and ensuring its longevity and long-term stability.

In the open spaces and below all the shrub and/or tree plantations, the project makes provision for polyphyte meadow, to be carried out by enhanced hydroseeding (adhesives and amendments) of a mixture of mainly graminacious and leguminous species.

Anti-erosion works and renaturalisation with perennial herbaceous species

These works consist of hydroseeding soil or rock surfaces (not integrated or compacted) treated with wall-mounted strengthening systems (anchored steel mesh) that do not require geomat, and filled with vegetable soil and classical hydroseeding.

They will be implemented by enhanced hydroseeding of particular perennial herbaceous plant species which, thanks to the combined action of their deep roots and lush foliage, will green the surface whilst blocking erosion of the soil and protecting it from weather events.

Shrub formations

Autochthonous shrub species represent a more developed dynamic stage of the meadow; the project gives them ample space so that their natural phytocoenoses can evolve towards more mature forms within the dynamic series to which they belong, and so that their root systems can consolidate the soil.

Arboreal-arbustive formations

The design and distribution of these species within the planting layouts attempt to borrow from natural forms, to favour their integration into the surrounding landscapes as much as possible and ensure that the new and the existing come together, re-naturalising the newly landscaped areas.

Tree rows

For the rows of trees screening the new works, thermophilic species are used that are ecologically consistent with the vegetation in the existing context.

Ivy

In compliance with MIBACT Provision B2, *Hedera helix* will be planted as masking for exposed fairfaced concrete walls.

2.7 *Specific tunnel mouths*

2.7.1 *The Delle Grazie Tunnel*

This tunnel mouth changed significantly at the DD stage, mainly because of a radical modification of the excavation site to preserve the morphology of the hill behind and the important buildings that stand on it, including parts of the romantic Park of Villa Duchessa di Galliera.

These modifications led to a revision of this stretch of the road on plan and in section, a significant reduction of the levelling, a different organisation of the tunnel excavation (Figure 4) and a reduction in the extent of the site roadways and preparatory works close to the abutment of the viaduct. Following a lengthy collaborative process with MIBACT, all of these aspects were developed in depth and are confirmed in the IP.

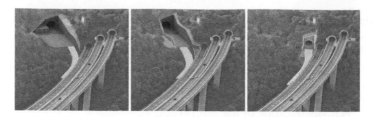

Figure 4.   Delle Grazie Tunnel mouth: the contraction of the excavation site over time, from left to right.

2.7.2 *The tunnel mouth at Monterosso Est with Palazzo Pareto and the Genoa Viaduct*

For the duration of the works, this tunnel mouth will be the departure point for the two TBMs. It was modified at the DD stage to preserve the historic complex of Palazzo Pareto and its outbuildings. Palazzo Pareto is an important testimony of the suburban residences that the Genoese nobility contructed along the valley of the Polcevera. Conserving, making safe, (Figure 5), and restoring the palace are integral parts of the Gronda bypass project, which when completed will include a new tree-lined avenue leading to the palace.

For the architecture-light-colour project of the adjacent Genoa Viaduct, the specially designed noise barriers for the viaduct will continue as far as the tunnel portals on both sides of the valley, next to the historic complex near the west carriageway.

2.7.3 *The Genova Ovest tunnel*

The Promontorio Tunnel (built between the two wars) and the San Bartolomeo Tunnel (built after the Second World War, when the A7 motorway was doubled) converge at this site with the mouths of the new tunnels for the Gronda motorway bypass, which are inserted via an eastwards extension of the toll barrier apron. The architectural theme here is to harmonise works that belong to three different historical periods.

As was the case at the other tunnel mouths, constructing inclined portals proved not particularly effective, due to the difficulty of covering over the flank alongside the existing carriageways and the high walls retaining the excavation. The solution adopted extends the new retaining wall for the earth bank, left by the earthmoving works, towards the north until it meets the existing wall (Figure 6). After a point where the wall makes a 90° turn at a water

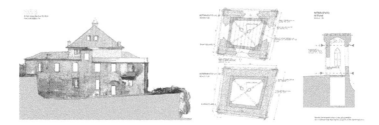

Figure 5.   Palazzo Pareto, laser-scan surveys of the main building, and the making-safe works.

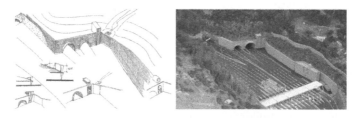

Figure 6.   Genova Ovest tunnel mouth: a study drawing and photosimulation of the architectural solution.

manhole, the portals of both tunnels, and the entrance and ventilation openings of the safety tunnels, emerge out of this wall; the three-storey technical machinery cabin is contained in a recess adjacent to the tollgate canopy.

3 THE LANDSCAPE REINSTATEMENT WORKS

3.1 *Sites and service roads*

The main objective of the project for reinstating the construction sites and service roads is to re-stitch together the smallest weave of the landscape, in some cases by the morphological and vegetational reinstatement of the planting and in other cases reinstating the tarmac road surface by demolishing the top layer (the binder and wearing layer) and re-laying a new surface.

3.2 *The landscape restoration of a transept of the Leiro Valley*

To meet the requirements of the Architectural and Landscape Heritage Superintendency of Liguria, the IP has developed a series of landscape restorations works that will compensate for some unmitigated residual impacts in the valley of the Leiro Stream and include reinstating the sites and the service roads that were used to construct the two viaducts and could be described as an operation of territorial reconnecting. (Figure 7).

These restorations works provide a new bridge for vehicles, a parking area serving the cemetery and the residential buildings, and reinstate the pedestrian walking route along the right bank of the stream, thereby connecting the two existing bridges upstream and downstream of the new bridge. These works are completed by reconstructing the dry stone walls, the new external stretch of pathway, and reinstating the urban paving, street furniture, and services.

Figure 7.   Landscape restoration in the Leiro Valley: general drawing and photosimulation.

4 THE REPRESENTATION OF THE PROJECT AT THE VARIOUS DESIGN STAGES

The way in which the project is set into the landscape, and the evolution of the works over time, are fully documented by aerial and ground-level CGIs showing the four stages of the operation: before the works, during the works, after completion, and the tree growth (Figure 8).

Detailed views showing the works from various angles were extracted from the 3D models to evaluate their constructional aspects. On completion of the DD stage a series of informative videos were made illustrating the whole of the works for the Gronda motorway bypass. All the 3D models were updated in the IP, increasing the resolution of the details and the background images. Small-scale models and samples of the finishing materials and the paint colours were prepared for the architectural and colour study of the Genoa Viaduct.

Figure 8. Photosimulations of the Monterosso Est tunnel mouth and the Genoa Viaduct, at the four stages.

5 CONCLUSIONS

The project for the mouths of the tunnels in the Gronda motorway bypass shows the design methodology and the multi-year approvals path required for a piece of major road infrastructure, integrating it into a complex urban and landscape context that is characterised by a historicised urban fabric, in conditions of considerable intervisibility generated by a variable topography, whilst keeping the site preparation requirements, which are essential for carrying out the works, within the time-frame of the programme.

The architectural and landscape solutions adopted give preference to integration with the context whilst maintaining a contemporary constructional vocabulary, coordinating all the works at all scales, from large-scale landscaping to construction details and finishing materials.

REFERENCES

Berta, M. 2002. Henri Coumoul e les Autoroutes du Sud de la France. *Architettura del Paesaggio* (8).
Crowe, S. 1960. *The Landscape of Roads*. London: The Architectural Press.
Iarrera, R. A. 2004. *Autostrade come progetto di paesaggio*. Rome: Gangemi.
Lassus, B. 1994. *Autoroutes et Paysages*. Paris: Ed. du Demi-Cercle.
Micheletti, C. & Ponticelli, L. 2003. *Nuove infrastrutture per nuovi paesaggi*. Milan: Skira.
Vallerini, L. 2009. *Il paesaggio attraversato. Inserimento paesaggistico di grandi infrastrutture lineari*. Florence: Edifir
Virano, M. 2002. *Parole sulla strada*. Turin: Piazza.
Maffioletti, S. & Rocchetto S. 2002. *Infrastrutture e paesaggi contemporanei*. Padua: Il Poligrafo.

Tunnels and Underground Cities: Engineering and Innovation meet Archaeology,
Architecture and Art, Volume 2: Environment sustainability in
underground construction – Peila, Viggiani & Celestino (Eds)
© 2020 Taylor & Francis Group, London, ISBN 978-0-367-46579-7

Transporting asbestos: Disposing of excavation waste in an urban context

S. Frisiani & L. Messina
Spea Engineering, Milano, Italy

M. Mazzola
Autostrade per l'Italia, Roma, Italy

ABSTRACT: Tunnels for the new Genoa bypass motorway system will pass through formations that potentially contain asbestos. The plan for disposing of such material is to confine it within the area of the seaport, by constructing a dedicated structure (in the form of sedimentation beds). An innovative solution has been adopted for transporting this material from the tunnel excavations to the sedimentation beds. The paper describes the rock formations where an asbestos risk exists, assesses the volume of material to be disposed of, describes the technology on which project is based - a pipeline that transports solid material mixed with bentonite sludge - and outlines the most important characteristics of the route for the line, which passes through built-up areas. Finally a description is given of the separation plant foreseen at the seaward end of the system, so that the bentonite fluid can be recovered and reused within a closed circuit.

1 INTRODUCTION

Some of the tunnels in the Genoa motorway node upgrading project will pass through formations potentially containing asbestos. In order to dispose of any excavated material of that type, the plan is to confine it at a land reclamation area adjacent to the "Canale di Calma", a body of water between the sea dock and Genoa airport, by constructing a dedicated confining facility (sedimentation beds) in that location.

The solid material from the excavations will be mixed with bentonite sludge and carried to the sedimentation beds using an innovative design solution (a slurry duct) that follows the line of the Polcevera Stream, passing through heavily urbanised areas.

The adoption of this technology will also require the installation of supporting systems, e.g. for separating the excavated spoil before disposal and for mixing the bentonite fluid.

2 EXCAVATIONS IN ROCK POTENTIALLY CONTAINING ASBESTOS AND THE FINAL DESTINATION FOR THE SPOIL

2.1 *The geological context*

The presence of asbestos in the rocks of Liguria is a constant and known fact. The Ligurian authorities have mapped, classified, and quantified the presence of asbestos in these rocks, whether in the form of breccias, blocks, or slabs, and have also determined the release index for assessing its hazardousness (in compliance with Ministerial Decree 06.09.1994).

In order to acquire information about the presence of asbestos within the lithologies that are flagged as being at risk in the areas to be tunnelled, specific studies were carried out at the design stage to define which geological and structural areas were likely to contain asbestos, the types of asbestoid minerals present in those areas, and the qualitative and quantitative levels of the asbestos fibres.

The rocky masses in the mountains west of the Polcevera Stream potentially contain asbestos because of their genesis and their mineral composition. The tunnel excavations for the Genoa bypass project will require more than 5,000,000 m³ of this hazardous material to be disposed of.

Petrographic studies have confirmed that some of the geological zones where tunnelling is to take place (such as metabasalts and metasediments like limestones, clay schists, and metapelites) would not be compatible with any presence of fibrous minerals and can therefore be considered, with complete certainty, to present no asbestos risk whatsoever. However in the case of rocks that are likely to contain asbestos (serpentinites, serpentinised peridotites, serpentinoschists, chlorite schists, and tremolites) a number of petrographic classes with an increasing asbestos risk (petro-structural facies) have been identified and the frequency and distribution of the veins of fibrous mineral, and their types, have been defined. These mineralogical and petrographical analyses also enable the fibre content in the various petro-structural facies to be estimated. Applying these estimates to the homogeneous stretches of the geological sections made it possible to define a zonisation along the route of the project that relates to the asbestos content of the rocks (see the example given in Figure 1), making allowance for the relative margins of uncertainty, which are

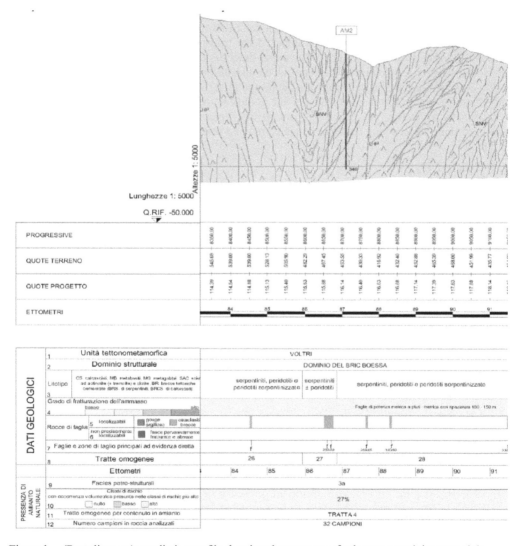

Figure 1. (Part diagram): predictive profile showing the presence of asbestos-containing materials.

inherent in the degree of reliability of the general geological predictions (the positions and typologies of the geological bodies), and for the problems associated with their density and areal distribution.

2.2 *Final disposal of potentially asbestos-containing spoil*

The spoil from the excavations will be deposited at a work site at Bolzaneto, where it will be stored temporarily in special silos that will prevent any contamination of the external environment.

In order to determine its final disposal and reuse it will be sampled, analysed, and classified at Bolzaneto in accordance with the regulatory requirements for managing asbestos-containing materials. Depending on how it is classified, three alternative destinations have been identified (see Figure 2):

– a land reclamation area (code green: asbestos content less than 1 g/kg): any material suitable for re-use as landfill will be used as part of the works to create a buffer zone alongside the runway of Genoa airport, which occupies reclaimed land at the seafront. This material will be pumped through the sealed slurry duct to its destination in the reclamation area at the airport area, which is adjacent to the main port of Genoa.
– tunnel inverts (code yellow: asbestos content greater than 1 g/kg): for any material with geotechnical characteristics that permit its re-use at the tunnelling sites, the technical project specification provides for it to be used as backfill for tunnel inverts on condition that these are appropriately protected by a layer of uncontaminated material and the new road surface;
– landfill (code red: asbestos content greater than 1 g/kg): this material cannot be reused because its geotechnical properties do not meet the technical project specifications, and will be sent to landfill for authorised disposal.

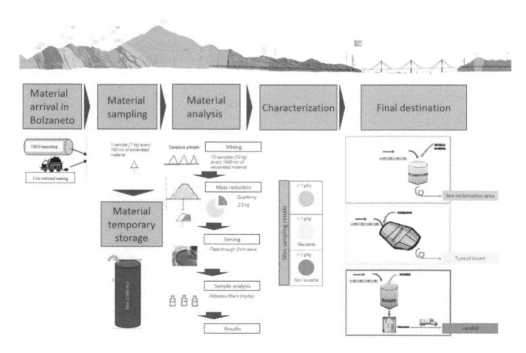

Figure 2.   The final destinations of the waste material.

3 THE DESIGN SOLUTION: TECHNOLOGICAL INNOVATION

After initial drying, the waste material that can be reused for land reclamation will be slurrified at Bolzaneto by adding a water and bentonite mix. This slurry will then be pumped to its final destination at the land reclamation site, adjacent to Genoa airport, via a 9 km sealed pipeline (see Figure 3) that follows the route of the Polcevera Stream. This innovative and specially designed hydraulic system will ensure that the material is transported in environmentally safe conditions.

3.1 *The final destination: the land reclamation site*

The tunnel excavations will generate more than 12,000,000 m³ of spoil, of which approximately 5,000,000 m³ will potentially contain asbestos. Long-term storage of this material

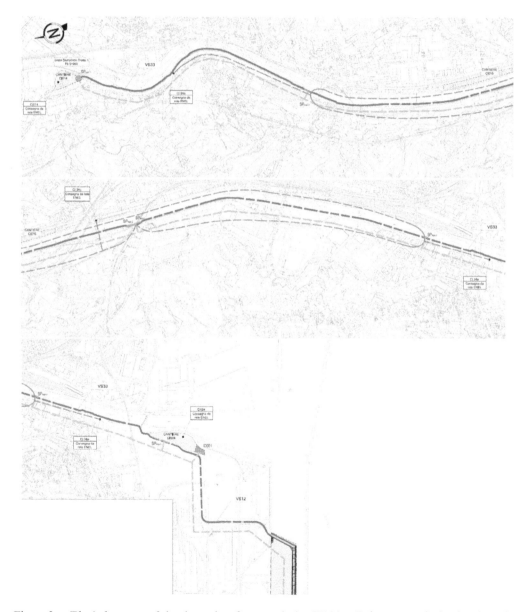

Figure 3. (Plan) the route of the slurry duct from work site CI014 at Bolzaneto to the land reclamation site.

anywhere in the surroundings of Genoa, where the territory is either strongly urbanised or is of high environmental value, would present almost insurmountable difficulties. The solution found consists of augmenting the width of the runway at Genoa airport, using the waste material to partly fill in the Canale di Calma (a body of water adjacent to the runway).

This land reclamation site addresses a number of technical and functional objectives:

– it widens the Genoa Airport runway strip (a buffer zone running alongside the runway) bringing it into line with the higher airport safety standards required by international regulations;
– it protects the airport from significant sea weather events;
– it improves the drainage system of the airport platform.

Figure 4 shows the location of the land reclamation area, which measures approximately 165 x 3600 m on plan, with the minimum widening strip (45 m) that would be required for the airport improvements.

The spoil that will be used to fill in the land reclamation area (see Figure 5) is of two types: excavated material originating from west of the Polcevera Stream and containing less than 1g/kg of asbestos, which will be used for the nucleus and will be delivered via the slurry duct, and the protective top surface layer of the reclamation area (the capping), which will use material from other contexts east of the Polcevera that do not contain asbestos.

3.2 *The work site at Bolzaneto*

The excavation waste disposal systems will be installed at work site CI014, which in hydrographic terms is situated left of the Burla Stream immediately upstream of the point where it flows into the Polcevera. This site, which from an operational point of view is a nodal point in the system, is where the material potentially containing asbestos will be sorted and treated. The following operations buildings will be constructed here (see Figure 6).

Figure 4. The reclamation area alongside the airport runway.

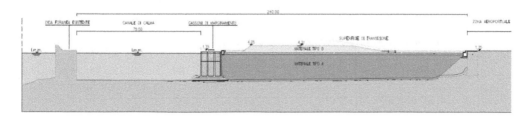

Figure 5. Cross-section of the reclaimed part of the Canale di Calma.

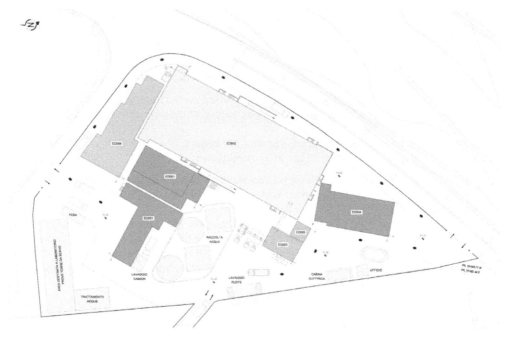

Figure 6.   The layout of work site CI014 showing the positions of the operations buildings.

- ED001 - Crushing: for receiving and breaking down the material excavated using traditional means;
- ED002 - Silos: for storing the material until it can be sorted;
- ED003 - Stabilisation, for processing material with an asbestos content greater than 1g/kg to make aggregate that can be reused in the construction work (tunnel inverts);
- ED004 - Slurrification: for mixing material containing less than 1g/kg of asbestos with the bentonite fluid and pumping it to the land reclamation area;
- ED005 - Bagging: for bagging other material in containers of "big bag" type and then preparing it to be transported to landfill;
- ED036 – Analysis Laboratory: for categorising the material for environmental purposes.

3.3   *The slurry duct: from the work site at Bolzaneto to the land reclamation area*

Even if the potentially asbestos-containing material falls within the limits indicated in Annex 5, Title V, Part IV, Table 1, Column B of Legislative Decree 152/2006, it nevertheless contains fibres that must not be dispersed in the air. Transporting it from work site CI014 to the land reclamation area will require construction of the slurry duct system.

This pipeline for transporting the slurry- an economical system for transporting minerals that has been in use for decades in the mining industry – consists of a hydraulic circuit (typically an enclosed steel conveyor) through which the material is pumped after it has been mixed with a liquid vector.

3.3.1 *The process diagram and performance requirements*

The spoil will be fluidified using a bentonite fluid (a mixture of water and bentonite) with rheological characteristics that optimise the granulometry of the material so that the coarsest parts can be transported without exceeding the permitted velocity of travel in the pipeline.

The slurry duct system consists of:

- The main line, for transporting the spoil from work site CI014 to the land reclamation area;

- The recirculation line, for returning bentonite fluid from the land reclamation area back to CI014;
- Operations building ED004 at CI014 (shown in Figure 5) where the mixture of bentonite fluid and waste material (the slurry) is prepared;
- Relaunching stations positioned along the line of the Polcevera Stream, for pumping the slurry and the returning bentonite fluid;
- Bentonite separation and recovery plant: situated at the land reclamation area, this facility returns the separated bentonite to CI014 via the recirculation line.

Belt conveyors take the sorted spoil, which has been temporarily stored in the sorting silos, to the slurry box, which also receives the returning bentonite fluid from the recirculation line. The mix of spoil and vector fluid is then conveyed to the first relaunching station. The mixed fluid, of maximum density 1.35 t/m³, is then launched into the pipeline by centrifugal slurry pumps that can handle solid granulometries up to 240 mm.

The slurry duct system has been dimensioned for a maximum capacity of 350 m³/h and must be able to transport a total of approximately 5,100,000 m³ of potentially asbestos-containing material. The sorting silos serve as "accumulation tanks" (24 silos each of capacity 900 m³ for a total volume of 21,600 m³) with a storage capacity equivalent to 4 days of production (considering a maximum permitted excavation rate for the TBMs of 15 m/day).

Because the slurry duct system is of strategic importance for the works as a whole, a high degree of reliability (95%) is required. Provision will therefore be made for regular planned maintenance, for the pipework of the main line and the recirculation line to be completely replaced as necessary, and for redundancy to be built into the critical elements of the system.

The maximum electricity demand of 13.5 MW must be compatible with the capacity of the local grid. The slurry duct system is therefore designed to minimise energy consumption, and because the bentonite fluid optimises the velocity of the mixture through the system, it requires less power to transport the material. Provision has also been made for an "unloaded" low-velocity, low-energy configuration for the bentonite fluid to keep it circulating during non-production times when completely shutting down the system would not be justifiable.

The transporting fluid is a very delicate and fundamentally important element of the system, and its operating parameters must be constantly monitored using density and flow meters installed along the route. Any requirement to resupply fresh bentonite fluid will be managed by checking these parameters, together with laboratory tests to measure the dynamic viscosity of the system and therefore its efficiency.

3.3.2 *The route taken by the slurry duct*
The total length of the route is approximately 9,100 m (see Figure 3). Most of it runs in the bed of the Polcevera Stream (6,800 m); the remainder runs on land, parallel to service road VS033 (which will be used for constructing, operating, and maintaining the slurry duct) and service road VS012 (which connects work site CI004 to the land reclamation area).

Beginning from work site CI014, the slurry duct passes at high level over the confluence of the Burla stream with the Polcevera and then runs along the right-hand side of the Polcevera, continuing almost to its mouth, where it deviates on to land and continues along service road VS033, passing below the "new Sea Road", crossing the entrance roundabout for work site CI004 at high level, and arriving at relaunching station 5, at the beginning of service road VS012. Along this stretch of the route the slurry duct is either supported on monopiles or bracketed off the riverbank walls.

Along road VS012 the pipes cross the ILVA steelworks sites, running first on a steel supporting structure and then in the former cooling channel of the ILVA works, crossing various roads and railways via a box section underpass, and arriving at the new land reclamation area, where they continue on the caissons of Holding Tank A1 until they reach the separation plant. Along this stretch of the route the slurry duct either runs at ground level or is hung from brackets within the underpass.

The presence of other pre-existing pipelines in the bed of the Polcevera was determinant for defining the route. These existing pipelines carry mainly oil and gas and are either currently in use or are being maintained in working order. They are partly fixed to the riverbank walls and partly buried below the riverbed. The positions of the subfoundations for the supports of the new slurry duct take into account these existing pipelines.

The level of the new slurry duct will be above the 200-year flood level of the Polcevera Stream. For bridges and other structures encountered *en route* the pipes will run below the decks of these structures, except for the Ratto Bridge, which, being already in a critical situation today from hydrological terms, will be overcome by the pipes.

3.3.3 *The supports for the slurry duct and the pumping stations*

Since most of the route runs along the Polcevera Stream, the slurry duct will be supported on steel elements, to minimise interference with the hydrological regime of the stream and to keep the pipes above the 200-year flood level.

These purpose-designed supports will consist of T-shaped steel elements bearing on large-diameter monopiles set at 12 m centres (see Figure 8). Wherever the 12 m distance between the piles cannot be maintained, provision has been made for steel lattice structures spanning up to 26–29 m (U-shaped so that the pipes can be installed and maintained from above).

In stretches where the slurry duct is to be anchored to the riverbank walls and there are no pre-existing pipes, a steel structure with three brackets was designed (one bracket per pipe), to be anchored to the wall at 12 m centres.

Where bridges have to be crossed, it will not be possible to construct subfoundations below them. At these points, "hanging" supports fabricated from steel profiles will therefore be fixed to the existing structure (whether this is steel or concrete).

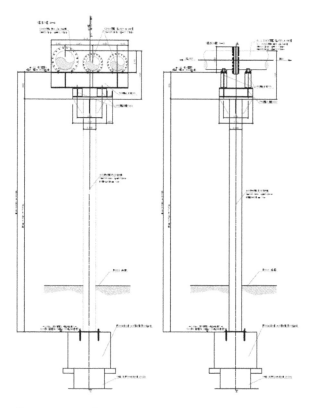

Figure 7. The mono-pile support typology.

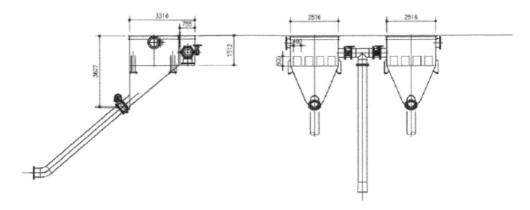

Figure 8.   Slurry box: detail.

3.3.4   *Operations building ED004 - Slurrification*

The asbestos content of much of the spoil material will be lower than 1 g/kg and will be sent to the land reclamation area. So that this material can be transported whilst reducing its impact on the urban environment, it will be carried by a transporting fluid. The solid material must therefore be mixed with the transporting fluid, to make a suspension that can be pumped.

The plant that will mix the suspension (the slurry), ready to be pumped to the land reclamation area, is within operations building ED004. Here, a belt conveyor coming from the sorting silos will feed the so-called "slurry box" (see Figure 8), which has a particular shape that enables the material to be mixed and guided towards the suction of the first relaunching pump. This integral part of the slurry circuit is fed from the bentonite fluid recirculation line (the feed line). There will be two slurry lines (one active and one in reserve) and the building will contain the slurry boxes, the hoppers for spreading the material and the relaunching pumps for the slurry pipeline, for both lines.

The building will be constructed from steel, with a continuously negative-pressurised internal space to prevent any dispersion of hazardous fibres.

3.3.5 *The pumping stations*

The pumping stations resemble shipping containers. Each is approximately 15 m long and 3 m wide (see Figure 9), They are positioned along the route of the slurry duct, where possible using existing car parking (pumping stations 2 and 3) or supported on a steel structure, similar to a system of piles, standing in the bed of the Polcevera Stream (pumping station 4).

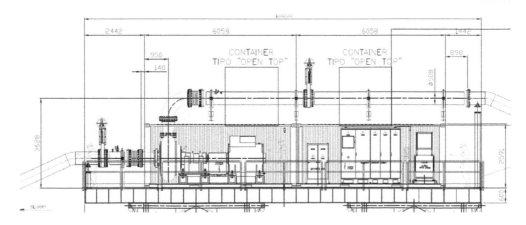

Figure 9.   Pumping station.

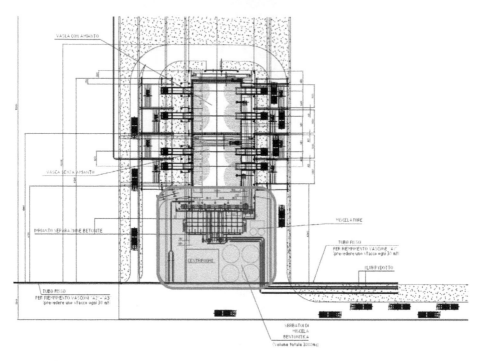

Figure 10. The separation plant at the land reclamation area.

3.3.6 *The separation plant at the land reclamation area*

The slurry duct technology requires a separation plant to separate the solid waste material from the liquid vector fluid with which it was mixed. The separation plant must be positioned close to the final disposal site so that the journey of the solid waste to its final destination is as short as possible.

The separation plant will therefore be installed on an embanked peninsula, which will be created by partially filling settling tank W1, which is confined by the caissons facing the Canale di Calma.

Settling tank W1 measures 40 m × 150 m overall with an average depth of 10 m. It will be partly filled with waste from the tunnel excavations to create a peninsula measuring 35 m × 35 m overall that will take the separation plant (see Figure 10), the tanks, the silos, and the mixing systems.

The first stage of the separation plant passes the waste through a sieve to separate out the largest material (D>20mm); the remaining suspension is then passed through two centrifuges to separate out the particles up to 30mm.

4 CONCLUSION

The decision to use a slurry duct to carry the solid waste (potentially containing asbestos) fluidified with bentonite to its final destination in the land reclamation area is an innovative design solution.

This paper describes the overall process from the mixing plant, where the bentonite and water are added to the solid material to form a sludge, through to the final separation plant, where the solid waste is separated from the bentonite, enabling the bentonite to be recovered via a closed recirculation system and used again.

The most significant innovative aspect of this system is that it takes an existing technology already in general use for tunnel excavation, and adapts it for use in an urbanised situation where other waste transportation systems would not be feasible.

A specially designed component of the system, the slurry box, mixes the solid waste/bentonite into a slurry with the requisite rheological properties that enable the slurry to be pumped.

This innovative system complies with the environmental requirement that no asbestos particles must be released when the waste material is transported from one place to another, and overcomes the spatial constraints associated with the problematic urban situation of Genoa.

This waste disposal system is a critical component of the tunnelling and excavation process for constructing the new Genoa bypass motorway system (and is included in the general critical path programme). Because it must therefore be completely reliable, its most important parts have been designed for 100% redundancy.

The key input data for designing the slurry duct system, and for meeting the energy consumption requirements, was to identify the correct solid transportation parameters for the vector fluid: its density, viscosity and flow rate. The flow rate in particular has a significant impact on the power requirements as well as on the wear and tear of the pump sets, elbows and bends.

REFERENCES

Capponi, Crispini, 2002. Structural and metamorphic signature of Alpine tectonics in the Voltri Massif (Ligurian Alps, North Western Italy). *Helv. Eclogae geol.*

Cortesogno, Haccard, 1984. Note illustrative alla carta geologica della zona Sestri-Voltaggio. *Mem. Soc. Geol. It.*

Maidl, Herrenknecht, Maidl, Wehrmeyer, 2011. Mechanised Shield Tunneling, 2nd edition. *Berlin: Ernst & Sohn.*

Praetorius, Schößer, 2017. Bentonite Handbook – Lubrification for Pipe Jacking. *Berlin: Ernst & Sohn.*

Stellini, 2014 Tesi di Laurea Magistrale, "La problematica ambientale delle fibre di amianto nei terreni e nelle rocce da scavo. Proposte analitiche innovative per affrontare casi reali complessi". *Torino: Univ. Studi Torino.*

Tunnels and Underground Cities: Engineering and Innovation meet Archaeology, Architecture and Art, Volume 2: Environment sustainability in underground construction – Peila, Viggiani & Celestino (Eds)
© 2020 Taylor & Francis Group, London, ISBN 978-0-367-46579-7

Passing through the valleys at Voltri: The technical challenges posed by asbestos and by the safeguard of the local territory

S. Frisiani & M. Pastorino
Spea Engineering, Milano, Italy

ABSTRACT: The Genoa Motorway Bypass is the new road system that connects the motorways going towards Genoa from Lombardy and Piedmont with the motorway going west and the motorway going east. Within this project, a fundamental role is played by the intersection at Voltri, connecting the Bypass to the existing motorways via a complex system of underground junctions, inserted into an environmental context potentially containing asbestos and subject to landscape protection constraints. The document describes the methods and criteria that were adopted and the technologies that were identified to control the asbestos risk during the excavation stages, in order to contain the fibres that might be released, confine the material produced during excavation and treat the site water before returning it to hydrographic system. The Bypass was properly inserted in the local context identifying morphological configurations consistent with that territory and studying specific materials and architectural characteristics for the final layout.

1 INTRODUCTION

The Genoa Motorway Bypass (the "Gronda") is the new road system connecting the motorways towards Genoa from Lombardy (A7) and Piedmont (A26) with the A10 motorway going west (to France and Spain) and the A12 going east (to central/southern Italy).

Within this project, the intersection at Voltri plays a fundamental role, by connecting the Bypass to the existing A10 and A26 motorways via a complex system of junctions, mostly underground.

The works in the Voltri area can be summed up as follows:

- construction of tunnels and tunnel entrances at Voltri Est, Voltri Ovest, Bric del Carmo, Ciocia and Delle Grazie;
- construction of viaducts at Cerusa Est, Cerusa Ovest, Leiro Est and Leiro Ovest;
- widening the existing Cerusa, Casanova and Leiro viaducts;
- realization of several service roads and operational sites and their interconnection with the public road system, to ensure that the works listed above can be carried out.

All these works had to be inserted into an environmental context potentially containing asbestos and which was subject to several landscape protection constraints. All of these aspects strongly conditioned the design solutions adopted, both in terms of the construction methods used and of the integration of the work into that context.

2 THE PROJECT IN THE VOLTRI AREA

The Voltri area includes the valleys of the Cerusa and Leiro streams, not far upstream from where the streams flow into the sea.

Within that area, the project for the main Bypass roadway provides for two viaducts crossing the Leiro and Cerusa valleys. These viaducts are separated by a short tunnelled stretch.

The Bypass then continues below ground towards west, via the twin-bore Borgonuovo tunnel, and towards east, via the twin-bore Amandola tunnel.

Figure 1 gives an overview of all the works in the Voltri area.

In addition to the two main carriageways of the Bypass (denominated motorway A10bis), new intersections will connect the new road with the existing A10 and A26 motorways. The sections of the motorway will be almost entirely in tunnels, except for brief sections where the new access ramps merge with the carriageways of the existing motorways; upgrading works will be carried out at those points to the existing Leiro and Cerusa viaducts on the existing A10 motorway.

2.1 *Works below ground*

The works in the Voltri area (see Figure 1) include 5 traditionally excavated tunnels; in fact these tunnels are not suitable for mechanical excavation either because they are short, have a

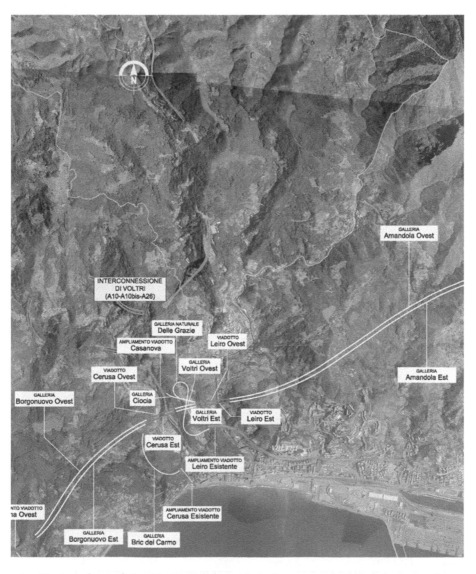

Figure 1.   Works to be carried out in the Voltri area (red: above ground; white: below ground).

reduced section, or because of their layout (due to underground connections with other branches, tight radii of curvature, etc).

The Bric del Carmo tunnel is a tunnelled ramp that connects the southern carriageway of the A26 motorway with the eastern carriageway of the A10bis going towards north and east. This ramp enables vehicles coming from the north to bypass the urban part of the A10 motorway and to access the new A10bis, going in an easterly direction.

The Ciocia tunnel is a tunnelled ramp that enables vehicles coming from east to connect with the A26 going towards north. Following the direction of travel, this tunnel encounters roadways with various cross-sections and passes through chamber number 8, which at its widest part has a maximum excavated width of 31 m.

Along with the Ciocia tunnel, the Delle Grazie tunnel is a ramp that enables vehicles coming from east to connect both with the A10 towards Genoa, and with the A26 towards north; this means that heavy vehicles going to the commercial port at Voltri will be obliged to use the A10bis because heavy traffic will be prohibited on the section between Genova Aeroporto and Genova Voltri.

The Voltri tunnels, West and East Carriageways, represent a short section of the new A10bis. In fact, the Bypass project includes two separate carriageways (the A10bis) that deviate from the route of the existing road, in the section between the Vesima interconnection and the intersection with the A7; these carriageways will be used respectively by vehicles coming from west and going east (the East Carriageway) and by vehicles coming from east and going west (the West Carriageway).

The characteristics of the road and the various tunnelled parts are shown in Table 1.

The illustrations below show:

• Two typical sections, one with an escape route and one without (Figure 2);
• The maximum excavation geometry, on plan and section, of chamber number 8 (Figure 3).

2.2 *The most important engineered works*

The works in the Voltri area (see Figure 1) include the construction of four new viaducts (Cerusa Est, Cerusa Ovest, Leio Est, and Leiro Ovest) and the widening of three existing viaducts (Cerusa, Leiro, Casanova).

2.2.1 *New viaducts*

The four new viaducts are all Category 1 bridges constructed as a steel/concrete mixed system using the finite element continuous beam method. The torsio-rigid open caisson structural typology was adopted. The viaducts are curvilinear on plan with a constant radius curve. The concrete slabs carry the running deck, with two lateral curbs supporting the safety barriers. The pylons vary in height and consist of a plinth in hollow-section r.c., with a coupled pair of

Table 1. Characteristics of the road cross-sections and where they apply.

Tunnel	Road cross-section code	Road type	Length [m]	Escape route	Platform dimensions [m]
Bric del Carmo	G9Tca	1 lane	842	Yes	6.00 ÷ 8.00
Ciocia	G3Tcc	2 lanes + E	118	Yes	11.00
	Chamber n. 8	-	87	No	~14.00 ÷ 24.00
	G7T	1 lane	69	No	6.00
	G8Ta	1 lane	25	No	6.00 ÷ 7.00
	G9Ta	1 lane	71	No	6.00 ÷ 8.00
Delle Grazie	G9Tca	1 lane	1311	Yes	6.00 ÷ 8.00
Voltri Ovest	G2Tvr	2 lanes + E	210	No	11.20
Voltri Est	G5TCPV	3 lanes	203	No	14.45

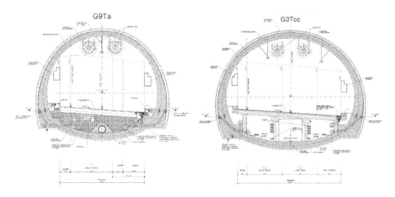

Figure 2.   Typical below-ground sections – with and without an escape route.

Figure 3.   Chamber number 8 – maximum plan and cross-section.

Table 2.   Characteristics of the new viaducts.

Viaduct	Number of spans	Length [m]	Deck size [m]	Deck height [m]	Noise barrier
Cerusa Est	5	367	14.45	4.45	Yes (integrated with H3 on one side)
Cerusa Ovest	4	290	11.20	4.45	No
Leiro Est	4	375	11.20	5.25	No
Leiro Ovest	4	353	18.20	5.25÷5.50	No

upper columns, of polygonal cross-section, connected transversally by a tubular steel lattice structure (see Figure 4).

The characteristics of the new viaducts are given in Table 2.

As an example, the longitudinal profile of the Cerusa Est viaduct is shown in Figure 4.

2.2.2 Increasing the width of existing viaducts

The structure of the three existing viaducts to be widened consists of precast concrete beams, and dates back to the second construction stage of the A10 motorway (1970s). The new structures will be connected at slab level to the existing structure by partly demolishing it and anchoring new reinforcing bars to it.

Widening the deck of the existing Leiro viaduct is necessary so that the exit ramp from the West carriageway of the Bypass can be connected to the A10 going towards the junction at Voltri. The widening is limited to span number 2.

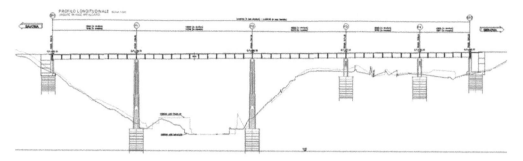

Figure 4.   Longitudinal profile of the Cerusa Est viaduct.

Widening the deck of the existing Cerusa viaduct is necessary so that the exit ramp from the West Carriageway of the A10 (coming from the junction at Voltri) can be connected to the East Carriageway of the Bypass (towards Genoa). The widening affects the full span of the viaduct.

Widening the deck of the existing Casanova viaduct is necessary so that the exit ramp from the West Carriageway of the Bypass can be connected to the A26 towards north. The widening is limited to the longest span only and the cantilever on the upstream side only, which is extended by a variable length.

2.3   Site preparation works

2.3.1   Service roads
A significant number of service roads are to be constructed in the Voltri area so that the new tunnel entrance sites can be connected to existing roads: both the ordinary roads running along the valleys of the Cerusa and Leiro streams and the A26 motorway. These connections will give access to the tunnel entrance construction sites and will enable excavated and drilled material to be transported via the existing motorways, thereby avoiding any increase of traffic on the ordinary roads in the area of Voltri.

Table 3 lists the service roads to be constructed.

Table 3.   Service roads

VS003	Road giving access to the site areas for the Borgonuovo and Bric del Carmo tunnels
VS004	Road giving access to the sites for constructing the Voltri and Ciocia/Delle Grazie tunnel entrances
VS005	Road to enable motorway access for works traffic from the sites for constructing the Ciocia and Voltri tunnel entrances
VS006	Road giving access to the west Amandola tunnel entrance construction site
VS011	Road giving access to the site areas for constructing the pylons and abutment of the widening of the Leiro viaduct and to the works for Delle Grazie tunnel entrance on the south side
VS015	Road giving access to the areas for constructing the pylons of the Leiro viaduct and for connecting between the sites for constructing the Voltri and Amandola tunnel entrances
VS019	Road giving access to the sites for constructing the pylons of the Cerusa Est and Ovest viaducts
VS020	Road giving access to the sites for constructing the Voltri tunnel entrance and the areas for constructing the pylons of the Leiro viaduct
VS021	Road giving access to the sites for constructing pylon 1 of the Cerusa viaduct
IN003	Deviation for roads in the Voltri area that have been interrupted

2.3.2   Operational construction sites
To support the works in the Voltri area an industrial operations field, two support areas for constructing the Leiro and Cerusa viaducts and seven tunnel entrance sites are to be constructed.

Industrial operations field CI003 is situated in the vicinity of the port and is accessed via the ordinary roads. This operations field has an overall surface area of approximately 10,500 sq. m, which will be subdivided into two zones: one zone will be dedicated for stockpiling materials and equipment and for installing standard equipment; the other will be dedicated for concrete manufacture and will therefore contain the mixing plant, the weighbridge, fuel tanks, the hopper washing tank for concrete trucks, and the aggregate storage areas.

Two support areas CO03L and CO03C for constructing the Leiro and Cerusa viaducts will be provided, with surface areas respectively of 3,340 sq.m and 4,300 sq.m. Access to the dedicated Leiro viaduct construction site will be from VS015 whilst access to the Cerusa viaduct construction site will be via the existing local roads.

At the tunnel entrance sites will be installed site offices, first aid post, WCs, wheel washing tanks, gas monitoring system, big bag stockpiling area, ventilation plant for the tunnel excavation, asbestos decontamination changing room, water tanks, etc.

Table 4 lists the tunnel entrance sites.

Table 4. Tunnel entrance sites

CO01E	To enable construction of the Borgonuovo and Bric del Carmo tunnels and the Cerusa Est and Ovest viaducts, and to be accessible from service road VS03B
CO02S	To enable construction of the tunnel Bric del Carmo, and to be accessible directly from motorway A10 towards Ventimiglia, taking advantage of the widening of the existing Cerusa viaduct
CO03E	To enable the excavation of the Voltri Est and Voltri Ovest tunnels, and to be accessible from service road VS020
CO03W	To enable the excavation of the Voltri Est and Voltri Ovest tunnels, and to be accessible from service road VS04B or directly from motorway A26 towards Gravellona Toce. Any material containing asbestos in percentages sufficiently high that would require it to be taken to an authorised landfill site will be transferred to big bags on a specially created dedicated site immediately adjacent to the works area
CO04N	To enable excavation of the Ciocia tunnel, and to be accessible directly from motorway A26 towards Gravellona Toce
CO05S	To enable construction of the Delle Grazie tunnel, and to be accessible directly from motorway A26 towards Genoa
CO06W	To enable construction of the Amandola tunnel and for commissioning the Leiro viaduct, and to be accessible from service road VS006

3 LOCAL ENVIRONMENTAL CONSTRAINTS AND DESIGN SOLUTIONS

3.1 *Natural presence of asbestos in rocks, and design solutions*

From a geological standpoint, the area involved in the construction of the Genoa Bypass is part of a region of great structural complexity where the Alpine orogenic domain is juxtaposed to the Apennine domain. Due to their genesis, the mineral composition of the rock mass in the west Polcevera mountains (where Voltri is situated) shows the potential presence of asbestos.

Since the potential presence of asbestos fibres released into the air presents a risk for human health, specific procedures and technologies were planned at the design stage for use during implementation, aimed at safeguarding the workers and the surrounding environment, and covering the whole "life-cycle": from initial excavation to categorising the material at the end, in order to identify its final destination.

3.1.1 *Operational prescriptions for excavations in the presence of asbestos*
Firstly and foremostly, the operational prescriptions to be adopted when excavating tunnels and/or surface excavations in ground potentially containing asbestos were identified.

The most significant impact, which acts directly on the atmosphere component and indirectly on other components, consists of dust emissions emanating from excavation and drilling activity and from the transportation and stockpiling of the excavated material; there may be a presence of dust at the site itself and along the route leading to the sites of final destination for the excavated material. For the purposes of environmental safeguarding and in respect of the provisions of prescription A13 in Environmental Impact Assessment Decree Dec/VIA 28/ 2014, various other measures were identified for containing these emissions in addition to the standard operational precautions:

– delimiting the site areas and using appropriate dust suppression systems of "fog cannon" type;
– continuous monitoring for asbestos fibres in the area surrounding the work sites;
– suspending open excavation works when the wind speed exceeds 5 m/s;
– covering the load beds of the vehicles used to transport the materials with sheets or similar equipment, to prevent the dispersion of dust.

3.1.2 *Technologies for excavating in the presence of asbestos*

Special excavation methods and other appropriate measures were also introduced.

In order to reduce asbestos fibres directly at the "source" as much as possible, for traditional tunnel excavation the material will be initially wetted using a spray system mounted directly on the demolition hammer; the rocky material deposited at the foot of the excavator will also be given a second wetting on the ground using fog cannons.

Conceptually, the tunnel can be divided into three parts:

– An uncontaminated zone, where all work not related to tunnel boring will be carried out, such as casting the linings for the second stage;
– a decontamination zone, where all decontamination of personnel will take place and the equipment that has been used in the contaminated area will be decontaminated;
– a contaminated zone, where all work relating to the excavations will be carried out.

To prevent the dispersion of fibres towards the outside and limit it to the contaminated area only, which is obviously where the most intense production of fibres will occur during excavation, a suction ventilation system will be installed to capture the fibres dispersed in the air as close as possible to the "source", to "reclaim" the environment in that zone as soon as possible after the excavation work has been completed, and to make safe the face by installing the pre-lining. The ventilation system must begin from the tunnel entrance and reach the suction point as close as possible to the excavation face, and must ensure a continuous flow of healthy air within the tunnel.

All water produced by the decontamination activity, by the wetting processes in the tunnel or on the external yard, and any water streams intercepted at the excavation face, will be collected and then conveyed to the purification plant, which will be include a section for the ultrafiltration of any water potentially containing asbestos fibres.

Two technologies have been identified for constructing structural wells (well foundations for the backfill to the tunnel entrance work areas and for stabilising slopes). Unlike the construction of a well of traditional type (where the machinery operates directly in the shaft and drives the excavation vertically downwards), these will enable the earth to be removed without any direct contact between the operators and the asbestos-containing material.

The first construction type consists of an outer ring of secant piles and piles constructed downwards from the surface ("well piles"). This solution will be adopted mainly for backfilling the work yards at the tunnel entrances and as the foundation for the pylons and abutments of the viaducts, in the absence of any significant pre-existing slope instability. "Well piles" will be used, for example, on a number of the pylons and abutments for the Cerusa and Leiro viaducts.

The second construction type will be used where there is a presence of significant landslide events, the stabilization of which will require internal inertia and strength that can only be provided by full-section reaction wells. These will be constructed using vertical cutter technology (VSM type: "vertical shaft sinking machines"), which enables the excavation to be carried out using a roadheader lowered directly into the excavation. The excavation is driven down vertically by using the excavation mud to remove detritus and as short-term support for the walls of the well; as the excavation proceeds to a greater depth this provisional support is provided by prefabricated concrete segments. When the lowest level of the well has been reached, the excavated mud will be dried out, the lean concrete for levelling the bottom of the excavation will be cast, the reinforcement will be placed, and the concrete will be cast to fill the well. This type of well will be used, for example, when backfilling the site work yards at the Voltri Est and Voltri Ovest tunnel entrances and as the deep foundations for a number of abutments and pylons of the Cerusa and Leiro viaducts.

3.1.3 *Managing potentially asbestos-containing excavations*

Finally, a system has been developed for sampling and subsequently managing the excavated potentially asbestos-containing material, based on the concentrations of asbestos that are expected in some parts of the project.

At the implementation project design stage, the asbestos risk parameters were defined for the full length of the geological section of the project. This forecasting analysis made it possible to estimate the total volume of rock in the zones (petrofacies) in which the concentration of asbestos could exceed the limit of 1,000 mg/kg.

Throughout the whole project, the material removed from the excavation will be sampled, analysed and classified in accordance with the regulatory requirements for managing asbestos-containing materials, to define its final disposal and reuse.

In general, deep and underground excavations within a rocky mass have chemical concentrations that can be attributed exclusively to the natural composition of the rock. However in the case of surface excavations it must be verified whether any anomalous value encountered is attributable to the nature of the material itself or is anthropogenic. In that sense, in the tunnel entrance areas, along the service roads, etc., the sampling plan provides for preliminary verification of the excavations. The direct on-site surveys and the analytical laboratory results will be accompanied by the information that has been developed at the various design stages; in this way it will be possible to confirm whether there is a presence of values higher than the Contaminant Threshold Value (CTV), as per Part 4, Annex 5, Table 1, columns A and B of Legislative Decree 152/06, based on the geological conditions in the territorial context to which it belongs, and to reuse of excavated material in those same settings and/or geological contexts, as per the approved Use Plan; on the other hand, anything that cannot be attributed to the natural background must be managed in accordance with the applicable regulations.

The excavated materials, including those that exceed the CTV limits mentioned above that can be attributed to background levels, can be re-used for backfills, embankments and earthworks for remodelling in the area of the construction works, since these works are similar to sites for industrial/commercial use as referred to in column B mentioned above.

3.2 *Landscape constraints and project solutions*

Specific project solutions had to be identified to comply with landscape protection constraints, particularly by providing for morphological figurations that are consistent with the local territory.

The landscaping and architectural design was developed on the basis of requirements introduced by prescriptions B1 (*"In relation to inserting the Delle Grazie tunnel of the so-called "Ramp 2" of the interconnection at Voltri the proposed solution must be agreed with the competent Superintendency...."*) and B2 (*"For the whole length of this section of the road, the solutions arrived at must be compatible with the context to which they belong, avoiding, in particular, the visual impact produced by the containing walls and proposing an arrangement for the landscaping that takes account of the pre-existing landscape, ..."*), contained in the Environmental Impact Assessment Decree Dec/VIA 28/2014.

Beginning from an analysis of the existing landscape that will be affected by the works, the project develops their architectural aspect and their insertion into the landscape, defining morphological remodelling works that are homogeneous and integrated into the landscape receiving them, paying particular attention to safeguarded areas or areas that in any case are characterised by landscape of high quality, not only because of the richness of their natural or man-made components, but also because of the extent to which are part of panoramic settings.

In line with these design solutions, the excavation fronts will be covered by graduated earth movements to form embankments and berms that make it possible to reconnect the areas that will have been altered by the presence of site works. This morphological remodelling will be carried out using banks of natural soil and on slopes whose surfaces will be suitable for the planting and growth of native species and will restore the continuity of the vegetation. In some areas, where the excavations are left open following the morphological remodelling, provision has been made for anti-erosion and re-naturalisation works with deep-rooted hardy perennial herbaceous seeding to guarantee the stability of the slope using only natural plant material.

In compliance with prescription B1 of Environmental Impact Assessment Decree Dec/VIA 28/2014 previously mentioned, the entrance of the Delle Grazie tunnel, which is within a national park, has been studied in depth; the project provides for the application of a natural stone facing to the reinforced concrete works to simulate the dry stone walls that are often used to construct terracing. In some areas where the excavations will remain open after completion of the morphological remodelling, provision has been made to restore the vegetation by spreading reinforced biomats and wire mesh on the excavation face, fixed with steel rods, followed by the final application of hydroseeding mixes: materials that are suitable for ensuring the stability of the slope.

Figure 5 is a photomontage showing the definitive morphological remodelling of the tunnel entrance at the south end of the Delle Grazie tunnel.

Considering the particularities of the valley of the Leiro stream, special care has also been taken in the design of the viaducts that cross the valley; significant environmental optimisations have been introduced (see Figure 6), to minimise the impact on the landscape by working on the following elements:

• regular alignment of the pylons of both viaducts;
• the particular design of the pylons (concrete plinth, surmounted by a trussed part);
• application of a paint finish to the solid parts of the pylons in a "grey" colour and construction of the trussed part in natural and painted corten.

Figure 5.   Photomontage: tunnel entrance at the south end of the Delle Grazie tunnel.

Figure 6.   Photomontage: Leiro Est and Ovest viaducts.

4   CONCLUSION

Since today's technology makes it possible to suitably manage the excavation of hazardous rocks, this is a critical issue that can now be overcome; it is now possible to operate in a healthy and safe environment during all the stages of construction, from initial excavation to final disposal of the material.

Specific technologies have been identified in order to control the asbestos risk during the excavation stages, containing the fibres that might be released, confining the material produced during excavation and treating site water before returning it to the hydro-graphic system.

The environmental context of the Voltri area is subject to various landscape protection constraints. For such reason, the Bypass had to be inserted into the local context by identifying morphological configurations consistent with that territory. In order to counterbalance some of the effects of the main works, particular attention was paid in selecting the materials and defining the architectural characteristics of the technical solutions adopted.

REFERENCES

Capponi, Crispini, 2002. Structural and metamorphic signature of Alpine tectonics in the Voltri Massif (Ligurian Alps, North Western Italy). *Helv. Eclogae geol.*

Cortesogno, Haccard, 1984. Note illustrative alla carta geologica della zona Sestri-Voltaggio. *Mem. Soc. Geol. It.*

Stellini, 2014 Tesi di Laurea Magistrale, "La problematica ambientale delle fibre di amianto nei terreni e nelle rocce da scavo. Proposte analitiche innovative per affrontare casi reali complessi". *Torino: Univ. Studi Torino.*

*Tunnels and Underground Cities: Engineering and Innovation meet Archaeology,
Architecture and Art, Volume 2: Environment sustainability in
underground construction – Peila, Viggiani & Celestino (Eds)*
© 2020 Taylor & Francis Group, London, ISBN 978-0-367-46579-7

Tunnel excavation material – waste or valuable mineral resource? – European research results on resource efficient tunnelling

R. Galler
Chair of Subsurface Engineering & Department ZaB-Zentrum am Berg, Montanuniversität Leoben, Austria

ABSTRACT: The amount of different minerals in tunnel excavation material varies in a large range. Therefor industrial clients who could be interested in raw materials extracted from underground construction sites were investigated. Parallel to the development of advanced online technologies for analysing the excavated materials research is looking for important information regarding requirements for raw materials used in industrial process-es, which are grain size distribution, mineralogical composition, geochemistry as well as different water content and water absorption properties.

1 INTRODUCTION

Europe 2020 is the European Union's ten-year growth strategy and is about delivering growth that is smart, through more effective investments in education, research and innovation and sustainable, thanks to a decisive move towards a low-carbon economy. Of relevance to the research project which is discussed here, were the targets related to climate change and energy sustainability. The ambitious targets that were set by the EU have been adopted into the *Europe 2020* strategy which are a reduction in EU greenhouse gas emissions of at least 20% below 1990 levels, 20% of EU energy consumption to come from renewable resources and a 20% reduction in primary energy use compared with projected levels, to be achieved by improving energy efficiency. Collectively these are known as the 20-20-20 targets. *A resource-efficient Europe* is one of seven flagship initiatives of the *Europe 2020 Strategy* and aims to support the shift towards a resource-efficient, low-carbon economy to achieve sustainable growth. This recognises that natural resources underpin our economy and our quality of life and that continuing our current patterns of resource use is not an option. Increasing resource efficiency is the key to secure growth and jobs for Europe and will potentially bring major economic opportunities, improve productivity, drive down costs and boost competitiveness. This flagship initiative has moved the extraction and use of natural resources into the centre of the political agenda of the European Commission. The strategy establishes resource efficiency as the guiding principle for EU policies on energy, transport, climate change, industry, commodities, agriculture, fisheries, biodiversity and regional development. The flagship initiative connects policies related to resources such as the *Roadmap for a resource efficient Europe* (EC 2011b) and the *Raw Materials Initiative* (EC 2009c).

2 OBJECTIVES OF THE RESEARCH

The research project – *Development of Resource-efficient and Advanced Underground Technologies* – intends to improve resource efficiency in tunnelling and other underground construction processes by providing the excavated material as a raw material for the construction site as well as for other industrial sectors which are using primary mineral ressources. Within the research project a system of automated chemical, physical and mineralogical online-analysis techniques all mounted on a bypass conveyor belt are developed. These analysing techniques are followed by a separation plant using recycling units like crushers, sieves, etc. (Figure 1). So the excavation material can be recycled and used directly onsite as construction material or by

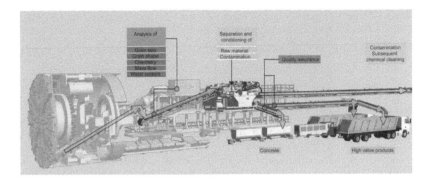

Figure 1. the concept of the research project.

being transported to the particular industry sector. One aim of the project is to place the whole process, starting from the characterisation of the excavated material to its classification and processing, completely underground.

The results of the project will help to reduce or even elimintate any material deposits, thus aiming the production of zero waste originated from underground construction sites; furthermore environmental impacts like production of noise and dust will significantly be reduced by recycling the excavated material onsite. The research project applies the methodology of Life Cycle Assessment (LCA) according to (ISO 14040/14044) in order to include life cycle thinking into the project. Additionally the mass flows are analyzed by the Mass Flow Analysis according to (Baccini and Brunner, 2004). The reason for applying these methods is to compare different scenarios of recycling or disposal of excavation material. This allows observation of the whole system in order to avoid any problem shifting into other parts of the environmental system. Substitution effects caused by replacement of primary material with excavation materials are identified and quantified. Considering these aspects, underground constructions in the future will much more contribute to resource efficiency and the reduction of CO_2- emissions compared to today's state of the art.

3 OVERALL STRATEGY AND GENERAL DESCRIPTION

The current legal framework for the utilisation of excavation material is not yet completely satisfying and requires a change of rules where geochemical, geotechnical, mineralogical parameters are clearly specified for being allowed to reuse the excavated material as mineral product. The declared goal is to reach zero waste from underground construction sites, at least when rock properties and the demand allow it. A legally binding rule that material extracted in tunnelling is preferred for use as long as its overall suitability can be demonstrated would not only create new raw materials potential. Companies in the mineral raw materials industry would obtain already crushed material at favourable rates, save their existing quarries and thus extend the lifetime of their companies. Following the results of the studies of (Kündig et.al., 1997) such a legally binding rule will get significantly important at least for European countries as mineable ressources get dramatically less in future (Figure 2).

The fear that insufficiently stringent statutory guidelines could lead to the propagation of hazardous substances is unfounded with regard to tunnel spoil as long as thresholds and tests are specified for the material, corresponding to the products of the mineral raw materials industry. A clear differentiation must be made between geogenous and anthropogenous contamination. In order to gain legal certainty, a product status is still lacking for tunnel spoil, which is allowed to be used as quality-assured recycling material. To go a step ahead Austrian engineers together with the Federal Ministry of Agriculture, Forestry, Environment and Water Management wrote a first guideline called *Verwendung von Tunnelausbruch* published by the ÖBV – Österreichische Bautechnik Vereinigung.

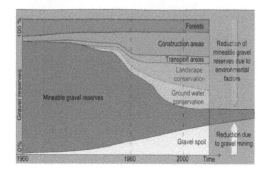

Figure 2.   The conflicting utilisation claims of the resource gravel (according to Kündig et al., 1997, and Jäckli & Schindler, 1986, modified).

4   IMPACT ON SOCIETY, ECONOMICAL AND ENVIRONMENTAL IMPACTS

The developed prototype could also be of interest for mining companies which focus on resources of critical raw materials like gold, platinum, rare earths and others. Extracting the minerals containing such elements in a selective manner could facilitate the separation from dead rock and could allow the exploitation of mineral deposits which are currently not attractive enough due to their relatively low concentration of valuable elements. One of the main objectives of the research project was to contribute to the natural resource conservation within the European Union. Depending on the geological composition of the material it is possible to recycle up to 100% of the excavated material. The reduction of transport ways, the diminution of pollutants as well as the recycling of the excavation material possess a large environmental protection potential. The main expected outcome of the LCA (life cycle analysis) is to provide scientific evidence that the recycling of excavated tunneling material will result in more resource-efficient and more closed-loop related systems (even in the industry-related economy) in Europe. One of the goals is to establish a close relationship to the external surroundings (national environmental authorities; standardisation bodies etc.) of the project in order to implement and integrate the project results and findings quite smoothly within the specific regional, national and/or international environment. Main target of the research project are to act as Best Practise case how underground excavation material can be recycled as valuable material in diverse industrial processes and sectors. The dependencies of the individual fields of interest which have to be considered in order to recycle excavation material from underground construction sites are shown in Figure 3.

4.1   *Economic considerations*

When tunnel spoil is recycled as a construction raw material, this has extensive economic effects in addition to the ecological effects. The economic benefit comes from the earnings that can be made by selling a certain material quality and also from savings, first from the substitution of purchased aggregates for the internal needs of the site with tunnel

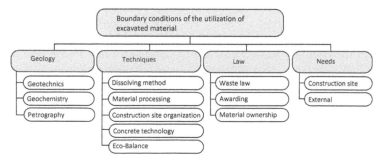

Figure 3.   Topics to be considered for utilization of excavation materials from construction sites.

spoil, and second through reduced landfill costs. As soon as the sum of earnings and savings exceeds the cost of additional materials handling, the recycling of tunnel spoil can be considered profitable.

4.2 *Recycling possibilities*

The recycling and marketing of excavated material can only be achieved when the material meets the requirements of the consumer companies and there is also a market for it. If the material meets the requirements for concrete aggregates, it should be recycled directly onsite during the construction phase of the project. If the production of high-quality and "problem-free" aggregates exceeds the demand onsite, markets to local producers and processors should be found as an alternative. If lithology is encountered, whose analysis has indicated suitability for recycling as industrial minerals during the site investigation, the aim is to reuse this rock in mineral-processing companies. Excavated material, which cannot be used for more high-quality applications can be used for landscaping inside and outside the site, as long as its suitability can be demonstrated and the legal framework is complied. If this kind of recycling is not possible, the excavation material has to be disposed. The intention of recycling is not to set up as a competitor to local raw materials companies but to make the material available at a reasonable price in order to save raw material deposits.

5 MATERIAL MANAGEMENT CONCEPT

The usability of excavated material mainly depends on the material properties and the composition of the material according to the respective raw material that is to be processed. In this connection the relevant requirements for the raw material are specified by the usability and demand at the jobsite as well as by the industrial companies in the surrounding area. This is why a material management concept that is coordinated with the expected masses of excavated material should already be prepared in the planning phase of a tunneling project. This concept includes the possibility of a self-supporting jobsite and a resource-efficient industrial utilization. Furthermore it considers the ecological intermediate storage of oversupply in excavated material to cover future demands, and the usage respectively disposal of less quality material. Additionally the transport routes, material specific processing, intermediate storage and landfilling are also taken into consideration in terms of CO_2-emission balancing. The material management concept is a basis to prepare delivery contracts with potential purchasers. The usability of excavated material mainly depends on the geological realities on job-site, and has to be essentially oriented to these facts. Generally it needs to be distinguished between hard rock and soft-/mixed ground. Figure 1 shows the overall material management concept for excavated material. The concept is based on an online database. This database contains a requirement matrix with integrated specific requirement lists for the relevant usage and further processing of raw material. The individual utilization scenarios can be derived from the requirement matrix.

Within the scope of a tunneling project, the constructor captures data of the subsoil via the exploration process. For example, probe drillings are made from the surface in order to obtain the essential information regarding complex geological situation. On the basis of the resulting geological profile, volume and properties of the future excavated material can be previously estimated. Using the database, the constructor can thus immediately determine the utilization potential of the subsoil. The subsoil data and the legal requirements are entered in this database in a project-specific way. Intermediate purchasers and material processing companies can also access this database and specifically store their demands and the requirements (material properties, time of demand, volume, maximum transport distances, etc.). The online analysis on the conveyor belt continuously transfers the measuring results to the control cabin. Using data processing software, the excavated material is classified by the directly comparison with the required specifications in the online database. Sorting the excavated material into different categories is thus already possible during the excavation process. The captured measured data for quality and quantity is transferred continuously to the online database. The relevant data can thus continuously be updated and provided to the intermediate purchasers via notification or inquiry. In parallel, the analysis data is used for the self-supply of

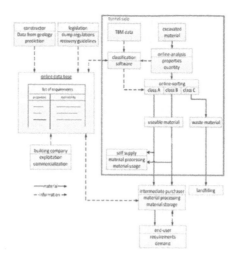

Figure 4. Material management concept.

the jobsite with excavated material. In case the properties of the excavated material do not comply with the quality requirements for the directly usability at the jobsite or in industry, then it is mainly used for embankments or landfilling. Therefor the relevant guidelines and regulations for landfilling need to be observed. In this context the waste avoidance has the topmost priority here. The constructor (owner) as well as the contractor (building company) are primarily responsible for the material usage and waste management and are the first who will realize material management concept. Marketing of the excavated material can have significant economically benefits for the constructor, even if there are at least no costs for landfilling, and ideally the material turns out to be a lucrative raw material as a by-product.

6 LATEST TECHNICAL DEVELOPMENTS

The main technological developments within the research project include fully automatic processes for fast detection of usable materials and the recycling the materials immediately on the conveyor belt. Conveyor belts offer great opportunities to automate analyses, classification and processing of the excavation material to ensure recycling in a proper way. Newly developed measuring gear for online-measurements need to be installed directly onto the hauling installations (Müller et.al, 2011). The main parameters of the material concerning its suitability as a resource are Chemistry and impurities, grain size distribution, water content, grain shape and mass flow.

6.1 DCLM – Disc Cutter Load Monitoring

The physical characterisation of excavated material can be best detected by measuring the disc force. (Entacher, 2013) has shown that geological conditions in front of the cutterhead can be described with the disc cutter force characteristics. These developments should lead to a better understanding of the relationship between pure rock parameters, rock mass strength, geology, cutting forces cutter wear.

For this purpose, Herrenknecht AG and the Montanuniversität Leoben developed various methodologies for measuring the cutter force on discs while the machine is boring.

6.2 Automated sampling from the conveyor belt

The sampling has to be done very close to the tunnel face. With automated sampling procedures new standards are set in underground constructions. The sampling system has to be available throughout the whole construction time with only short breaks for cleaning and maintenance during TBM standstill times.

Figure 5.   Assembled DCLM test rig built in the MTS 815 servo-hydraulic press.

6.3   *Automated measuring of grain size and grain shape*

For various reuses the grain size distribution is an important input data. Therefore the research project uses a technology which determines the particle size distribution of the material fully automated. The excavated material is analysed photo-optical and the resulting image is evaluated digitally. This measurement method is able to analyse grain sizes and shapes of dry, non-agglomerating particles in the range from 10 μm to 400 mm. The main advantages of photo-optical particle analysis to be used underground are the rapid detection of particle size, particle shape and particle number, the time savings in the measurements compared to a sieve analysis and the high level of automation

Figure 6.   Photo-optical system mounted along the prototype.

6.4   *Material analysis techniques by X-ray fluorescence analysis*

The X-ray fluorescence spectroscopy is one of several possible methods to find out the exact elementary composition qualitatively and quantitatively. The results are converted to the

Figure 7.   Online x-ray chemical element analyser OXEA (by INDUTECH).

oxides of the single elements and give an overview of the percentage of each component in the sample. This analyzing technology is installed immediately behind the particle size analyser.

7 USER SCENARIOS

7.1 Railroad projects in Germany

PORR is involved in major projects in southern Germany, where 26 km of tunnels, producing 4.5 million tons of excavation materials. A large amount of these materials have a high content of gypsum, salt and clay. Other layers are mainly formed of sandstone. The goal is to clearly separate materials which can be used for dams, noise protection walls, concrete production and diaphragm walls from such which can only be used for landfilling. Some of the materials also have the potential to be used in the ceramic industries for bricks and other coarse ceramic products. The sandstones could mainly be used in the construction sector. It can be clearly stated that the addressed industries are only willing to accept the materials if the quality is near or equivalent to the raw materials they are used to procure. In any case an early detection of the chemical and morphological characteristics would help to increase the amount of usable materials and would be a precondition for utilization in some industries. On the other hand the detection and separation of some expected metals like Nickel, Chromium, Cadmium could either lead to an industrial utilization or at least to a reduction of disposal areas required for hazardous waste. In both cases it is mandatory not to mix up excavation materials containing these metals in a relatively high content with others because that would lead to a dilution and so to an unfavorable increase of the masses which have to be processed or disposed.

7.2 Tunnel Lyon – Turin connecting France and Italy

The new railway line from Lyon to Torino is part of the West-East European corridor linking Barcelona to Kiev. Some 250 km of tunnels will be built on the section including the French and the Italian parts. The key part of this new route is the Maurienne-Ambin Base tunnel, about 57 km long, crossing various and difficult geological formations. The valorisation of the excavated material is a major issue for this project which will produce some 35 million tons of rocks, of which an aggregate for a total quantity of about 12 million tons could be produced for the 6 million m^3 of concrete for the project. On the central section, sulphated materials are expected, and they must be identified on the transportation system, into the tunnel, before the stockpiling and before the transformation into aggregates.

7.3 Les Farettes – Romande Energie Hydropower Project in Switzerland

This case study is of major interest for the research project because it is showing that it is possible to process excavation material with a light dry process and to produce concrete on a simple way, inside the tunnel, if the diameter of the tunnel is big enough to implement the required processing equipment.

7.4 Nant de Drance Hydropower Project in Switzerland

This case study is representing the most beautiful case for the *on the job site* excavation material monitoring and processing organization. The Owner is supported by a dedicated staff managing the excavation material classification, handling, storing and the aggregates production. The aggregates processing plant is fitted with the latest technology equipment. It is possible to feed a large range of concrete types with TBM and NATM excavation materials. Furthermore the quality of the aggregates is completely in accordance with the special requirements for crushed aggregates and high quality concrete for underground works.

The analysis of Reference Projects is showing that the objectives of the research project are definitely on the right track in order to develop online excavation materials analysis methods for both chemical and physical characterization.

8 CONCLUSION

Using the analyses techniques developed in the described research project allows an immediate analyses and recognition of valuable materials close to the tunnel face and thus forces the direct utilization of the excavated material. The new developed technologies in order to recycle the excavation material will have a strategic impact on sustainable management of limited mineral resources, higher resource efficiency through a recycling process and related decrease of EU dependency on resource imports, lower negative environmental impact, more competitiveness of all underground construction related companies and organisations and new resource-efficient environmental technologies. The rate of utilization of excavated materials from tunnelling will be significantly increased. Most of this expertise is already based on previous funded national projects and initiatives, in which some of the partners have been involved dealing with the recycling of tunnelling excavation material and the development of an innovative separation technology for bentonite.

The new and innovative technologies within the research project can also be used for the control of a tunnel boring machine by the fragmentation of the excavated rock which leads to an optimisation of the machine operation. Such a concept has never been realised before. However, previous fundamental research work at the Montanuniversität Leoben has shown that such a concept is feasible.

The exploitation of minerals is usually affected by conflicts between the economic interests of the extractive industries or the construction sector and environmental protection concerns, but also declining sizes of natural stocks have an influence. To overcome such problems new ways of making minerals available on a regional and local level through tunnel excavation projects is a possible solution. The excavation close to the users is very important as it will help to reduce the consumption of fossil fuels for production, processing and transport.

ACKNOWLEDGEMENT

This research has received funding from the European Union's Seventh Programme for research, technological development and demonstration under grant agreement No 308389.

REFERENCES

Barwart, S., Dissertation, Montanuniversität Leoben, "Entwicklung eines Schneidkraftmesssystems für diskenbestückte Tunnelvortriebsmaschinen", 2016

Brunner, P.H. & Rechberger, H. 2004. Practical Handbook of Material Flow Analysis, in CRC Press LLC, Boca Raton, Florida.

Baccini, P & Brunner, P.H. 1991. Metabolism of the Anthrosphere, in Springer, Berlin, Heidelberg, New York, 1991, Page 157

Entacher, M. & Lorenz, S. & Galler, R., 2014. Tunnel boring machine performance prediction with scaled rock cutting tests. Int. Journal of Rock Mechanics & Mining Sciences 70, pp. 450–459, 2014.

Erben, H., Dissertation, Montanuniversität Leoben, "Real-Time Material Analysis and Development of a Collaboration and Trading Platform for Mineral Resources from Underground Construction Projects", 2016

Erben, H. & Galler, R. 2014. Tunnel spoil – New technologies on the way from waste to raw material, Geomechanics and Tunnelling 7 (2014), No. 5, pp. 402–410.

Galler R. & Voit K., 2014. Tunnelausbruch – wertvoller mineralischer Rohstoff, in Betonkalender 2014, 103. Jahrgang, Verlag Ernst & Sohn, pp. 423–468

ISO 14040: Environmental management - Life cycle assessment - Principles and framework/ISO 14044: Environmental management - Life cycle assessment - Requirements and guidelines.

Kündig, R., 1997. Die mineralischen Rohstoffe der Schweiz, Zürich, Schweizerische Geotechnische Kommission, 1997.

Müller, S. B. & Zwicky, C. N. & Blahous, L., 2011. Kalibrierung eines NIR-Online-Analysators mittels Bohrmehlproben zur Optimierung des Zement- Rohmaterials, BHM 156. Jg. 2011, Heft 6, pp. 225–226.

Resch, D. & Lassnig, K. & Galler, R. & Ebner, F., 2009. Tunnel excavation material – high value raw material, Geomechanics and Tunnelling 2 (2009), No. 5, pp. 612–618

Tunnels and Underground Cities: Engineering and Innovation meet Archaeology,
Architecture and Art, Volume 2: Environment sustainability in
underground construction – Peila, Viggiani & Celestino (Eds)
© 2020 Taylor & Francis Group, London, ISBN 978-0-367-46579-7

City Rail Link, New Zealand – reduction of land acquisition time and risk on the City Rail Link Project

R. Galli
City Rail Link Limited, Auckland, New Zealand

T. Ireland & C. Howard
Aurecon, Auckland, New Zealand

ABSTRACT: The City Rail Link is New Zealand's first underground passenger railway and largest transportation project undertaken in New Zealand to date. The $3.4b project comprises 3.4km of twin tunnels and two new underground stations. Commercial property subsurface land acquisition can be complex and inefficient, and a major program risk. A portfolio of 120 sites were assessed for future development loads with screening tools based on international best practice. Each site assessment included a geotechnical profile, potential foundation schemes, a load check and cost advice where modified foundations will be necessary above the tunnels. Investment in early landowner engagement and focused communication processes minimised the fear of the unknown for landowners, enhanced project relationships and improved City Rail Link Ltd's (CRLL) reputation.

1 INTRODUCTION

Auckland's City Rail Link (CRL) comprises the construction, operation, and maintenance of a 3.4km underground passenger railway running between the existing Britomart Station and North Auckland Line (NAL). The CRL will provide more frequent trains with more direct services to the city centre and allow a train every five to ten minutes from most Auckland stations.

Construction of the CRL will double the train capacity through Britomart and will provide significant network capacity improvements with up to 24tph operating in each direction. CRL will introduce 3.4km of underground railway including three new underground stations running between the existing underground Britomart Station and the NAL. The alignment and long section is shown in Figure 1 and 2. The new stations comprise two cut and cover stations at Aotea and Mt Eden and one mined station at Karangahape Road. With a maximum design line speed of 70km/h, the travel times between stations will be approximately 2 to 2.5 minutes.

CRL will be the largest NZ infrastructure project to date. One of the key risk elements for any major underground project in the CBD is the property acquisition workstream. CRLL completed acquisition for the entire title properties (including surface stations, cut and cover sections) in 2016, involving 68 freehold and 14 commercial lease interests. A further 91 negotiations are required for the 160 substrata interests from private landowners. This paper describes a new approach to compensation assessment and engagement with property owners to de-risk the process.

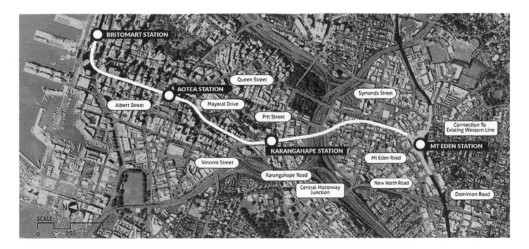

Figure 1.   Aerial view of Auckland CBD showing the CRL alignment.

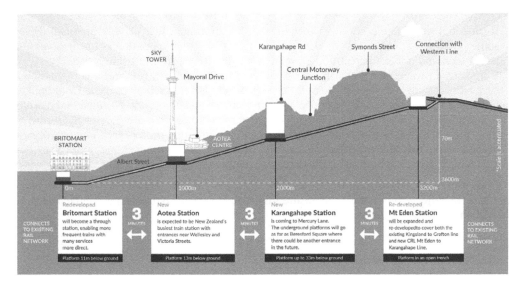

Figure 2.   Illustrated long section of the Auckland City Rail Link.

2   BACKGROUND

The property acquisition team considered domestic and international experience as well as issues encountered by New Zealand Transport Agency (NZTA) during the recent Waterview State Highway tunnels substrata acquisition programme. The key issues relating to the acquisition of substrata for the CRL are:

– Definition of required substrata land for the CRL and protection requirements
– A robust compensation methodology, supported by the NZ Valuer General
– Consideration of the existing built environment and potential redevelopment within the Auckland Operative District Plan
– Communication with landowners and high-quality communication materials
– A well-coordinated negotiation processes programme to acquire interests by agreement
– Excellent process for reviewing and approving future developments above the CRL

Acquiring substrata land for transport infrastructure in built-up urban environments is commonplace in major cities throughout the world; however, it is new to New Zealand. The NZTA's recent experience in the state highway Waterview tunnels primarily involved low-density residential zoned land (much of it owned by a New Zealand Government agency), NZ Railway land and Council land. Only a few commercial interests were affected by the Waterview tunnels.

The Waterview project was very helpful for several reasons:

– It set a 'first project' range on compensation for acquiring substrata from low-density residential zoned land. We would expect commercial land to see less before and after value difference
– The project actual ground settlement that drove initial fear of value loss was far less than predicted (and in some cases immeasurable). There have been no claims for damages to property
– There were no instances of the detectable operational or tunneling construction noise and vibration that caused considerable initial public fear

International precedent was investigated including mid to high-rise experience with substrata acquisition in California and Seattle (USA), Melbourne and Brisbane (Australia) to learn about the experiences of acquiring and compensating commercial owners for subterranean land. These municipalities have similar land ownership to New Zealand. While the land ownership, acquisition processes and local legislation and rights vary, there were common themes in terms of landowner engagement that are generalised below:

– Residential single-family dwelling owners: can be emotional, fearful, non-financial driven, some physically affected (loss of sleep) by change. Courts are empathetic, but it is value to the market and not lost perception of value to the landowner that prevails.
– Commercial owners: are primarily economic, used to dealing with risk and a combination of variables, 'cash in and expenditure out' that is generally not affected by tunnels. Land banking for future is important. Capable of dealing with change and complexity. Normally take professional advice including geotech, planning and structural advice, the presence of the tunnel for most properties will be a minor further consideration. Unit title owners are more accepting of dealing with a raft of issues affecting common property, tunnels are just one more. They have a structured group support and body corporate that moderates impact to any one-unit owner.

Given this, the CRL acquisition programme has been shaped to respond to the commercial nature of the affected landowners to ensure that their individual circumstances have been considered and reflected in the processes.

3 SUBSURFACE ACQUISITION

As is common for urban railway projects, the route alignment was originally developed to minimise private property, direct impacts and acquisition as summarised in hectares and as a percentage of the overall CRL substrata requirement as shown in Figure 3 below.

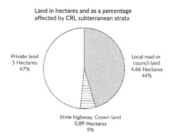

Figure 3. Substrata land required for the CRL by landowner type.

There is a mix of large residential developments with body corporate managers, residential properties, commercial development and retail development. There are a number of sites where redevelopment is likely. There are no significant land holdings by a single private landowner.

A few sites within the designation have recently applied to Auckland Council (AC) for consent to develop above the CRL. AC provides the applications to CRLL to review in terms of the CRL prior to granting consent. This process will be ongoing.

Confirming substrata volumetric requirements

The substrata acquisition programme encompasses an envelope, located between [20–40 metres] below the existing ground level to the center of the earth. In New Zealand every landowner (unless indicated otherwise by survey) owns their land to the center of the earth.

As the design has progressed significantly since the land requirements planning processes were initiated, the project has been optimised and design refinement is underway. The land designation footprint has been reviewed to confirm the acquisition boundaries for the current alignment and minimise the impact of land both surface and substrata for the Project.

A set of criteria with respect to the proximity of the tunnel structures to property boundaries has been developed to provide a minimum distance to protect the CRL tunnels. The dimensions are based on the influence zone around a tunnel which is the equivalent to one tunnel (bored or mined) diameter as illustrated in Figure 4 below.

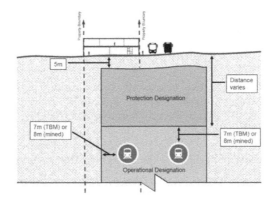

Figure 4. Typical cross section of substrata volumetric requirements.

Optimisation and reference design impacts on substrata land requirements

The CRL tunnel design has been undertaken using standard design criteria consisting of a future development surcharge working load of +/- 50 KPa and excavation of up to 5m from the ground surface for basement construction. The standard development load was checked against a more detailed load for the 'highest and best' development arrangement based on an individual actual urban planning analysis for each site. Where loads exceeded the standard criteria, a detailed assessment was undertaken to determine the load impact on the tunnels. This approach gives a more accurate understanding of what redevelopment loads we would expect in the future.

Programme and timelines

The programme for the substrata acquisition involves:

– finalising the valuations for each property
– preparing a compensation package to be provided to each landowner including planning and development assessments, and
– preparing a communications strategy for open days

The acquisition programme is well advanced and is expected to be completed for construction entry by December 2019. Resolving compensation claims may extend beyond that date.

Protection of assets

Considerations in protecting the underground infrastructure include:

- Direct impact to the infrastructure
- Disturbance of the soil supporting the CRL, and
- Significant load change, positive or negative that exceeds the design criteria the CRL is required to withstand

These threats are most likely to be brought on by future development over the life span of the CRL and not by the existing built environment that the CRL is designed to accommodate (within a 50KPa +/- variance). The designation width and upper acquisition levels proposed currently provides an 8-metre fee simple ownership offset from the edge of any subsurface CRL infrastructure to the boundary of the adjacent landowner. The restrictive covenants registered on each title within the designation protect the subsoil above the tunnel in terms of load change. In New Zealand the Resource Management Act 1991 (RMA), Section 176 provides that no person may, without the prior written consent of that requiring authority, do anything in relation to the land that is subject to the designation that would prevent or hinder a public work or project or work to which the designation relates.

Guidelines and approvals for future developments above the CRL private land

Two new developments have already been submitted and approved as part of the S176 approval process:

- a 20 storey development with two basements, foundation design already proposed piles loading below the CRL that took the load path out of CRL interaction
- a 7 storey building with full basement incorporating car stackers. Risk (unrelated to CRL) to damage of adjacent developments caused the developer to consider shallow foundations and limited excavation that placed the pile loading well above the CRL and provided adequate distance separation to disperse the loads prior to reaching the CRL

Future approvals may not be as straight forward as these. Information learned from the engagement process will assist in the review of guidelines to make it easy for developers, Council and future CRL staff to undertake assessments with predictability and consistency.

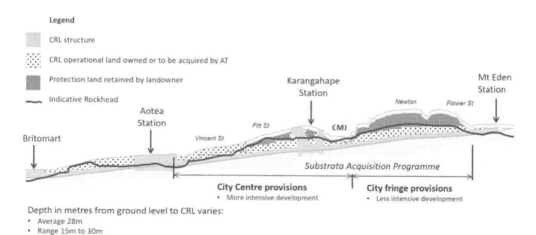

Figure 5.   Long section of substrata volumetric requirements.

Subsurface acquisition process (the How)

A team of specialists were engaged to assist in the technical aspects of the negotiations.

Table 1.   Expert advisors providing advice on compensation assessments.

Discipline	Advice/outputs
Cadastral surveyor	Land requirement plans and new survey plans for subterranean land
Geotechnical engineer	Advises on underlying geology/foundation systems for overhead development
Urban planner	Advises on permissible development under the Operative District Plan
Registered valuer	Advises on highest and best use of land in the present market and prepares a before and after compensation assessment
Structural engineer	Profiles loads for 'highest and best use' future development and advises pile loads
CRL – Principal Technical Advisor (PTA)	Assesses any interaction between future development loads and the CRL
Quantity surveyors	Assesses any cost differential to accommodate the CRL
Independent negotiators	Suppliers that facilitate negotiations between CRL and the landowner

A total of 91 highly customised landowner engagement assessment packages were prepared following the same structured process that yielded:

- Consistency
- Speed
- Necessary flexibility
- Services and output that delivered greater value and transparency

Each landowner received a detailed planning report with multiple development scenarios.

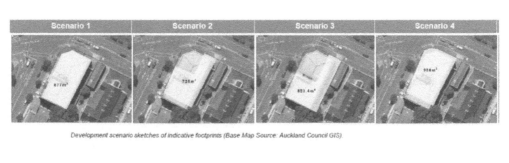

Development scenario sketches of indicative footprints (Base Map Source: Auckland Council GIS)

Figure 6.   Development scenarios.

CRLL's advisors undertook prototype structural assessments for each site based upon the valuer's opinion of the highest and best use of the land. The prototypes comprised a site specific structural grid and a likely foundation plan to ascertain likely building loads, as shown in Figure 7 below. The prototype building engineering assessments have been undertaken in two stages.

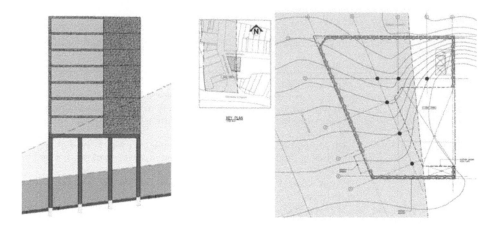

Figure 7.   Prototype development cross section and foundation plan.

– Stage 1 – This is the prototype as if CRL is not present.
– Stage 2 – Is undertaken if CRLL advises that the foundation or loads cannot be accommodated by the presence of the CRL. This may result in modified foundations or reduction of building floors or loads.

CRLL's advisors analysed each of the prototype developments and the first pass was a Numerical Analysis. Finite Element Analysis was undertaken where the prototype loads were of a concern. These are shown in Figure 8 below.

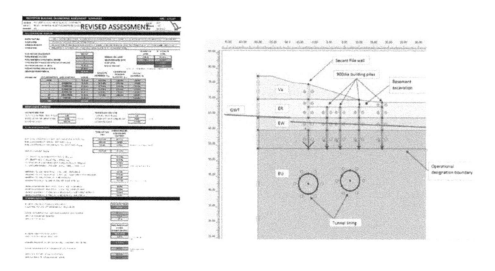

Figure 8.   Numerical analysis and finite element analysis of prototype development.

Results of portfolio assessments

CRLL assessed 106 prototype developments and the results of the foundation impacts are summarized below. Only portions of 3% of the sites had significant development constraints imposed by the CRL.

Table 2.   Potential future development foundation impacts caused by the presence of the CRL.

Foundation type	Category	Quantity	Subtotals	%
Any	Piles or shallow or raft	32		
	Grooved piles or shallow	31		
	De-bonded piles	15		
	Pile array or shallow	2	80	78%
Shallow only	Shallow or raft	20		
	Shallow or raft potentially screw piles	3	23	22%
None allowed	Development exclusion zones		3	3%
	Total assessed	103		

Communication and engagement to mitigate risk

As previously stated, subsurface acquisition is relatively new to New Zealand with very little experience of it among property owners. It was determined at the project outset that investment in communication and engagement with affected landowners was key to mitigating risk in terms of organisational reputation and programme, and could also build ongoing relationships for the project. It was also important to build trust and understanding of the technical requirements and acquisition processes among landowners and their advisors.

Initially this was through targeted briefings and individual meetings when the project was funded for property acquisition, but not for construction. As the project gained political and funding momentum, so did engagement until eventually a comprehensive and individualised portfolio of information was developed for each site and each owner. A personal approach was also taken in how it was communicated, with individual meetings with landowners where project staff talked through every section explaining its content and purpose.

Figure 9.   Pages from the Landowner Engagement Information folder.

Thanks to the relationships built up during this process, the portfolio was able to be tested with independent peer review experts and trusted owners to ensure it was comprehensive, open and transparent, fit for purpose, and easy to use and understand. This resulted in a portfolio of information that has been praised, uncontested in some instances and robust enough to withstand interrogation.

Compensation

The subterranean land to be acquired is identified on a survey office plan, and will be assessed by valuers completing a "before and after" valuation to estimate compensation payable.

Shallow tunnels - have the potential to create a variety of physical impacts on the surface of the affected property. The tunnels may be below or adjacent to existing underground structures. These tunnels may limit the use or development of the remainder of the property, segregate or isolate certain portions of the property for purposes of development.

Deep tunnels – have less overall physical impacts on the surface of the affected property. Any damage to the remainder of the property is non-physical in nature. The remainder of the property may or may not have a decreased value because it is encumbered by the tunnel acquisition.

CRLL sought advice from overseas transport authorities in Australia and USA on the method of assessing compensation and likely quantum of compensation for subterranean land.

After comparing compensation assessment methodologies in New Zealand, Australia and Seattle, the Seattle approach was the most sophisticated. Having been used in more than 1,000 valuations for subsoil, it was developed in an environment where compensation disputes are heard in jury trials. The approach clearly sets out the assumptions in a manner that non-professional people can follow, draws measurable conclusions, and responds to a variety of scenarios that can be easily checked and compared for consistency. The NZ Valuer General, who is the chair of the Valuers Registration Board, supported the Seattle methodology as a preferred basis for the CRL.

The compensation is presented to each landowner in a packaged format including structured framework that apportions the compensation, shown on Figure 10. The benefit of this compensation frame work is that it allows for each aspect of the claim to be interrogated by the appropriate expert, and for adjustments to be considered on a property by property basis.

Figure 10.   CRL compensation package.

Technical concerns encountered in negotiations for subsurface land

Two primary concern issues were raised by landowners in terms of their ability to maximise underground future parking and undertake a seismic upgrade to their property.

Underground car parking

CRL reviewed 35 recent developments in the Auckland centre. In the mid-rise and low-rise zones, buildings of 1 to 5 levels had one basement, buildings of 6 to 20 levels had up to two basements. No evidence from this study suggested the presence of the CRL would have any impact on the market value of the land if up to two basements could be accommodated.

Ability to undertake a seismic upgrade

CRL reviewed 15 seismic retrofit projects in the Auckland region and discussed those with AC. The conclusions from the representative retrofit case studies are that:

– The majority of work occurred at the roof level or within the above ground structure
– 9 of the 15 buildings had no foundation work undertaken, 5 had foundation work within the first 3m below ground, and
– 1 had piles extending 8m below ground level and into the CRL covenant area - this was undertaken in 1998 prior to understanding of more effective techniques

No evidence from this study suggested the presence of the CRL would have any impact on the methodology, techniques or ability for landowners to undertake a seismic upgrade.

4 CONCLUSIONS

The CRL subsurface acquisition programme sought to strike the balance between protecting the asset and not unduly impacting development over or adjacent to the project.

The Seattle valuation methodology as developed by Mr B. McKee provides a robust and adaptable framework and rationale to assess compensation where a large range of variables exist.

CRL developed all the experts' reports in consultation with urban planners and communication (written and visual) professionals that ensured the information was concise, but also accessible to lay people and that it was then well communicated.

CRL put considerable effort into the layout, content and standardisation of experts reports, thus allowing the technical advice involving more than 500 custom reports to be prepared in a way that they are easy to follow by experts as well as lay people.

Regardless how well the acquisition programme has been prepared, property developers do not like the idea of the uncertainty of another approval. The only way to deal with this is to commit the time and resources to work through their individual concerns.

ACs' focus, away from mandatory car parking requirements to maximum allowable parking to reduce vehicles entering the city, has also made way for the use of higher density car stackers. This has been a comfort to property developers that are used to costly multilevel conventional (deep) basements, that could also be problematic for the CRL to accommodate.

The international collaboration of subsurface land valuation and acquisition has lagged behind other aspects of tunneling expertise.

CRLL developed a subsurface acquisition programme that resulted in the before and after effect that could be represented in the range of 0.1%–2% of the capital value of the property.

Tunnels and Underground Cities: Engineering and Innovation meet Archaeology,
Architecture and Art, Volume 2: Environment sustainability in
underground construction – Peila, Viggiani & Celestino (Eds)
© 2020 Taylor & Francis Group, London, ISBN 978-0-367-46579-7

Site-specific protocols for evaluating environmental compatibility of spoil materials produced by EPB-TBMs

P. Grenni, A. Barra Caracciolo & L. Patrolecco
National Research Council, Water Research Institute, Rome, Italy

ABSTRACT: The EPB-TBM performance relies on the selection of the appropriate condition-ing additives. The anionic surfactant sodium lauryl ether sulphate (SLES) is the main component of several foaming agents. Consequently, tunneling spoil material can contain SLES residual con-centrations. Owing to the lack of SLES threshold limits in soil in both European and Italian legis-lation, it is necessary to apply an ecological approach to produce a site-specific Protocol to be used during the excavation phase for classifying spoil material as a by-product. The approach includes: preliminary environmental risk assessment based on the ecotoxicity of each compound inside the conditioning additives; microcosm/mesocosms studies for evaluating the SLES bio-degradability in the spoil materials during their temporary deposit and, finally, the evaluation of the potential ecotoxicity on test organisms selected on the basis of the possible scenarios of soil and water exposure to SLES residual concentrations in the final destination site.

1 INTRODUCTION

The Earth Pressure Balance-Tunnel Boring Machines (EPB-TBM) is a mechanized tunneling technology alternative to the widely-used drill and blast method. The great amounts of spoil pro-duced (excavated material) can be utilized in different ways such as rock filling material in aban-doned quarries, as filling and paving material in road construction, as construction material and as a soil replacement for covering or foresting stony/rocky areas (Gertsch et al., 2000; Tokgöz, 2013; Oggeri et al., 2014). It is calculated that in a 50 km long and 100 m^2 cross section tunnel, about 8 million cubic meters of excavated material is produced (Oreste and Castellano, 2012). The reuse of tunnel excavation material can save soil, which is a non-renewable resource, avoiding waste deposits, limiting negative effects on the population near a construction site, and reducing material transportation. The EPB-TBM machine generally requires the use of lubricant products (foaming agents and additives), which change the mechanical and hydraulic behaviour of a soil, changing it into a plastic paste, thus permitting soil pressure applications at the tunnel face (Vinai et al., 2008; Peila et al., 2016). The selection of the foaming product to be used depends mainly on the excavated material characteristics and on the TBM performances (Borio and Peila, 2011; Peila, 2014; Talebi et al., 2015). Additives can include special anti-clogging agents and anti-abra-sion additives for the cutter head and its tools for the extraction screw (EFNARC, 2005). The amount of foaming agents to be applied in the tunneling process (L/m^3) is determined by the Foam Injection Ratio (FIR, which can be 10 – 80%, in most cases around 30 – 60%), Foam Expansion Ratio (FER, typically 5 – 30) and the concentration of foaming agent in water (CF, typically in the 0.5 – 5.0% range), (EFNARC, 2005).

Planning the handling of the excavated materials for a tunnel construction project avoids waste deposits and reduces project costs. In accordance with the Italian legislation, the re-use of spoil material from excavation processes as a by-product in industrial or green areas is pos-sible if the chemical thresholds for organic and inorganic contaminants (e.g. heavy metals, hydrocarbons C>12; Italian Decree No. 152/2006 and No. 120/2017) are not exceeded.

Most soil foaming agents used for tunneling are water-based solutions of anionic surfactants in a variable percentage. Among anionic surfactants, SLES is one of the main compounds of the commercial products currently used in Italy, at a typical concentration between 5 and 50% (Barra Caracciolo et al., 2016; 2017; 2018). The absence of soil threshold limits for these anionic surfactants in the Italian and EU legislation does not facilitate the management of the spoil material containing residual concentration of SLES as a by-product (Mininni et al., 2018). However, this issue can be overcome through an "ecological approach" aimed at preparing a site-specific Operative Protocol to be used during a mechanized tunnel excavation. The Protocol has to be shared with National and local control authorities.

2 MATERIAL AND METHODS

In the example here reported, some commercial products containing different percentages of the anionic surfactant sodium lauryl ether sulfate (SLES) as the main substance were selected for soil conditioning. The conditioning parameters used in these studies were the same applied for the subsequent tunneling excavation.

The approach described here includes three main aspects to be taken into consideration and evaluated such as:

- preliminary environmental risk assessment based on the ecotoxicity of each compound inside each commercial product, considering both the aquatic and terrestrial compartment;
- estimation of the real concentration of the main component of each foaming agents in the spoil material; the SLES amount in soils depends on the conditioning parameters (e.g. initial treatment ratio, TR, L/m^3 soil), but above all on the possible biodegradation of this chemical in the temporary deposit areas. In order to assess SLED biodegradation, micro- or mesocosms experiments are performed for 28 days;
- the evaluation of the potential ecotoxicity on test organisms selected on the basis of the possible scenarios of soil and water exposure to SLES residual concentrations in the final destination site, using soil samples and water extracts from the biodegradation experiment at different times.

2.1 *Preliminary ecotoxicological risk assessment and ecotoxicological characteristics of commercial products*

In a tunnel project various foaming agents and additives can be used, with different soil conditioning parameters (Concentration Factor, Cf, vol%; Foam Expansion Ratio, FER; Foam Injection Ratio, FIR vol%; Treatment ratio, TR, L/m^3). In order to select the products with the best environmental characteristics, taking into consideration the quantity used in different soils (which depends on the geological characteristics), two different aspects should be considered: the preliminary environmental risk analyses and the ecotoxicological characteristics of the specific foaming agent used.

The environmental risk assessment of a chemical is performed following the European methodology for chemical risk assessment described in the Technical Guidance Documents-TGDs (European Commission, 2003), to authorize the marketing of a chemical. This approach is based on the evaluation and comparison between the exposure to contamination (i.e. concentrations of the substance in a given environmental compartment) and the effects (e.g. ecotoxicity) that the substance exhibits on organisms or, more generally, ecosystems. The exposure can be measured experimentally, through environmental monitoring (MEC: Measured Environmental Concentration) or, as in our case, theoretically estimated by means of predictive models (PEC: Predicted Environmental Concentration) for the soil and water compartments. In the case of commercial foaming agents containing SLES as the main component, the PEC value for the soil compartment is the amount of SLES in the soil calculated on the basis of its percentage in the commercial product and the real TR used. The PEC for the water compartment was estimated in accordance with the TGD and European VICH guidelines (VICH, 2000; 2004; Kolar and Finizio, 2017; Finizio et al., 2018).

To have a risk quotient, PEC is compared to PNEC (Predicted No Effect Concentration). The latter is the concentration in each environment below which adverse effects will most likely not occur during long-term or short-term exposure (Finizio et al., 2014). It is, in general, assessed for each compartment through laboratory tests or derived from literature data on acute and chronic toxicity, taking into account appropriate safety factors (AF: Assessment Factor). In accordance with the European TGDs, the Equilibrium Partitioning method is used to calculate the PNEC value if no ecotoxicological data are available for a compartment. The risk analysis is carried out for each substance in the commercial products (specified in their Safety Data Sheet), taking also into consideration their relative percentage (Finizio et al., 2018). The results of the preliminary risk analysis performed on different commercial products make it possible to compare their potential environmental risk and define the "most environmental friendly" commercial products.

In some cases, an additional ecotoxicological evaluation can be performed to test the toxicity of the overall commercial products; in this case the *Vibrio fischeri* test can be used (ISO 11348-3:2007 Water quality - Determination of the inhibitory effect of water samples on the light emission of Vibrio fischeri luminescent bacteria test -Part 3: Method using freeze-dried bacteria). The ecotoxicity of the different commercial products is reported as EC_{50} (median effective concentration). In this test, the endpoint is the inhibition of the bacterium light emission. Higher values of EC_{50} correspond to a lower ecotoxicity of the specific product.

2.2 *Biodegradation experiments performed in microcosms/mesocosms*

Based on the preliminary ecotoxicological risk assessment results, the most environmentally-friendly commercial products are chosen for the microcosm/mesocosms experiments to test SLES concentration in the spoil material at different time from the conditioning, simulating the soil temporary deposit in a construction site. Soil samples representative of the main geo-pedological formations faced in the excavated area are to be considered for the experiments. Each soil is conditioned with the selected foaming agent (eventually in the presence/absence of a polymer, if needed, Figure 1C) using the real excavation parameters (TR, FIR, FER etc.). In the case of microcosms, beakers (2 L) with a glass cap containing about 2 kg of soil (conditioned or non-conditioned with the commercial products) are maintained for 28 days under controlled conditions in the laboratory (Figure 1A; Barra Caracciolo et al., 2018). In the case of mesocosms, containers (1 m^3) with conditioned and non-conditioned soils are maintained in the construction site at open air (Figure 1B).

Figure 1. A1 and A2: Photos of mesocosms at open air. B: Photos of microcosm experiments in laboratory. C: soil samples conditioned with foaming agents used for microcosm experiments.

The degradation of the main component of the foaming agents (SLES) is assessed by analyzing its concentration in soil during the 28 day-experiments. For this purpose, soil samples are collected at fixed times (0, 7, 14, 21 and 28 days). SLES concentration is evaluated both in soil and soil water extracts (1:10, soil:distilled water) using the MBAS (Methylene Blue Active Substances) spectrophotometric method as described in detail in Rauseo et al. (2017) and Grenni et al. (2018). The abundance and activity of the natural microbial populations can be determined (Grenni et al. 2009; 2012; Barra Caracciolo et al., 2018) in order to assess if the disappearance of the anionic surfactant is due to biodegradation. In fact, microbes are among the most important biological agents in removing and degrading waste materials, enabling their recycling in the environment.

2.3 Ecotoxicological tests

Aliquots of soil and soil water extract (elutriates) are used for performing ecotoxicological tests at the same sampling times of the biodegradation experiment above described (0, 7, 14, 21 and 28 days). The selection of the tests depends on the possible scenarios of soil and water exposure to SLES residual concentrations in the final destination site. A minimum of three tests to be performed is required.

The most common tests for soil compartment are the growth test using a plant (e.g. *Lepidium sativum*, OECD 208, 2006), the survival and reproduction test with an earthworm (e.g. *Eisenia foetida*, OECD 222, 2004) and the toxicity to Crustacea (*Heterocypris incongruens*, ISO 14371:2012). The most common tests for water compartment to be performed on water extracts are the acute toxicity tests with the bioluminescent bacterium *Vibrio fischeri* (ISO 11348–3:2007), the plant germination test (e.g. *Lepidium sativum* seed germination test, US EPA OPPTS 850.4200 guideline, 1996), the FET-Fish Embryo acute toxicity test (using embryos of the fish *Danio rerio*, OECD 236, 2013), the algal growth inhibition test with *Pseudokirchneriella subcapitata* (OECD 201, 2011) and the acute immobilization test with the crustacean *Daphnia magna* (OECD 202, 2004).

The overall ecotoxicological results of our experiments conducted so far show that SLES can have more detrimental effects on aquatic organisms (effective concentrations in the range of few mg/L to dozen of mg/L) than terrestrial ones. Its ecotoxicity depends on the sensitivity of the organisms tested (Barra Caracciolo et al., 2017; Grenni et al., 2018; Barra Caracciolo et al., 2018) and on its residual concentrations, which in turn depend on its degradation time.

3 RESULTS AND DISCUSSION

In Table 1 an example of four different products containing SLES as the main component but in different concentrations (SLES %) is reported. The Predicted Environmental concentration (PEC) of SLES used for the preliminary environmental risk assessment in the soil when similar TRs are used depends on the SLES % in the commercial product.

The preliminary ecotoxicological risk assessment provided the PEC/PNEC values. They were >1 for each foaming agent considered. However, it should be clarified that degradability was not taken into account in this risk assessment. It referred, in fact, to a worst case situation (worst case assumption); as a result, the realistic level of risk for aquatic and/or terrestrial

Table 1. Example of Predicted Environmental concentration values of SLES in soil (PEC$_{soil}$) considering four different commercial products and their relative soil conditioning parameters (Cf, FER, FIR, TR).

	SLES %	C_f %	FER	FIR %	TR L/m^3	PEC$_{soil}$
Foaming agent 1	<30	2.2	15	40	0.59	35-80
Foaming agent 2	10-30	2.0	9	80	1.77	160-200
Foaming agent 3	5-10	1.7	11	60	1.6	36-72

C_f: Concentration factor of the foaming agent within the foaming liquid; FER: Foam expansion ratio; FIR: Foam injection ratio; TR: Treatment ratio, amount of foaming agent used to treat 1 m^3 of soil.

environments may be lower than calculated. In any case, the commercial products with the lowest PEC/PNEC values were considered that with the lowest environmental impact.

The results of the ecotoxicological tests performed on different commercial products provided EC_{50} values for the evaluation of the impact on a test organism. An example of different EC_{50} values in the *Vibrio fischeri* test for three foaming agents and one polymer containing SLES as the main component in different percentages are reported in Table 2.

In this example, the foaming agent 4 and foaming agent 4+Polymer had the highest EC_{50} values. This means that they have the lowest ecotoxicity.

Regarding the biodegradation, it is reported that SLES is readily biodegradable both in aerobic and anaerobic conditions (Barra Caracciolo et al., 2017), does not produce metabolites persistent and/or toxic in the aquatic environment and is classifiable as a substance non-bioaccumulative in organisms. In the different biodegradation studies, the time required for SLES concentration to decline to half of the initial value (DT_{50}) was calculated in soil and water extracts in microcosm experiments (where abiotic conditions such as temperature, light and humidity were similar to the soil deposit area) or in mesocosms kept on the excavation site. Under the same abiotic conditions, the DT_{50} value depends on the type and depth of soil, the natural microbial community presence and the initial amount of SLES; the latter depends on its percentage in the commercial product and on soil conditioning parameters (mainly TR and water content).

Regarding the tests performed in soil or water extracts (to evaluate the potential ecotoxicity of the samples treated with foaming agents coming from the biodegradation experiments), different organisms could display diverse sensitivities. For this reason, the results of each individual assay could be combined in an ecotoxicological test battery integrated index to assess the overall risk (Grenni et al., 2018).

The assessment of variations in microbial community is essential for the evaluation of the impact of an environmental stressor. In fact, the contaminant occurrence can significantly affect the abundance and activity of microbial populations and harm some crucial ecological functioning. In particular, it is possible to observe changes in community structure (in terms of dominance and/or disappearance of some bacterial groups with consequent loss of important ecosystem functions) and/or occurrence of some microbial populations adapted to xenobiotics and able to use it as a source of carbon and/or nitrogen. In our case, the results of the natural microbial populations in terms of total number, cell viability and activity provided useful information to assess the possibility that a contaminant (in this case SLES) can be removed thanks to the presence of microbial populations able to use it as a substrate for their growth (Barra Caracciolo et al., 2018).

The results of the biodegradation studies together with those of the ecotoxicological tests are fundamental to evaluate the residence time in the temporary deposit and for choosing the best ecotoxicological test to be used during the tunnel excavation.

The overall results were useful to produce the site-specific Operative Protocol for the excavation phase. In particular, the microcosm/mesocosm experiments provided information on both the time potentially required (from 0 to 28 days) for considering the spoil material a by-product

Table 2. EC_{50} values estimated for three foaming agents at 5, 15 and 30 min of the *Vibrio fischeri* test.

Commercial product	SLES (%)	EC_{50} (mg/L) ± S.E.	EC_{50} (mg/L) ± S.E.	EC_{50} (mg/L) ± S.E.
		5 min	15 min	30 min
Foaming agent 1	<30	19.15 ± 1.7	11.31 ± 0.9	6.96 ± 0.9
Foaming agent 2	10-30	20.68 ± 3.0	10.34 ± 1.5	6.89 ± 0.9
Foaming agent 4	10-20	51.88 ± 5.9	26.36 ± 0.9	17.86 ± 1.5
Foaming agent 4 + Polymer (26.7: 1 v/v)	10-20 (Foaming agent) and 25-50 (Polymer)	34.02 ± 3.1	18.71 ± 0.9	12.75 ± 0.0
Polymer	25-50	19.19 ± 1.7	9.59 ± 0.9	6.11 ± 0.9

and the most sensitive test to be used for environmental monitoring in the construction site. The same commercial product can be degraded in different times in different construction sites and detected in the soil extract depending on the soil texture and mineralogy. Ecotoxicological and degradation tests performed in our research showed that the residual concentration of SLES found in conditioned soils were generally not effective concentrations for the soil target species. On the other hand, the residual SLES concentration in water extracts from soil can be in some cases potentially toxic for aquatic organisms, which are the most sensitive to anionic surfactants (Grenni et al., 2018). This means that in specific site-conditions in which the spoil material use implies a potential contact with a water ecosystem (e.g. green areas close to water bodies), SLES needs to be degraded until a concentration which has no effect on aquatic organisms is reached.

Therefore, an Operative Protocol for the excavation phase needs to assess SLES persistence in the real conditioned soil and perform tests on soil or water extract at different times in the deposit area before to move the spoil material in the final destination area.

4   CONCLUSIONS

The ecological approach here described has been successfully applied for producing a "Site-specific Operative Protocol" for supporting engineering contractors and stakeholders (e.g. Railway and Motorway operators) to classify as a by-product the spoil material before its allocation in the final destination site.

The protocol is the result of several theoretical considerations and lab experiments. It involves analyses of the effects that the real conditioned soils can have on soil and aquatic test organisms. These effects depend on the residual concentrations of the anionic surfactant (e.g. SLES), on the lithological and mineralogical characteristics of each soil (which determine chemical release of SLES into the aqueous phase) and on soil residence time in the temporary deposit at the construction site. The selection of an ecotoxicological test for the protocol used during the tunnel excavation phase depends on its sensitivity and repeatability and on the final destination use of the spoil material, taking into particular account possible contacts with water bodies. For this reason, site-specific studies and protocols are necessary for each tunnel project.

ACKNOWLEDGMENTS

These multidisciplinary studies were performed thanks to the collaboration among IRSA-CNR (Water Research Institute of the National Research Council), other CNR Institutes (IMC-Chemical Methodologies Institute; IBAF-Institute of Agro-Environmental and Forest Biology; IGAG-Institute of Environmental Geology and Geoengineering), Universities (Bicocca University of Milan, Sapienza University of Rome, Polytechnic University of Turin) and the Italian Institute of Health (ISS).

REFERENCES

Barra Caracciolo, A., Grenni, P., Patrolecco, L. & Mininni, G. 2016. Qualificazione ambientale e recupero delle terre e rocce da scavo prodotte nella realizzazione di grandi opere ingegneristiche. In E. Brugnoli & V.F. Uricchio (eds), *La ricerca sulle Acque e le nuove prospettive di valorizzazione dei risultati in abito pubblico e private*. Caciucci Editore, 299-304.
Barra Caracciolo, A., Cardoni, M., Pescatore, T. & Patrolecco, L. 2017. Characteristics and environmental fate of the anionic surfactant sodium lauryl ether sulphate (SLES) used as the main component in foaming agents for mechanized tunnelling. *Environmental Pollution* 226: 94-103.
Barra Caracciolo, A., Ademollo, N., Cardoni, M., Di Giulio, A., Grenni, P., Pescatore, T., Rauseo, J. & Patrolecco L. 2018. Assessment of biodegradation of the anionic surfactant sodium lauryl ether sulphate used in two foaming agents for mechanized tunnelling excavation. *Journal of Hazardous Materials* 365: 538-545

Borio, L. & Peila D. 2011. Laboratory test for EPB tunnelling assessment: results of test campaign on two different granular soils. *Gospodarka Surowcami Mineralnymi* 27 (1): 85-100.

EFNARC 2005. *Specification and Guidelines for the use of specialist products for Mechanized Tunnelling (TBM) in Soft Ground and Hard Rock.* EFNARC, Association House, 99 West Street, Farnham, Surrey GU9 7EN, UK. Available at http://www.efnarc.org/pdf/TBMGuidelinesApril05.pdf

European Commission 2003. Technical Guidance Document in support of Commission Directive 93/67/ EEC on Risk assessment for new notified substances and Commission Regulation (EC) No 1488/94 on Risk assessment for existing substances and Commission Directive (EC) 98/8 on biocides, second ed. European Commission, Luxembourg, Part 1, 2 and 3, 760 pp.

Finizio, A. & Vighi, M. 2014. Predicted No Effect Concentration (PNEC). In: *Encyclopedia of toxicology* (third edition) pp. 1061-1065. Elsevier.

Finizio, A., Grenni, P., Patrolecco, L., Galli, E., Donati, E., Lacchetti, I., Gucci, P. & Barra Caracciolo, A. 2018. Preliminary environmental risk assessment of sodium lauryl ether sulphate contained in foaming agents used in mechanized tunnelling. In SETAC Europe (ed.) *SETAC Europe 28th Annual Meeting Abstract Book*, Society of Environmental Toxicology and Chemistry Europe, p. 338. Available at https://www.setac.org/store/ViewProduct.aspx?id=11710512

Gertsch, L., Fjeld, A., Nilsen, B. & Gertsch, R. 2000. Use of TBM muck as construction material. *Tunnelling and Underground Space Technology* 15 (4): 379-402

Grenni, P., Barra Caracciolo, A., Rodríguez-Cruz, M.S. & Sánchez-Martín, M.J. 2009. Changes in the microbial activity in a soil amended with oak and pine residues and treated with linuron herbicide. *Applied Soil Ecology*, 41: 2-7.

Grenni, P., Rodríguez-Cruz, M.S., Herrero-Hernández, E., Marín-Benito, J.M., Sánchez-Martín, M.J. & Barra Caracciolo, A. 2012. Effects of wood-amendments on the degradation of terbuthylazine and on soil microbial community activity in a clay loam soil. *Water, Air, & Soil Pollution* 223: 5401–5412.

Grenni, P., Barra Caracciolo, A., Patrolecco, L., Ademollo, N., Rauseo, J., Saccà, M.L., Mingazzini, M., Palumbo, M.T., Galli, E., Muzzini, V., Polcaro, C.M., Donati, E., Lacchetti, I., Di Giulio, A., Gucci, P., Beccaloni, E. & Mininni, G. 2018. A bioassay battery for the ecotoxicity assessment of soils conditioned with two different commercial foaming products. *Ecotoxicology & Environmental Safety* 148: 1067–1077.

Kolar, B. & Finizio, A. 2017. Assessment of environmental risks to groundwater ecosystems related to use of veterinary medicinal products. *Regulatory Toxicology and Pharmacology* 88: 303-309.

Mininni, G., Sciotti, A. & Martelli, F. 2018. Characterization and management of excavated soil and rock. In: SETAC Europe (ed.) *SETAC Europe 28th Annual Meeting Abstract Book*, Society of Environmental Toxicology and Chemistry Europe, p. 96. Available at https://www.setac.org/store/ViewProduct.aspx?id=11710512

Oggeri, C., Fenoglio, T.M. & Vinai, R. 2014. Muck classification: raw material or waste in tunnelling operation. *Revista Minelor/Min Revue* 20 (4): 240-49.

Oreste, P. & Castellano, M. 2012. An Applied Study on the Debris Recycling in Tunnelling. *American Journal of Environmental Sciences* 8 (2), 179-184.

Peila, D. 2014. Soil Conditioning for EPB Shield Tunnelling. *KSCE Journal of Civil Engineering* 18 (3): 831–836.

Peila, D., Picchio, A., Martinelli, D. & Dal Negro, E. 2016. Laboratory tests on soil conditioning of clayey soil. *Acta Geotechnica* 11(5): 1061–1074

Rauseo, J., Ademollo, N., Pescatore, T. & Patrolecco, L., 2017. Determinazione di tensioattivi anionici in terreni provenienti dallo scavo in sotterraneo mediante estrazione liquida pressurizzata (PLE) e metodo MBAS (sostanze attive al blu di metilene). *Notiziario dei Metodi Analitici e IRSA* news 1, 15-22.

Talebi, K., Memarian, H., Rostami, J. & Alavi Gharahbagh, E. 2015. Modeling of soil movement in the screw conveyor of the earth pressure balance machines (EPBM) using computational fluid dynamics. *Tunnelling and Underground Space Technology* 47: 136-142

Tokgöz, N. 2013. Use of TBM excavated materials as rock filling material in an abandoned quarry pit designed for water storage. *Engineering Geology* 153: 152-162

VICH 2000. International Cooperation on Harmonisation of Technical Requirements for Registration of Veterinary Products. Topic GL6: Environmental Impact Assessments (EIAs) for Veterinary Medicinal Products (VMPs) – Phase I. VICH, London, UK. CVMP/VICH/592/98.

VICH 2004 International Cooperation on Harmonisation of Technical Requirements for Registration of Veterinary Products. Topic GL38: Environmental Impact Assessment (EIAs) for Veterinary Medicinal Products (VMPs) – Phase II. VICH, London, UK. CVMP/VICH/790/03.

Vinai, R., Oggeri, C. & Peila, D. 2008. Soil conditioning of sand for EPB applications: A laboratory research. *Tunnelling and Underground Space Technology* 23 (3): 308-317.

Tunnels and Underground Cities: Engineering and Innovation meet Archaeology, Architecture and Art, Volume 2: Environment sustainability in underground construction – Peila, Viggiani & Celestino (Eds)
© 2020 Taylor & Francis Group, London, ISBN 978-0-367-46579-7

Monitoring of surface water resources and buildings due to tunneling, Uma Oya Project, Sri Lanka

A.H. Hosseini & A. Rahbar Farshbar
Farab Co., Tehran, Iran

R.M.P.G.L.S. Gunapala
CECB, Colombo, Sri Lanka

ABSTRACT: A comprehensive baseline building condition survey and periodical monitoring program are being carried out in the areas along the entire Head Race Tunnel and Link Tunnel traces to monitor the present and future status of structures due to tunneling within the stipulated area in Uma Oya Multipurpose Development Project. Further, the variation of surface water resources in major streams should be monitored by establishing V-notches or similar and also monitoring of ground water level in domestic wells, deep wells and other surface water sources of minor streams. Information collected from such monitoring are submitted to assess the impact of water ingress into the tunnel and surface settlement due to tunneling which affect the natives' buildings and surface water resources. This paper tries to describe the monitoring procedure and results.

1   INTRODUCTION

Uma Oya Multipurpose Development Project (UOMDP) as a hydro mechanical project in Sri Lanka has been targeted generating hydro-electric power, transferring water for irrigation purposes and controlling seasonal devastating flood. This project lies in the south-eastern part of the central highland region of Sri Lanka (Figure 1). The main part of the project is located in the south-western part of the Badulla district in the province of Uva. The south-eastern end of the project site is in the central west of the Moneragala district (province of Uva). Bandarawela is the largest urban setting within the project area, located halfway between the upstream and the downstream ends of the planned hydropower plant and combination of dams, tunnels, and downstream facilities.

The Multi-Purpose Project Of Uma Oya is planned to convey water from Uma Oya to Kirindi Oya Basin for hydroelectric energy generation includes Puhulpola RCC Dam on the Uma Oya river, Dyrabaa RCC Dam on the Mahatotilla Oya river, a link tunnel from Uma Oya to Mahatotilla Oya (L=3.8Km), a long feeder tunnel from Mahatotilla Oya to Kirindi Oya. The feeder (power) tunnel Lot consists of a headrace tunnel (L=15.2Km), the surge shaft (L=200m) and tank on the upstream of the vertical shaft, a vertical shaft with steel lining (L=650m), an underground powerhouse and a tailrace tunnel (L=3.87Km).

The underground powerhouse with a capacity of 120 MW is located at the end of a 2 km main access tunnel that has been mined by drill and blast method to allow for excavation of turbine and transformer caverns. The Tailrace tunnel with a length of 3.5 Km will transfer water to downstream of project from the underground hydropower facilities. The main access tunnel to the power house (MAT) is the principal means of accessing the underground space through construction period and for the operation and maintenance of the facilities in the future.

Head race tunnel (HRT) being excavated by a Tunnel Boring Machine (TBM) is a major component of the Project whereas link tunnel (LT) excavation has been fulfilled by using

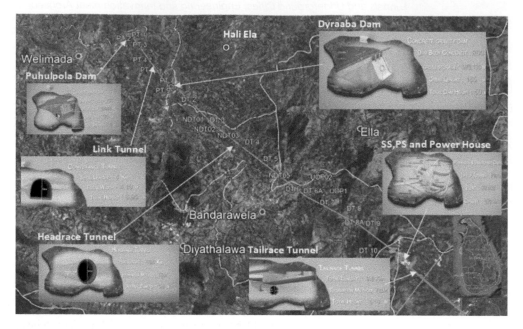

Figure 1.   Overview of the Sri Lankan island with project area.

conventional drilling and blasting techniques. A comprehensive baseline building condition survey and periodical monitoring program are being carried out in the areas along the entire HRT/LT trace to monitor the present and future status of structures within stipulated area. Further, the variation of surface water flows in major streams should be monitored by establishing V-notches or similar and also monitoring of ground water level in domestic wells, deep wells and other surface water sources. Information collected from such monitoring are submitted to assess the impact of water ingress into the tunnel and surface settlement due to tunneling which affect the natives' buildings and surface water resources. Further, the geological studies suggested that groundwater levels could be lowered during construction of the tunnel, but provided no data on the distribution of groundwater or the numbers of wells likely to be affected at present. These factors therefore need to be investigated to allow a more informed assessment of these impacts.

2   GEOLOGICAL CONDITIONS

From geological point of view, the project area is located in Highland complex area forming the rugged high grounds, covering the central part of Sri Lanka. This complex strongly tectonized, comprising mixed ortho and para gneissic granulite units which structurally overlies amphibolite grade rocks of Vijayan complex to the east and southeast. The Uma Oya-Kirindi Oya project lies in the Uva basin of the highest penplain and in the edge of the southeastern side of the southern platform of the middle peneplain. The Uva basin is a depression surrounded on nearly all sides by a rim of mountains. In the center of the basin the general elevation is 1000 m.s.l. Two rivers, Baduluoya and Uma Oya drain the Uva basin. The escarpment in the project area has some basement structural control (essentially by principal shear zones, major joint systems and the orientation of large-scale folds) that its uplift is a recent phenomenon.

The Highland series rocks cover the whole project area. They include pre-Cambrian rocks formed under high grade metamorphic conditions and are composed of two main types of rocks namely metasediments and charnockites or Charnochitic-gneisses. The metasedimentary rocks are metamorphosed sedimentary rocks and consist of garnet sillimanite gneisses or

Khondalites, quartzite, quartz feldspar granulites, garnet gneisses, marble and impure crystalline limestone. Charnochitic gneisses are the most common rock types of the Highland series. Quartzite's are very hard rocks and difficult to drill through. Marble and impure crystalline limestone may ran into a problem when encountered in tunnels, shafts and dam foundations or across reservoir periphery, especially when associated with faults and karstic zones. The project area is folded into a series of large domes, basins, anticline and syncline structures (Rahbar, A. & Rostami, J., 2016).

3 MONITORING

The scope of the monitoring is execution and completion of condition survey, monitoring and reporting of structures, infrastructures, and water resources, which are affected or might affect due to the excavation of Reservoir LT and HRT and also sampling water resources, including the items as follows:

– Pre-condition and post condition of buildings affected,
– Measuring the water level of domestic wells, deep wells and also the lakes, and
– Measuring the water discharge of streams and springs.

The boundary of the monitoring surveying within a corridor of 600m width at ground level along the tunnels alignment, which is 300m each side of the tunnels alignment in normal condition.

4 HRT MONITORING

Headrace tunnel designed to excavate as mechanized method by a 4.3m diameter Double Shield (DS) TBM in the basic feature as follows (Figure 2):

– Elevation Intake (invert): 960masl
– Length: ~15.2km
– Slope: 0.01015 & 0.002
– Internal diameter (ID): 3.4m
– Design Discharge (OD): 9.5
– Flow Condition: Pressure Flow

The vulnerability of the environment, especially related to changes in the groundwater table caused by the tunnel construction, is evaluated with the aim to develop methods to quantify accepted levels of leakage into a tunnel. Procedures and guidelines for various conditions are studied. The tunnel is running through the 16 sub divisional in five divisional secretariats in Sri Lanka. It was selected as monitoring area as 300m each side of the tunnel.

The monitoring has been done in forested area including deep valleys, which made lots of challenges for the monitoring team. There were 2811 buildings covered under the pre-construction conditions survey (CECB 2015-2018A) and also 846 domestic wells monitor (CECB 2015-2018B) in monthly basis. The pre-construction survey required a

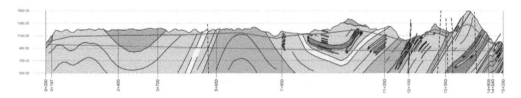

Figure 2.   The geological profile of HRT.

detailed report of every well within the 600m wide corridor. Whilst there is a record of the tube wells (CECB 2015-2018C), which have been constructed by the Water Resources Board, there is no register of the domestic hand dug wells in the local library or with the local district council. The property owners dig the wells without license and the list was compiled simply by asking local people in the area. Figure 3 shows the spread of the monitored building and Figure 4 shows the spread of the monitored Domestic well in HRT monitoring corridor.

After finishing the pre-construction monitoring of the buildings and the domestic wells, post monitoring of buildings (CECB 2015-2018D) and re-monitoring of the domestic wells has been continuously in progress based on the daily and weekly or monthly.

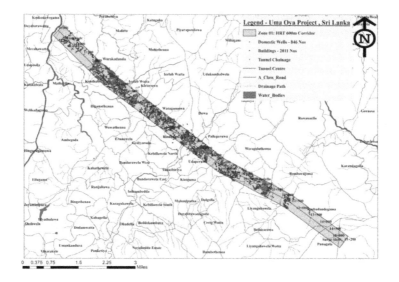

Figure 3.   Pre-construction Building Condition Surveyed Building and Water Resources Layout.

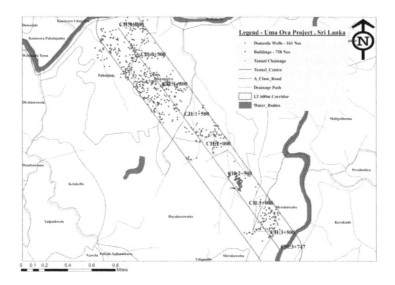

Figure 4.   Pre-construction Building Condition Surveyed Buildings and Water Resources Layout.

5 LINK TUNNEL MONITORING

Link tunnel designed to excavate by Drill & Blast method by a 4.7 m x 5.10 m (W×H) diameter in the basic feature as follows:

– Elevation Intake (invert): 981masl
– Length: 3.87km
– Slope: 0.0003
– Flow Condition: Free Flow

Scope of monitoring covered 758 properties within the 600m corridor of LT (CECB 2015-2018E). In addition, 161 domestic wells monitored within the corridor (CECB 2015-2018F). Figure 5 shows the spread of the monitored building & Figure 6 shows the spread of the monitored Domestic well in HRT.

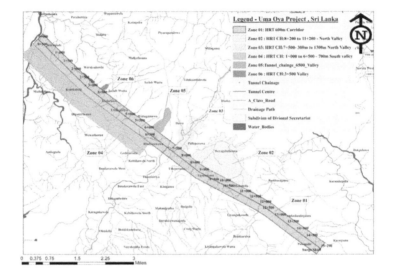

Figure 5.   Pre construction Building Condition Surveyed Zones along Headrace Tunnel.

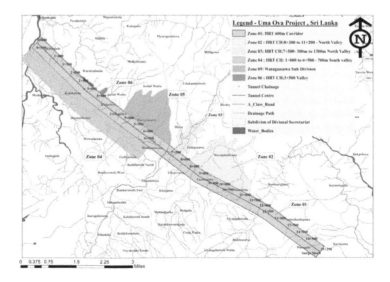

Figure 6.   Water Resource Monitoring Zones along Headrace Tunnel.

6 BOREHOLE

There were 19 borehole constructed along the headrace tunnel and all boreholes monitor biweekly. Anyhow, between the tunnel excavation progressive sections in selected boreholes monitors daily basis. The boreholes were driven into the earth deeper than the level of the tunnel. These boreholes directly read the fluctuation of deep aquifer with the tunnel inflow. Most of the sections of the tunnel are grouting and these reading has been collaborated the success of grouting process.

The reservoir link tunnel has nine boreholes and there is a collapse zone between the tunnel chainage 1+500 and 2+500. The between the collapse zone has three borehole and which monitor weekly basis to check the success of the grouting.

7 TUBE WELLS

There are 49 tube well in the 600m corridor of the tunnel and 19 tube wells monitor monthly. These reading directly show that, the connection in between the deep and shallow aquifer and check the effect of tunnel excavation. Some section of doubtful excavation progressive closely monitor with the excavation. These tube wells are property of the Water Recourse Board and were constructed in problematic drinking water areas.

8 STEAMS AND SPRINGS

There are lots of streams cross the headrace tunnel. Therefore, there are 8 Parshall flumes constructed to monitor the discharges. These monitoring has done on biweekly basis. Some streams monitoring has done visually because most of the people, who are living in this area, are farmers and they directly rose to the project in the dry period because inadequate water for farming (CECB 2015-2018G).

9 RAIN GAUGES

There are 4 local rain gauges installed as cover the entire tunnel and which are monitored daily because success of recharging of ground water compare with rainfall. According to the last records, station-to-station has shown considerable changes of received rainfall.

10 LAKES

There are two lakes in the both tunnel. The lake closer to the headrace tunnel, monitor weekly because it was a huge complaint in last drought period, lake was dried due to headrace tunnel. Recently, the TBM machine excavated more the 50% of the lake area. Same as the in the headrace tunnel has complaint closed lake in link tunnel and which has been monitored biweekly basis (CECB 2015-2018H).

11 EXTRA MONITORING

Whilst this has not been able to determine the depths of water in these wells before the major water ingress event in December 2014 it has enables us to monitor later events and monitor the changes due to the seasonal rainfall, and in areas where increased ingress has been intercepted in the tunnel excavation. Typically, little change has occurred. However, drying out, probably due to seasonal changes, has also been recorded.

The wells along the valley up to a distance of about 3 km from the point of the tunnel heading where the water ingress occurred in December 2014 have been surveyed. The soil overburden observed in many dug wells in this area is over 10 m deep. The bedrock observed was often weathered, highly jointed gneiss. The majority of these domestic water supply wells became unusable and it is reported that other points at which the groundwater emerges as small springs also became dry

The surface water resources monitoring and building condition survey was started after the first water ingress occurred at the tunnel chainage 11+169 on the date of 24 December 2014 with the amount of 400 liter per second. In this incident, there were 2061 building affected complaints with 340 dried domestic well complaints, which tunnel chainage between the 11+169 and 8+300. This area is shown in Figures 5 & 6 as Zone 02 (CECB 2017J). In this tunnel section, the north alignment of the tunnel has two lineaments, which has intersected to the tunnel and caused for significant water drain the tunnel. Affected area is belongs to the four subdivisions of divisional secretariat and in these subdivisions have 2746 properties. In zone 02, all the complaints wells have been seasonally measuring for monitoring the progress of recharging ground water after the post grouting of tunnel with the rainfall.

Another four potential zones additionally identified according to the geological condition of surface. Basically, which five of the zones were selected the rock joints intersected to the tunnel alignment and one zone, lineament running parallel to the tunnel trace. Some areas only water resource monitored in different frequencies and which are shown in Figures 5 and 6 separately.

One of the major intersecting lineaments has identified at the tunnel chainage 7+500 and it is shown as Zone 03 in Figures 5 & 6 (CECB 2017K). This is deep valley and monitoring corridor was selected according the geology. Period since the 06th August 2017 to 22nd October 2017, there are 303 properties pre-construction surveyed and 34 domestic wells monitor seasonally. These wells were constructed on the shallow aquifer and which are simply answered for short dry periods. Anyhow, the tunnel boring machine (TBM) was hit the aquifer at the tunnel chainage 7+317 with the inflow of more than 1000 liter per second. In this incident, reported 844 building affected and more than 500 domestic wells becoming drying complaint. Some of the complaints was received more than 3km away from the tunnel excavation face.

In the incident of tunnel chainage 7+317, one of the lineament which is running parallel to the tunnel chainage between the 1+000 and 6+500. Then the buffer zone 300m was extended to 1000m in south alignment of the tunnel. The monitoring area is shown Figures 5 & 6 as Zone 04 (CECB 2017M). Then the monitoring was conducted period between 27th April 2017 and 30th August 2018. There are 1285 properties done the pre-construction survey and 534 domestic wells have monitored in different frequencies with the advance of TBM.

In the meantime, some of the domestic wells complaints was raised tunnel chainage 5+000 and 7+000 in North alignment. The area is namely the Watagamuwa sub division of divisional secretariat. Only domestic wells has monitored with the different frequencies. The area is shown in Figure 6 as Zone 05 (CECB 2017-2018N). There are 334 domestic wells identified and initial monitoring was done period since the May 2017 to July 2018.

Then the major intersected lineament was identified at the tunnel chainage 6+500. Therefore, the pre-construction building condition survey was done and it is shown as Zone 05 in Figure 5. The monitoring was conducted period between the 26th February 2018 to 07th March 2018 and 154 properties covered. This area is belongs to the Watagamuwa subdivision and domestic wells covered that area.

Lastly identified lineament, which is intersected at the tunnel chainage 3+500. Pre-construction survey for building was done in this area. The survey was conducted period between the April 2018 and May 2018 and 160 building covered. The monitoring corridor is shown in Figure 5 as Zone 06 (CECB 2017P). For more detail, see Table 1.

Table 1. Monitoring activities along HRT and LT.

	Pre-Construction Survey		Post Construction Survey		Tube Wells	Deep Boreholes
	Buildings	Domestic Wells	Buildings	Domestic Wells		
HRT	4,573	2,016	2,913	789	18	19
Link Tunnel	758	161	758	161	03	09
Zone 1	2,811	846	852	522	18	19
Zone 2	-	267	2,061	267	-	-
Zone 3	303	34	-	-	-	-
Zone 4	1,588	534	-	-	-	-
Zone 5	154	334	-	-	-	-
Zone 6	160	-	-	-	-	-

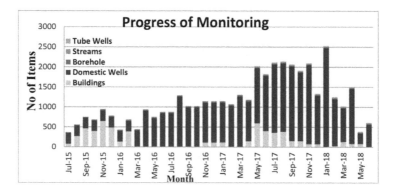

Figure 7. Progress of Monitoring of Water Resources and Building Surveying.

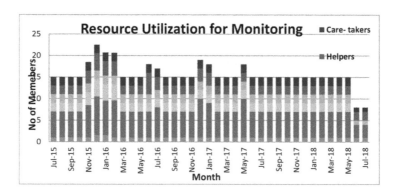

Figure 8. Resource Utilization for Surface Monitoring.

12 RESOURCES UTILIZATION

Covering more than 15Km length of HRT area and 4Km Link Tunnel in 600m corridor was a huge job in local road area with spread location of the buildings and domestic wells. Totally 5,471 buildings, 2,177 domestic wells, 18 tube wells and 19 deep bore holes monitored with the frequencies of daily, Weekly, biweekly or monthly during 1,551 days with 582 personnel.

13 CONCLUSION

A comprehensive baseline building condition survey and periodical monitoring program are being carried out in the areas along the entire Head Race Tunnel and Link Tunnel traces to monitor the present and future status of surface structures and water resources due to tunneling within the stipulated area in Uma Oya Multipurpose Development Project. Surface water resources in major streams are being monitored by establishing V-notches or similar and also monitoring of ground water level in domestic wells, deep wells and other surface water sources. Information collected from such monitoring are submitted to assess the impact of water ingress into the tunnel and surface settlement due to tunneling which affect the natives' buildings on the ground and surface water resources.

Monitoring of the properties as a pre-construction and post construction survey is one of the challenging and managerial specifications of the work in this project. In this paper, all the monitoring area and also extra effectible monitored zones and personnel planning for such a long (HRT) and medium tunnel (Link) were reviewed.

Covering more than 15Km length of HRT area and 4Km Link Tunnel in 600m corridor was a huge job in local road area with spread location of the buildings and domestic wells. Totally 5,471 buildings, 2,177 domestic wells, 18 tube wells and 19 deep bore holes monitored with 582 personnel in 1,551 days.

REFERENCES

Rahbar, A. & Rostami, J. 2016. Construction of Headrace Tunnel of Uma Oya Water Conveyance Project Sri Lanka, World Tunneling Congress 2016; Proc. intern. symp. San Francisco, USA, 22-28 April 2016.
CECB 2015-2018A, Central Engineering Consultancy Bureau (CECB), HRT Pre Construction Building Condition Report, Volume 1 to 30.
CECB 2015-2018B, Central Engineering Consultancy Bureau (CECB), HRT Monitoring and re-monitoring of Domestic wells, Volume 1 to 80 & Monitoring of Domestic Wells Monthly reports.
CECB 2015-2018C, Central Engineering Consultancy Bureau (CECB), Tube Wells Monitoring and re-monitoring of Domestic wells, Monthly Reports.
CECB 2015-2018D, Central Engineering Consultancy Bureau (CECB), HRT Post Construction Building Condition Report, Volume 1 to 29.
CECB 2015-2018E, Central Engineering Consultancy Bureau (CECB), LT Pre Construction Building Condition Report, Volume 1 to 9.
CECB 2015-2018F, Central Engineering Consultancy Bureau (CECB), LT Post Construction Building Condition Report, Volume 1 to 17.
CECB 2015-2018G, Central Engineering Consultancy Bureau (CECB), Monitoring and re-monitoring of Tube wells, Volume 1 to 5 & Monthly reports of Tube Wells.
CECB 2015-2018H, Central Engineering Consultancy Bureau (CECB), Monitoring and re-monitoring of Lakes, Volume 1 to 3 & Monthly reports of Lakes Monitoring.
CECB 2017J, Central Engineering Consultancy Bureau (CECB), HRT Zone 2 Monitoring of Domestic Wells Report.
CECB 2017K, Central Engineering Consultancy Bureau (CECB), HRT Zone 3 Pre Construction Building Condition Reports and Monitoring of Domestic Wells Reports, Volume 1 to 3.
CECB 2017M, Central Engineering Consultancy Bureau (CECB), HRT Zone 4 Pre Construction Building Condition Reports and Monitoring of Domestic Wells Reports, Volume 1 to 14.
CECB 2017N, Central Engineering Consultancy Bureau (CECB), HRT Zone 5 Pre Construction Building Condition Reports and Monitoring of Domestic Wells Reports, Volume 1.
CECB 2017P, Central Engineering Consultancy Bureau (CECB), HRT Zone 6 Pre Construction Building Condition Reports and Monitoring of Domestic Wells Reports, Volume 1.

*Tunnels and Underground Cities: Engineering and Innovation meet Archaeology,
Architecture and Art, Volume 2: Environment sustainability in
underground construction – Peila, Viggiani & Celestino (Eds)*
© 2020 Taylor & Francis Group, London, ISBN 978-0-367-46579-7

Hydrogeological investigation and environmental impact assessment during a large water inflow tunnel construction in Taiwan

F.Y. Hsiao & H.C. Kao
Sinotech Engineering Consultants, INC., Taipei City, Taiwan (ROC)

H.J. Shau
Directorate General of Highways, Ministry of Transportation and Communications, Taiwan (ROC)

ABSTRACT: Groundwater inflow during tunneling not only causes construction schedule delay but also long-term groundwater depletion. A highway tunnel excavated through high-permeability formations is illustrated in the paper. During the tunnel construction, a large accumulation inflow of more than 10 T/min was measured. In order to assess the impact of tunnel construction on groundwater resource, various hydrogeological investigations were performed, including tunnel inflow measurement, groundwater level observation and 3D hydrogeological modeling. Even though groundwater level was found showing a temporary drawdown during tunneling, the recovery of groundwater level was observed after the completion of tunnel construction. The assessment results provide a scientific reference for understanding the relationship between tunnel construction and groundwater resource protection.

1 INTRODUCTION

Taiwan is located on the convergent plate boundary between the Asiatic Continental Plate and the Philippine Oceanic Plate, strongly folded, faulted and highly fractured rocks are commonly observed. The existence of fissures and voids in fractured rocks is deemed to provide preferential flow path and storage space for groundwater. Taiwan has abundant rainfall throughout the year, and the annual average rainfall is approximately 2500mm, which is about 2.6 times of the global average. In certain areas, particularly to the windward mountainous regions, localized heavy rainfall with 6,000 to 7,000 mm/year was once recorded. Massive rainfall associated with favorable geological condition allows excess surface runoff to quickly infiltrate and replenish groundwater table. As already known, severe climate change is impacting the world. In Taiwan, even though the annual rainfall has no obviously changed, it has been confirmed that the number of raining days is significantly reduced and the days of heavy rain (the daily rainfall of more than 200mm) are gradually increased (Hsu et al. 2011). The conditions imply that the total annual rainfall in Taiwan has no apparently changed, but the rainfall is concentrated in a short period of time and the rainfall intensity is increased. In other words, heavy rainfall and short-term drought seem to become common.

Tunnel construction encountered large amount of water inflow is often occurred in Taiwan. Some tunnel construction projects were even delayed or halted by the unexpected large water inflow. The most well-known case in Taiwan is the construction of New Yungchuen Tunnel, which original tunnel route was abandoned by the disastrous water ingresses (Wang et al. 2011). Not to mention the occurrence of shortage of water supply during dry season is somehow easily attributed to neighboring tunnel construction. Especially when a long-term water outflow is encountered during tunnel construction, that is very possible under the suspicion of main cause of the insufficiency of water supply. Therefore, the variation of the regional groundwater resources induced by tunnel construction is necessary to be understood to verify the actual reason for the shortage of water supply.

In the past, the geological survey of tunnels mostly focused on investigation of mechanical properties for tunneling stability. With the awareness of groundwater resource protection, the science of hydrogeology that emphasized on the investigation of occurrence, distribution, movement and properties of groundwater is progressively important and practical adopted in tunnel investigation. Hsiao et al (2014) proposed the hydrogeological investigation technologies commonly used in tunnel engineering, including hydrogeological remote sensing, hydrogeological geophysical exploration, hydrogeological mapping, borehole hydrogeological survey, hydrogeological field test, and hydrogeological dynamic measurement. Although hydrogeological investigation is progressively applied in tunnel design, the problem of unexpected water gush of still often occurred during tunnel construction. It may be caused by complicated hydrogeological condition in a long tunnel. Under a limited survey budget and limited investigation schedule, it is impossible to accurately realize the hydrogeological conditions along a tunnel of several kilometers long. Therefore, it is necessary to carry out hydrogeological investigation and assessment during construction for a tunnel with high potential water inflow. The related measures and required adjustments should be made based on the in-situ hydrogeological conditions for minize the risk of tunnel inflow.

A high potential water inflow tunnel in eastern Taiwan is introduced to illustrate the hydrogeological investigations performed during construction. The regional groundwater resources variation induced by tunnel construction is estimated according to the field measurement and the hydrogeological modelling. The background and the hydrogeological investigation of the case tunnel are described briefly as flows.

2 TUNNEL BACKGROUND

In Taiwan, Highway plays a pivotal role in the passengers and materials transportation. Among numerous highways, Suhua Highway along the steep cliff by Pacific Ocean is an important road connecting between eastern and northern Taiwan. Unfortunately, there are many natural disasters, such as typhoon, heavy rainfall and landslide, regularly attacked this road in the past two decades. The transportation of the highway was often blocked and even led to casualty accident. To provide a safe and reliable highway for people to travel between the eastern and northern Taiwan, the mountainous section improvement project for Suhua Highway was launched in 2011.

Dangao Tunnel, one of the tunnels in Suhua Highway improvement project, is adopted as the case tunnel in the paper. The length of Dangao Tunnel is about 3.3 km. Tunnel north portal is near Yongle Rail Station, and south through Dangao Ridge to the valley of Dangao Creek, as shown in Figure 1. Due to the occurrence of multi-stage metamorphism, the rock strata are characterized by complex folds and sheared structures. There are two major thrust faults, Houishan Fault and Hsiaomaoshan Fault, intersect the tunnel north section. These thrust faults are nearly parallel southerly inclined, and trend generally east-west direction. In addition, several transverse faults with generally south-north trend appeared in the tunnel both sides. These transverse faults often form a series of well-developed tensile cracks that offer a well flow path of groundwater. Tunneling in fractured rock has a high potential of large water inflow. Furthermore, two completed rail tunnels, Yungcheun Tunnel and New Yungcheun Tunnel, excavated in the area underwent the difficulty of tunnel inflow during construction. New Yungcheun Tunnel even met a huge water inflow of 83 T/min, which seriously delayed the tunnel construction and finally caused the modification of partial tunnel route. Up to now, water is continuously outflow from both rail tunnel portals after construction. Dangao Tunnel is closed to the rail tunnels, the distance between them is only approximate hundreds of meters, as illustrated in Figure 1. Therefore, groundwater flow into tunnel during construction is anticipated for Dangao Tunnel. The assessment of the neighboring groundwater resources variation during tunneling is necessary.

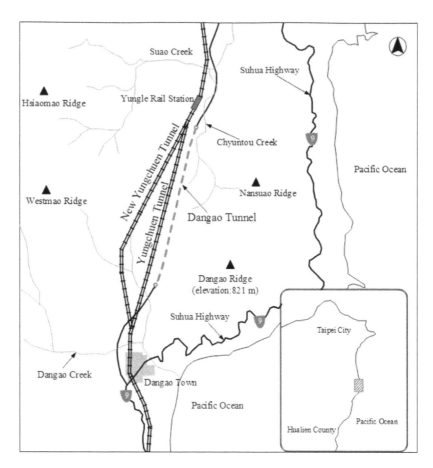

Figure 1. Location of Dangao Tunnel.

2.1 *Hydrogeological condition of the case tunnel*

There are several ridges, including Hsiaomao Ridge, Westmao Ridge, Nansuao Ridge, and Dangao Ridge, around Dangao tunnel. The topography in this area is rugged and two main watersheds can be divided, Suao Creek and Dangao Creek watersheds. Furthermore, this area is located in the windward side of monsoon. The rainfall is heavy and concentrating during a small period from November to next February. According to the long-term meteorological rainfall data collected by weather stations, the precipitation in varied from 4,000 to 6,500 mm/ year. And an extremely heavy rainfall of 7,300 mm/year was recorded at Dangao Ridge rainfall station in 2017. The precipitation in the area is overall higher than the average rainfall of Taiwan. Groundwater recharge potential is, therefore, high.

Complicated metamorphic rocks can be found around the tunnel. The main rock types in the tunnel alignment consist of slate, metamorphic sandstone, diabase, schist, marble and amphibolite (Figure 2). The contact area between different metamorphic rocks with apparently difference in rock strength is easily to form a slip plane or fractured zone during vigorous tectonic activity process. The permeability of intact metamorphic rock, such as amphibolite and schist, is low. However, the permeability for fractured rocks on the contact area between different rocks is high, that is preferential flow paths of groundwater. The hydraulic conductivity in this region would change drastically. According to the results of double packer hydraulic test (Shau et al. 2015), the hydraulic conductivity for the intact schist is very low, approximate in the order of 10^{-10} m/s. The hydraulic conductivity for the well-developed schistose is slightly increased in the order of $10^{-7} \sim 10^{-8}$ m/s. And the permeability capability would increase obviously for fractured

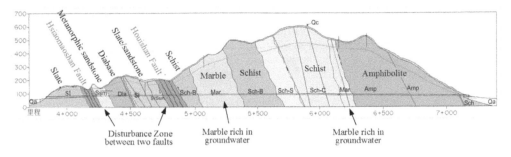

Figure 2. Geologic profile of Dangao Tunnel.

schist, which hydraulic conductivity reaches in the order of 10^{-6} m/s. In addition, the permeability of fractured marble is considerably high, which hydraulic conductivity is in the order of 10^{-5} m/s. The test results demonstrated that the hydraulic property of the rock mass in the region is dominated by the fractured development conditions.

2.2 Hydrogeological model in design stage

The hydrogeological characteristics around Dangao Tunnel are extremely complicated, dramatically varied permeability for metamorphic rocks and geological structures affects regional groundwater flow behavior. Simplified theoretical solution would not be suitable to estimate the complex groundwater flow in this region. Therefore, groundwater modelling software package Groundwater Modelling System (GMS), which supports groundwater numerical codes MODFLOW (McDonald & Harbaugh 1988), was adopted to assess the tunnel inflow risk and possible regional hydrogeological impact in the tunnel design stage (Sinotech 2012). The MODFLOW model has been improved and verified by academia and engineers in many countries and has been widely used for the past two decades (Yang et al., 2008). Through the categorization and stratification of aquifers, the complicated hydrogeological environment in the neighboring area of Dangao Tunnel is converted into a simplified hydrogeological conceptual model, as shown in Figure 3a. The model consists of several hydrogeological stratum units; groundwater inflow and outflow in the model are based on water balance reaching a steady state. Groundwater flow field can be obtained by using 3D groundwater flow control equations with known initial and boundary conditions and groundwater loads.

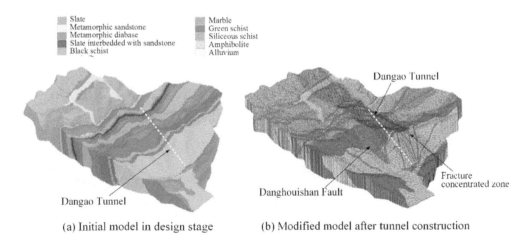

(a) Initial model in design stage   (b) Modified model after tunnel construction

Figure 3. 3D hydrogeological conceptual model of Dangao Tunnel.

The modeling results in the design stage indicate that the major variation of groundwater flow field after tunnel excavation is restricted to the region of 1~2 km along Dangao tunnel. Tunnel inflow is anticipated to occur in fault zone and the interlayer fractured zone. Furthermore, a large water inflow of about 1,500 L/min is expected for the excavation of marble due to its high permeability.

3 HYDROGEOLOGICAL INVESTIGATION DURING CONSTRUCTION

The construction of Dangao Tunnel began in 2013 and completed by the end of 2016. During the tunnel construction, a number of hydrogeological surveys, including tunnel inflow measurement, groundwater level observation, rainfall record, and river flow observation, were carried out to understand the possible variation of the regional groundwater resources. The representative results of the investigations are presented below.

3.1 Tunnel inflow measurement

Dangao Tunnel is a twin-tube tunnel with 3.3 km. The tunnel was excavated from both sides of north portal and south portal. There are four tunnel advancing faces during construction. The weirs were installed both at the tunnel portals and the site near tunnel advancing face to measure the amount of tunnel inflow. The measurement results near tunnel advancing face for the northbound tunnel and the southbound tunnel are illustrated in Figure 4–5. The main results are summarized as follows.

(1) In the north portal excavation section, a variety of rocks, including slate, metamorphic sandstone, diabase, schist and marble, were encountered. The rock masses between Hsiaomaoshan Fault and Houishan Fault are broken and partially infilled with soft materials. Tunnel excavation was very difficult and several tunnel collapses were occurred. Meanwhile, different amounts of water inflow were encountered. The instantaneous maximum water inflow was approximately 2,050 L/min. Water inflows were reduced soon after waterproofing as expected.

(2) In the south portal excavation section, amphibolite of several hundred meters long was encountered. The permeability of amphibolite is very low. No obvious water inflow occurred during tunnel excavation.

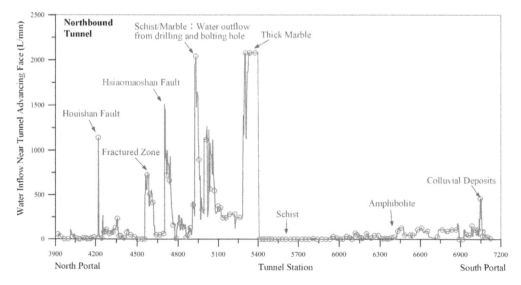

Figure 4.   Measurement of water inflow near advancing face on the northbound tunnel of Dangao Tunnel.

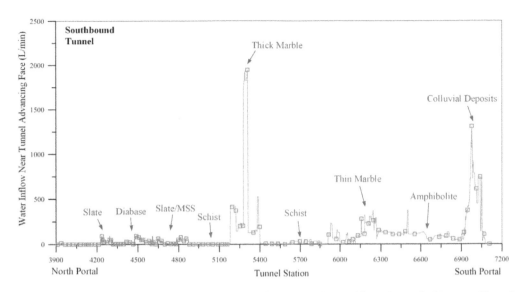

Figure 5.   Measurement of water inflow near advancing face on the southbound tunnel of Dangao Tunnel.

(3) Except for the broken rocks between faults, the major water inflow event was occurred in the thick layer of marble at the tunnel station of about 5K+300~5K+400. Large amount of groundwater flowed out of the marble fractures. And a lot of water continuously flowed out of the drilling boreholes during rockbolt installation. Even huge amount of water jetted out in some area, as shown in Figure 6. The total tunnel inflow was obviously increased. The maximum water outflow measured at the north portal of northbound tunnel reached to approximate 10 T/min. In addition, the tunnel inflow in the southern marble of thin layer at the tunnel station of about 6K+250 was not significant. The amount of water inflow was smaller than 500 L/min, while the water flow was reduced soon in a short period of time.

3.2   Tunnel inflow characteristic

In order to understand tunnel inflow behavior, the relationship between the tunnel excavation length and the total water discharge measured at tunnel portal are discussed and plotted in Figure 7. It is discovered that the water discharge would not cumulatively increase with the increasing of tunnel excavation length. An obvious increase of water discharge for the

Figure 6.   Large amount of groundwater flowed out of the boreholes during rockbolt drilling.

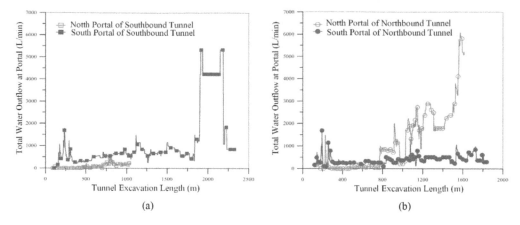

Figure 7. Relationship between tunnel excavation length and total water outflow at tunnel portal. (a) Southbound tunnel (b) Northbound tunnel

southbound tunnel was occurred as the excavation length exceeding about 1800 m, where is the thick layer of marble at the tunnel station of about 5K+300. A lot of water flowed out of the marble fractures and the drilling boreholes of rockbolts. The similar condition was found in the construction of the northbound tunnel. Overall, there is no obvious correlation between the tunnel excavation length and the water discharge. The amount of tunnel inflow is in the control of the permeability of stratum encountered.

The thick marble in the north section is anticipated to be the main groundwater flow stratum. During the actual construction, the southbound tunnel with faster excavation progress encountered the thick marble first and of course a large amount of water inflow occurred. Then the northbound tunnel was excavated to the thick marble, a large amount of groundwater inflow also occurred. A very interesting condition was observed, that is the amount of water inflow in the southbound tunnel was dramatically reduced to almost dry condition. At the same time, large amount of water continuously flowed into the northbound tunnel. It is implied that the main source of groundwater flow is from the eastern mountainous area, the Dangao Ridge. Groundwater flow from the west side is very less. This condition is consistent with the scenario of regional groundwater system proposed in the tunnel design stage (CECI 2010). The scenario is that a transverse fault, named Danghouishan Fault, with general south-north trend exists between Yungcheun Tunnel and New Yungcheun Tunnel, as shown in Figure 8. The transverse fault would block groundwater movement and separate the east groundwater system from the west groundwater system in the region. Dangao Tunnel is located at the east side of Danghouishan Fault, that is, in the east groundwater system. The west groundwater system connected far with the abundant groundwater resources within Central Mountain of Taiwan would not affect the tunnel construction. In other words, the possible water inflow during tunneling would mainly come from groundwater under Dangao Ridge. A wide fracture concentration zone with general south-north trend discovered in Dangao Ridge, as portrayed in Figure 8, would transmit the large amount of rainfall into underground.

4 IMPACT ASSESSMENT OF REGIONAL GROUNDWATER RESOURCE

4.1 Groundwater level measurement

There are four groundwater level observation wells, including the Well BH-01 at the upstream mountain of Chyuntou Creek, the Well BH-02 at the upstream mountain of the north tributary of Dangao Creek, the Well BT-01 at the thick marble above the tunnel, and the Well BT-02 in the east side of tunnel at Dangao Ridge, were installed around the tunnel. These wells measure

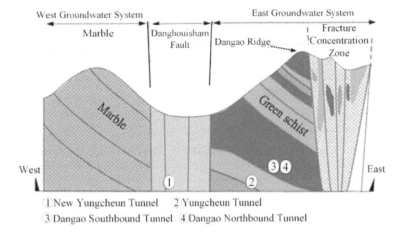

Figure 8.   Schematic profile of groundwater system in the region of Dangao Ridge.

groundwater level every 10 minutes. The variation of groundwater level during tunnel construction can therefore be analyzed and the results are illustrated in Figure 9.

It is found that the change of groundwater levels for wells BH-01 and BH-02 was not obvious during tunnel construction. And the variation of groundwater levels for wells BT-01 and BT-02 was not significant but seems to be varied with the rainfall in the initial period of tunnel construction. A clear drawdown was occurred in early 2016. In addition to the influence of the rainfall shortage during the year, groundwater drawdown could be mainly caused by the large amount of water outflow as the tunnel excavated through the thick marble, resulting in a significant drawdown of groundwater level in Well BT-01. As groundwater continuously flowed into the tunnel and the drawdown area progressively has expanded, groundwater level of Well BT-02 eventually decline. Fortunately, the regional groundwater level is

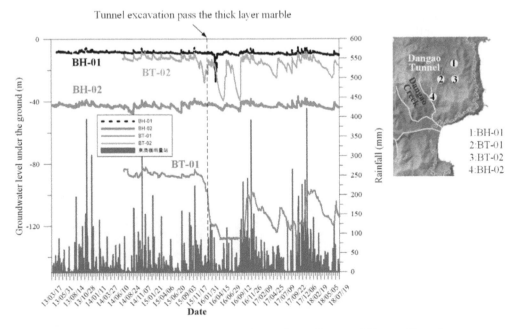

Figure 9.   Variation of groundwater level of the observation wells around Dangao Tunnel.

gradually rebounded with the installation of concrete lining and the abundant recharge of massive rainfall in Dabgao Ridge. Groundwater level in the well of BT-02 has approximately restored to the initial state before tunnel construction. Although groundwater level in Well BT-01 is not fully recovered, it has been rebounded obviously. It demonstrates that the regional groundwater resources were indeed affected by tunnel construction, while the influence is reduced to a small area adjacent to the tunnel up to now.

4.2 *Hydrogeological modeling assessment*

Field measurement data are regularly collected for understanding the variation of groundwater level during construction. The assessment of regional groundwater resources should be initiated from the perspective of water balance in a catchment area, which often involves a very wide area. In the past decades, in-situ measurement data was usually insufficient, and as such it is difficult to exactly represent the groundwater variation of a wide area. In recent decades, many theoretical solutions and numerical modeling have been developed for the assessment of hydraulic flow through rock masses. The theoretical solution method is not suitable for the case tunnel because of the complicated hydrogeological conditions. Therefore, the 3D numerical modeling was used to assess the variation of the regional groundwater environment. The initial hydrogeological model installed in the tunnel design stage, as shown in Figure 3a, was adopted as the fundamental framework. During the construction of tunnel, the model was continuously modified and calibrated based on the real hydrogeological conditions revealed and field measurement data. The hydrogeological model was revised several times and the final version is shown in Figure 3b, which is apparently different from the initial proposed model.

The regional groundwater flow field before tunnel construction presents the gravity flow pattern varied with the topography. The changes in the regional groundwater flow mainly occur in high groundwater head area (high rock cover) and the area with high permeability stratum (such as fault or fractured zone). Tunnel excavation may cause changes in the surrounding hydraulic gradient and result in variation in groundwater flow field. The variation of the regional groundwater flow field caused by the construction of Dangao Tunnel was examined from the hydrogeological modeling. The major change of groundwater flow field during construction is occurred in the thick marble at the tunnel station of about 5K+300~5K +400. The maximum groundwater drawdown is about 8.73 m, as shown in Figure 10. Groundwater inflow is reducing after the completion of tunnel construction. And there was a massive rainfall of about 7,315 mm in Dangao Ridge in 2017. The modeling results show that the drawdown area of groundwater is significantly reduced and groundwater level is gradually recovered.

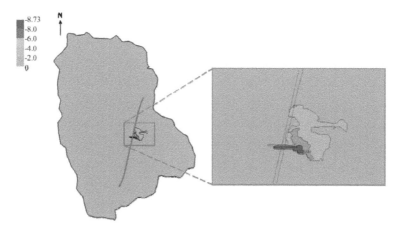

Figure 10.   Groundwater drawdown assessment for the construction of Dangao Ridge.

5 SUMMARY AND CONCLUSIONS

During tunnel construction, the occurrence of unexpected gush of water inflow is a very common problem and always making serious troubles. In Taiwan, such problems would appear to be more intractable since the hydrogeological conditions in the mountainous area are extremely complex. In addition, under the conditions of severe climate change, it is a crucial issue to secure the stability of water supply, and as such the hydrogeological investigation and environmental impact assessment are progressively important for tunnel construction.

A number of hydrogeological investigations were carried out in our study, including tunnel inflow measurement, groundwater level observation, double packer hydraulic test, and so on during the case tunnel construction. The amount of tunnel inflow was found mainly in the control of the permeability of stratum encountered. The major water inflow was occurred in the thick layer of marble. Large amount of groundwater inflow from the east side continuously flowed into the tunnel. Groundwater flowing from the west side, stopped by Danghouishan Fault, is less affected the tunnel construction. Groundwater level was found showing a temporary drawdown during the excavation of the thick marble. According to the field observation and the 3D hydrogeological modeling, the recovery of groundwater levels was observed after the completion of tunnel construction. The further restoration of groundwater in the area is anticipated due to the reduction of tunnel inflow and the abundant rainfall recharge of more than 6,000 mm/year in the region.

It is suggested that the hydrogeological investigation should be carried out in tunnel design stage and during tunnel construction to identify the variation of regional groundwater resource. Limited investigation data are available and therefore it is difficult to represent groundwater variation of the entire catchment. We proposed a method combined 3D numerical modeling with field hydrogeological surveys can be used to assess the impact of tunnel construction on regional groundwater resource in complicated hydrogeological conditions. The assessment results provide a valuable scientific reference for understanding the relationship between tunnel construction and groundwater resource protection.

REFERENCES

CECI Engineering Consultants, Inc. 2010. *The report of geotechnical investigation and integration assessment in the feasibility study and priority route design for the Suhua Highway mountainous section improvement project (in Chinese)*. Taipei City, Suhua Improvement Office, Directorate General of Highway, Minostry of Transportation and Communication.

Hsiao, F.Y., Peng S.J., Kao, H.C., & Chi, S.Y. 2014. Application of hydrogeological survey and numerical modeling in tunnel construction, *Tunnel Construction*, 34:6–14.

Hsu, H.H., Wu, I.C., Chou, C., Chen, C.T., Chen, Y.M. & Lu, M.M. 2011. *Science report on climate change in Taiwan*, Department of Natural Sciences and Sustainable Development, National Science Council.

McDonald, M.C. & Harbaugh, A.W. 1988. *A modular three-dimensional finite-difference groundwater flow model*. USGS, U.S.A.

Shau, H.J., Hsiao, F.Y. & Kao, H.C. 2015. Preliminary assessment of environmental impact for a high potential water inflow tunnel in Taiwan, *ITA WTC 2015 Congress*, Dubrovnik, Croatia.

Sinotech Engineering Consultants, Ltd. 2012. *The report of geotechnical investigation for the Suoao-Dangao section of Suhua Highway mountain area improvement project (in Chinese)*. Taipei City, Suhua Improvement Office, Directorate General of Highway, Minostry of Transportation and Communication.

Singhal, B.B.S. & Gupta, R.P. 1999. *Applied Hydrogeology of Fractured Rocks*. Kluwer Academic Publishers.

Wang T.T., Jeng, F.S., & Wei, L. 2011. Mitigating large water ingresses into the New Yungchuen Tunnel, Taiwan. *Bull Eng Geol Environ*, 70:173–186.

Yang, F.R., Lee, C.H., Kung, W.J. & Yeh, H.F. 2008. The impact of tunneling construction on the hydrogeological environment of Tseng-Wen Reservoir Transbasin Diversion Project in Taiwan. *Engineering Geology*, 103:39–58.

Tunnels and Underground Cities: Engineering and Innovation meet Archaeology,
Architecture and Art, Volume 2: Environment sustainability in
underground construction – Peila, Viggiani & Celestino (Eds)
© 2020 Taylor & Francis Group, London, ISBN 978-0-367-46579-7

Use and management of sulphated excavation material from the Montcenis Base Tunnel

E. Hugot
TELT geologist, Le Bourget-du-lac, France

J. Burdin
TELT Engineering Consultant, La Féclaz, France

L. Brino
TELT Director Engineering, Turin, Italy

P. Schriqui
TELT Construction manager opérative work excavation materials, Le Bourget-du-lac, France

M.E. Parisi
TELT Function manager Geology, Turin, Italy

ABSTRACT: The 57.5 km long Montcenis Base Tunnel will link up Saint-Jean-de-Maurienne in France to Bussoleno in Italy, with some 45 km being excavated on the French side producing almost 27 million tons of material.The exploratory works already carried out have shown the presence of sulphates within the excavated material, which could not allow the production of concrete aggregates according to the dedicated standards. TELT has launched tests as part of a research and development programme focused on developing concretes produced with sulphated aggregates by using specific cements. The results are very encouraging. However, taking into account the boring progress and the small available space at the Villarodin-Bourget/Modane jobsite, the implementation of the chemical analysis of the excavated material is being critical, and especially for the SO_3 content, with the convenient accuracy and within a very short notice. The only reliable way to do this is to set up a chemical characterisation unit on site directly on the conveyor belt, for example a Prompt Gamma Neutron Activation Analysis (PGNAA)-type unit. Tests have been implemented on behalf of the French atomic energy commission (CEA) in order to certify the use of this kind of equipment. This paper is presenting all the information regarding the Villarodin-Bourget/Modane jobsite, the quality and the quantity of the dedicated excavated material, the need of aggregates and the details of all the tests carried out.

1 INTRODUCTION

The Montcenis Base Tunnel is the main part of the international section of the new Lyon-Turin rail link. This 57.5 km long twin-tube will link up Saint-Jean-de-Maurienne in France to Bussoleno in Italy, some 45 km being excavated on the French side from six main tunnelling faces. The French side will produce some 27 million tons of material, able to produce 7.7 million tons of aggregates for the concrete tunnel lining and ancillary works. Tunnelling from the Villarodin-Bourget/Modane adit will produce over 12.7 million tons of material.

The latest 2017 estimation is showing the aggregate needs for the whole cross-border section of about 11.4 million tons, including 8.6 million tons for the French side alone. It means that TELT would have to face a potential shortage on the French side as the production of aggregates is not enough to cover the needs. Explorations and other works already carried out have shown the presence of sulphates impacting 1.7 million tons of materials, with a knock-on effect equivalent to approximately 1 million tons of aggregates, i.e. almost 24 per cent of the material required to produce concrete for this sector. When boring the Villarodin-Bourget/ Modane access tunnel, some of the excavated material could not be used as concrete aggregates, as planned, because their sulphate content was higher than the standard requirements. The NF EN12-620 and NF EN 206 standards limit sulphate content in aggregates to 0.2% in order to protect the concrete from deterioration caused by sulphated reactions.

In order to comply with the EU environmental rules by maximizing the use of excavated material, TELT undertook a research and development programme in 2009 to check the sulphate content of this material and to find out a way of use. The first laboratory results and on-site studies are presented here. Furthermore, taking into account the forecasted excavation progress and the small available space at the Villarodin-Bourget/Modane site, the implementation of the chemical composition analysis of the excavated material is being critical, and especially their SO_3 content, with the convenient accuracy and within a very short notice. One reliable on-site chemical characterisation process today identified is the one provided by the Prompt Gamma Neutron Activation Analyser (PGNAA). TELT therefore asked the CEA to carry out an expert appraisal of the equipment used and of the material submitted to radiation, in order to demonstrate that no residual radioactivity is remaining within the tested material. The details and results of the protocol, and the tests performed are presented here.

2 CONTEXT

In the Villarodin-Bourget/Modane sector, the material to be extracted over a length of approximately 10 km could have a significant sulphate content, in the form of gypsum and/or anhydrite in blocks or inclusions in the rock mass. Regarding the standards requirements, this material cannot be used as aggregates for civil engineering concrete applications, as their SO_3 content is $\geq 0.2\%$.

A research and development programme was initiated in 2009 by LTF and was carried on by TELT with the partnership of IFSTTAR (leadership), VICAT, HOLCIM and LERM. An initial laboratory study was concluded by the Jeremy Colas' thesis in 2012. It delivered some very promising results for the use of this excavation material in the future Base Tunnel, providing the use of special cements (Monin et al., 2013).

The tests were performed with aggregates produced by the experimental processing plant that had been set up to make the concrete of the access gallery. The SO_3 content divided by grain-size band of this material is given on table 1.

The results showed that calcium sulphates are encountered in the form of gypsum (40%) and anhydrite (60%) and are most accessible in grains smaller than 1 mm. Indeed, they are mostly found in the form of independent grains. In the case of grains larger than 1 mm, they are located at the periphery, or in inclusions into the aggregates that limit accessibility. Given these findings, two solutions stand out and could be considered to use this excavation material:

1/Extended washing time. Washing the excavated sand with water at 20°C would enable sulphate content to be reduced from approximately 3.5% down to 2.5%.

2/Fine particle fraction of the sand removing. Indeed, the 0/315 μm band has a sulphate content up to 6 times greater than other grain-size bands. By eliminating the grains smaller than 315 μm, sulphate content into the excavated sand can be lowered down to 1.9%.

In both cases, this content is still over the standard, which limits the sulphate content to 0.2% in concrete aggregates

The IFSTTAR studies also showed that the different cement types analysed did not have the same behaviour in terms of sulphate reaction with the formation of ettringite or thaumasite. This

Table 1. Sulphate content by grain-size band.

Grain-size band (mm)	0/0.315	0.315/1	1/4	4/8	8/16
Sulphate content (as a SO_3%)	7.2%	2.8%	1.6%	1.4%	1.2%

leads to another possible use for this excavation material, i.e. using a special cement settled for aggregates having a high sulphate content. The current findings are showing that two cements could be suitable: Super Sulphated Cement (SSC) and Portland Cement with a very low C3A content, referred as CEM I 52.5 N PM SR 0.

A more in-depth study on an industrial scale should help to clarify the following issues:

– identification of the mechanisms responsible of the good behaviour of the SSC and CEM-I 52.5 N PM SR0 cements and the effects of using a correcting sand;
– long-term behaviour of sulphate in excess into the concrete;
– feasibility of using such concretes on an industrial scale;
– durability of these concretes.

3 INDUSTRIAL TEST PROGRAMME

The research partnership between TELT, IFSTTAR, VICAT and LAFARGE/HOLCIM was going on with the development of concrete blocks for on-site tests. The goal of this in-depth study is to check the durability of concretes in existing tunnel curing conditions and especially how they react to the internal sulphate attack when using super sulphated cement (CSS) and Ultimat cement (CEM-I 52.5 N PM SR0), which showed very good behaviour in the previous phase of the study (Colas, 2012).

Industrial-scale testing was performed in order to confirm the initial very encouraging results. These tests consisted of producing 11 demonstrators using 11 different concrete mixes.

3.1 Concrete mix design

It is the same as the one used during the first test phase and it is close to the mix design made with aggregates processed from the excavation material from the Villarodin-Bourget/Modane access tunnel, which is for 1 m^3 of concrete:

– Cement: 400 kg (reference 300kg of CEM I + 100 kg of fly ash)
– Sand 0/4 mm: 770 kg
– Grit 4/8 mm: 280 kg
– Gravel 8/16 mm: 700 kg
– Crushed gypsum: depending on the SO_3 content of the aggregates

3.1.1 TELT aggregates

Aggregates produced with the material excavated from the Villarodin-Bourget/Modane access tunnel were used and are referred as "TELT aggregates" in the remainder of this article. Each grain-size band was first homogenised and sampled (NF EN 932-1 standard) for chemical analysis and to check its SO_3 content (table 2).

3.1.2 External aggregates

Some mix designs were developed by replacing the finest grains of the TELT 0/4mm sand, with a silico-limestone sand from external origin that is referred in the remainder of this article as "Barraux sand".

Table 2. SO_3 content of grain-size brackets.

	sand 0/4	aggregate 4/8	aggregate 8/16
SO_3 by gravimetry	2.03	0.89	0.91
Uncertainties	0.43	0.20	10.20

3.1.3 *Gypsum*

The TELT aggregates have been stored outside since 2008 and have therefore been exposed to the effects of the environmental agents. Consequently, the leaching of the aggregates has modified their sulphate content. In order to make concrete mixes with aggregates having sulphate content of up to 4%, crushed gypsum was added to some mix designs. Mixes with 2% of SO_3 are only produced with TELT aggregates without gypsum addition.

3.2 *List of the 11 concrete compositions*

No. 1) Control concrete made from external aggregates and CEM I PM ES cement + 25% fly ash

No. 2) Concrete Mix with external aggregates + Ultimat cement

No. 3) Concrete Mix with TELT aggregates as they are (2% SO_3) + Ultimat cement

No. 4) Concrete Mix with TELT aggregates as they are + SO_3 (4%) + Ultimat cement

No. 5) Concrete Mix with TELT aggregates with the 0/4mm bracket being replaced by Barraux sand + Ultimat cement

No. 6) Concrete Mix with external aggregates + SSC cement

No. 7) Concrete Mix with TELT aggregates as they are (2% SO_3) + SSC cement

No. 8) Concrete Mix with TELT aggregates as they are + SO_3 (4%) + SSC cement

No. 9) Concrete Mix with TELT aggregates with the 0/4mm bracket being replaced by Barraux sand + SSC cement

No. 10) Concrete Mix with TELT aggregates as they are (2% SO_3) + 320 kg/m3 of Ultimat cement + 80 kg/m3 of fly ash

No. 11) Concrete Mix with TELT aggregates as they are + SO_3 (4%) + 320 kg/m3 of Ultimat cement + 80 kg/m3 of fly ash

The concrete slump consistence class is S5 (NF EN 206): virtually self-compacting concrete with good flow characteristics and requiring very little vibration when used. The concrete strength development class is C30/37.

An initial laboratory optimisation study and was carried out by VICAT and LAFARGE/ HOLCIM especially to fix the quality and the quantity of admixtures.

The concretes were produced by the Saint-Michel-de-Maurienne "Bétons VICAT" unit and cast in the Villarodin-Bourget/Modane access tunnel. Test Concrete samples for durability trials were cast at the same time as the test blocks, then stored on site with the same curing conditions as the test blocks, with a watering system in place until completion of the test period (i.e., for 3 months). They were then transported to IFSTTAR Laboratory to run the tests (Figure 1).

3.3 *Reference grain size distribution curve for the concrete*

Figure 2 shows the reference grain size distribution curve of the concrete.

Figure 1. The samples in the access tunnel.

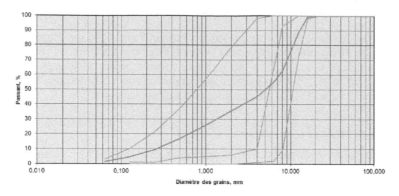

Figure 2. Reference grain size distribution curve for the TELT 0/16 mm concrete without binder and without gypsum addition.

3.4 *The test blocks*

3.4.1 *Description*

Eleven L-shaped test blocks with the following dimensions were on-site cast:

– Slab: length = 3.3 m, width = 2.5 m, variable thickness; slab reinforced with a welded mesh 11 x 10 m², ø = 8 mm
– Sidewall: length = 3.3m, height = 2m, thickness = 0.40m; precast non-reinforced sidewall.

370 samples were produced under the guidance of a representative from the IFSTTAR research laboratory.

3.4.2 *Pouring and placing method*

The following tests to control the fresh concrete were carried out before leaving the production unit and again on the site: Flow-Test, air temperature, concrete temperature, air content. These data enabled to characterise the rheology of the concrete and to determine how it varies during transport.

The concrete was then cast. The slab was cast first and a welded mesh was put in place halfway through the casting process (Figure 3).

Once the casting of the slab was completed, a formwork panel was placed over the slab to cast the sidewall. The sidewall was cast by discharging the concrete between the formwork panel and the wall of the tunnel. The strength developed during the casting of the sidewall led to the blocks being adapted. Only the control block (F1) was cast in place. The sidewall elements were finally cast on an horizontal formwork, then placed upright and sealed to the side wall of the access tunnel (Figures 4 a, b, 5).

Figure 3. End of casting of the slab after placing the welded mesh.

(a)   (b)

Figure 4.   Slabs in place after formwork removal and sidewall with rods used to seal the block in the siding of the access tunnel.

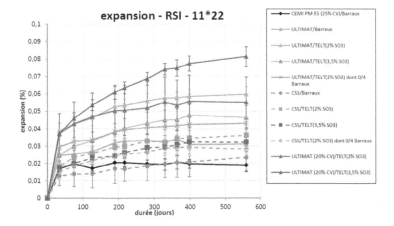

Figure 5.   Test blocks in place with the sealed tacheometric monitoring marks.

4   LABORATORY TESTS

Laboratory tests enable to assess the risk of internal swelling reactions for the different concrete type vs reactions resulting in the formation of ettringite and thaumasite. The first results are given on figures 6 and 7 below.

After monitoring during 550 days, 6 concretes (including the reference) presented swelling less than or equal to 0.04%. No damage was found, in light of the slow kinetics of the reactions and the small swelling value.

Figure 6.   ISA (Internal Sulphate Attack) expansion test results.

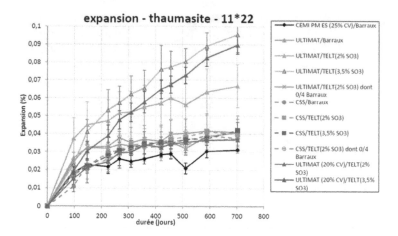

Figure 7.   Thaumasite expansion test results.

After 700 days monitoring, expansion measurements showed that 3 concretes were quite reactive (>0.05%). No damage was found, in light of the slow kinetics of the reactions and the small swelling value.

5   ON-SITE DEFORMATIONS MONITORING

5.1   *Equipment*

Each block was fitted out with 10 sealed bases, which will allow periodic monitoring readings to highlight and measure deformations and abnormal movements due to internal strains (Figure 8).

The brass marks were sealed into the blocks one month after the blocks casting and the first measurement was made 15 days later. Taking into account the accuracy required (uncertainty less than 0.07mm), the measurements were carried out using a Laser Tracker of the Leica AT402 type.

5.2   *Operating mode*

Two sets of independent measurements were performed and combined in order to ensure consistency and thus determine the final coordinates of the points observed. The results are presented in a three-dimensional reference specific framework to each block, free of tunnel movements and by separating the sidewall block from the slab.

5.3   *Results*

The deformations observed are weak on most of the blocks (Figure 9).

Figure 8.   Names of the points by block and brass mark.

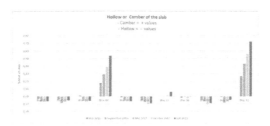

Figure 9.   Monitoring of the deformations observed on the horizontal blocks.

Only blocks 4 and 11 are presenting more deformations, showing a swelling phenomenon of the vertical and horizontal blocks.

6   RADIATION PROTECTION EXPERT APPRAISAL

The main problem will then come from the need to perform quickly chemical characterisation of the excavated material as it comes out of the tunnel. To do this, our research focused on continuous chemical analysis processes, used commonly and successfully by the time being in cement plants when preparing the raw mix, as well as in the mining industry.

6.1   Description of the equipment

The basic principle is to submit the material to neutron radiation and analyse the gamma radiation recovered above the bed of material to produce an elementary chemical analysis. This device is called a Prompt Gamma Neutron Activation Analyser (PGNAA) and is fixed on the conveyor belt.

It operates with a ^{252}Cf radioactive source that requires special authorisation from DGPR (French Risks Prevention General Directorate) and ASN (French Nuclear Safety Authority). In order to speed up the preparation of this authorisation, TELT asked the CEA in Paris-Saclay to perform a radiation protection expert appraisal, the goal was to prove that no residual radioactivity is remaining within the material analysed. It also aimed to demonstrate that this type of device fully meets TELT's needs in term of chemical characterisation and accuracy of the percentages measured.

6.2   Test protocol according to the CEA's requirements

In order to get the necessary administrative authorisations, the CEA was asked to check that the material did not have any residual radioactivity after going through the neutron analyser. TELT therefore got in touch with the CEA who accepted to provide TELT with this expert appraisal request.

LAFARGE, who is already a partner in the research program on sulphated materials presented above, gave the chance to TELT to carry out this expert appraisal by using the operating equipment of their Saint-Pierre-La-Cour plant.

TELT brought 60t of material from SMP4 (silico-limestone 0/40mm SMP4) and 60t of material stored in Villarodin-Bourget/Modane (0/4mm VBM sand) from the tunnelling works at this same access tunnel. These two families of material had chemical characterisation performed in the laboratory on the main component elements before delivery to Saint-Pierre-La-Cour, the site on which the tests would be carried out. Sulphate contents measured were: 2% in the VBM sand and 2.30% in the 0/40 SMP4.

The radiation protection expert appraisal of the material passed through the neutron analyser was carried out in the normal way, with a belt speed of 2.3 m/s and a flow rate of approximately 1000 t/h, and with an accident simulation that shut down the conveyor belt

Figure 10. Measuring equipment in the on-site laboratory measuring the radioactivity of the sample (0/40 SMP4).

and left the material exposed during 2h to radiation emitted by the neutron analyser. Each family of material was analysed individually.

The samples to determine the radioactivity level of the material before and after passing through the neutron analyser in these two exposure conditions were performed on-site and in the CEA's laboratory in Paris-Saclay (Figure 10).

6.3 Results

The results of the analyses carried out by the analyser and those carried out for comparison in the laboratory are shown on the following figure:

%	Sable 0/4 mm VBM		0/40 mm SMP 4	
	LERM [*]	PGNAA [**]	LERM [*]	PGNAA [**]
LOI	3,12	10,46	35,20	33,24
SiO$_2$	83,06	70,23	10,78	16,16
Al$_2$O$_3$	4,03	4,17	2,83	4,68
Fe$_2$O$_3$	0,67	1,06	1,31	1,67
CaO	3,64	9,82	39,49	34,82
MgO	0,92	2,57	6,97	5,91
TiO$_2$	0,13		0,12	0,19
MnO	0,02	0,07	0,05	0,03
P$_2$O$_5$	0,03		0,06	
Cr$_2$O$_3$	<0,01		<0,01	
SrO	0,03		0,08	
Na$_2$O	0,18	0,75	0,23	0,23
K$_2$O	2,09	0,93	0,69	1,29
SO$_3$	2,00	2,57	2,30	1,87
Total	99,92	102,63	100,11	100,09

[*]	Dosage des éléments majeurs et mineurs par ICP
	Dosage SO$_3$ par chromatographie ionique
	Dosage des alcalins par Spectromètre à flamme
[**]	Exposition statique PGNAA 2 heures

Figure 11. The results of the PGNAA chemical analysis vs laboratory results.

The PGNAA SO$_3$ measurement results are showing a quite good accuracy. Some other elements show some discrepancy; this is probably due to the fact that the analyser used for these tests could not be calibrated before the test (too the short time available, there was no question to stop the cement production), TELT material is quite different from cement raw-mix (The analyser calibration must be achieved with the dedicated "reference" elements to get the convenient accuracy).

Furthermore, the readings also show that the neutron activation of the material required for chemical characterisation created almost zero residual radioactivity, in the form of radionuclides having a very short radioactive period of around a few hours. This radioactivity will therefore disappear by the time the material is used. The use of a neutron analyser does not therefore lead to the presence of residual radioactivity that can be differentiated from the natural radioactivity in the tunnel construction material.

6.4 Perspectives in the framework of Future Operational Work Site CO 5a

The contract N° CO 5a for Villarodin/-Bourget/Modane will be very shortly awarded by TELT. It will be dedicated to the preparatory works, to the construction of ventilation shafts and a range of different underground galleries, and the TBM assembly caverns. These works will be performed from the Villarodin-Bourget/Modane access tunnel and they will provide a possible testing site for managing, classifying, storing and processing the excavation material. On-site chemical analysis such above described will be used to check on site the reliability of the analysis method on an industrial scale and to develop the corresponding working procedures.

7 CONCLUSION

TELT, Alpetunnel and LTF, who were involved upstream in preparing the project, are committed and determined to apply a policy of using excavated materials for the construction of the Montcenis base tunnel. The above discribed work is giving a clear proof of its feasibility.

This policy has taken them on a path to innovation, innovation in a unique research programme conducted under the umbrella of qualified state agencies and innovation in using equipment never used before on such construction sites. The quality of the work carried out will hopefully become a reference in this field and will provide major results to prepare the improvement of the national standards. This step is on the way with the AFNOR Commission P 18 B in order to agree upon a Performance Test which will be able to check the durability of a concrete made with aggregates having SO_3 content over the requirements. If this NF standard improvement proceed correctly, the Villarodin-Bourget/Modane site would then be able to work in total self-efficiency to produce the concrete for the tunnel.

ACKNOWLEDGEMENTS

TELT would like to thank all the partners for the interest they have shown in this research and development programme and especially IFSTTAR for running the tests and LAFARGE/ HOLCIM and VICAT for their involvement.

REFERENCES

AFGC 2004. Conception des bétons pour une durée de vie donnée des ouvrages.
Colas, J. 2012. *Etude de la valorisation des déblais du chantier et des tunnels riches en sulfates en granulats à béton*. Thesis defended on 4 December 2012. Université de Paris-Est.
Darmendrail, X., Rimey, J., Brino, L. & Burdin, J. 2003. *Nouvelle liaison ferroviaire Lyon-Turin – Une approche d'étude originale pour la valorisation des déblais des tunnels. Tunnel et Ouvrages souterrains* n. 176: 55-61.
Darmendrail, X., Brino, L. & Burdin, J. 2011. *Bilancio e risultati degli studi sulla gestione e valorizzazione dei materiali di scavo del Tunnel di Base della Nuova Linea Torino Lione*. In Convegno SIG 2011, Verona.
Divet, L., Colas, J., Chaussadent, T., Lavaud, S. & Desrues, B. 2015. *Perspectives de valorisation de granulats riches en sulfates dans les bétons*. In Congrès AFGC, Paris.
Monin, N., Burdin, J., Brino, L., Colas, J., Divet, L. & Chaussadent, T. 2013. *Preliminary tests on concrete aggregates with high sulphate content for tunnel lining*. In AITES-ITA 2013 World Tunnel Congress, Geneva, 31 May - 7 June 2013: 1459-1466. CRC Press.
Thalmann, C., Carron, C., Brino, L. & Burdin, J. 2005 *Gestion et valorisation des matériaux d'excavation de tunnels – Analyse comparative de trois grands projets*. In Congrès International AFTES 2005, Chambéry, 10-12 Octobre 2005: 237-248. Edition Spécifique.

*Tunnels and Underground Cities: Engineering and Innovation meet Archaeology,
Architecture and Art, Volume 2: Environment sustainability in
underground construction – Peila, Viggiani & Celestino (Eds)*
© *2020 Taylor & Francis Group, London, ISBN 978-0-367-46579-7*

Blasting method to reduce the adverse effect on the living environment around tunnel portals using electronic detonators

Y. Kitamura, Y. Tezuka & K. Fuchisaki
Kajima Corporation, Tokyo, Japan

Y. Maeda
NEXCO Research Institute, Tokyo, Japan

ABSTRACT: When residential areas are close to a tunnel portal, noise, low frequency sound and ground vibration generated by blasting are factors that cause adverse impact to the neighboring environment. Noise can be reduced by using soundproof equipment and the vibration attenuates as the distance between the blasting location and the residential area increases, however the low frequency sound would not attenuate sufficiently even if blocked by soundproof equipment nor with increasing distance between the blasting location and the residential area. In the Minoh tunnel where tunnel portal and the houses are separated by 100 m, hard rocks were found close to the portal location. The result from the analysis of the data measured during the blasting demonstrates that vibration velocity and low frequency sound can be effectively attenuated and rattling of windows in houses due to low frequency sound can also be eliminated by electronic detonators.

1 INTRODUCTION

The Minoh tunnel is a two-lane two-way road tunnel with a total length of about 5 km, which will be a part of the Shin-Meishin expressway connecting Nagoya and Kobe. As shown in Figure 1, at the portal on the western end of this tunnel, private houses are sprawled right up to the portal, and since nearby residents requested that impact on the environment from blasting be reduced as much as possible. For the first time in a road tunnel in Japan, an electronically-controlled detonator was used to reduce impact on the residential environment. Moreover, results from the measurement of vibration and noise in the residential area at the time of blasting shows that not only was it possible to effectively reduce ground vibrations and low frequency sounds by appropriately controlling the delay time interval, but also suppress rattling in neighboring houses. In this paper, we report these results.

Figure 1. Separation between tunnel portal and houses.

2 VERIFICATION OF CHARACTERISTICS OF ELECTRONIC DETONATOR WITH EXPERIMENTAL BLASTING

Ground vibration velocity, noise and low frequency sound level were measured during the experimental blasting which was conducted with several delay time intervals and the relationship between the interval and the impact on the residential environment was studied.

2.1 Features of the electronic detonator used

Table 1 shows the features of the electronic detonator (Figure 2) used in Minoh-Tunnel West construction.

2.2 Overview of experimental blasting

Experimental blasting was conducted inside the tunnel about 120 m from the entry portal. A total of 11 measuring points was set during the experimental blasting: 6 in the yard (A – F), and 5

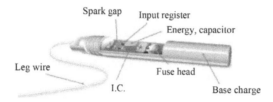

Figure 2.   Structure of electronic detonator.

Table 1.   Features of high-accuracy electronic detonator.

Setting of Delay time interval	Ability to set the interval in ms increments from 0 to 20 seconds
Maximum number of delay blasts	500 stages
Blasting order	With a dedicated scanner to read an assigned ID to the detonator, it is possible to set an arbitrary delay and the order of blasts arbitrarily.
Wiring	Wiring can be completed by pinching the plastic connector of the detonator leg wire terminal to a harness wire.
Detonation signal	Responds only to signals from special explosive devices No erroneous detonation due to stray current or static electricity

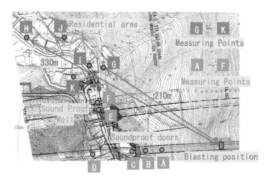

Figure 3.   Blasting positions and measuring points.

in the residential areas (G – K) as shown in the Figure 3. Noise, low frequency sound, and vibration velocity of the ground were recorded at each point. During the experimental blasting, we compared 3 different delay time intervals. The specifications for the experimental blasting are shown in Table 2.

2.3 Overview of experimental blasting

Figure 4 shows the relationship between the distance from the blasting position and the vibration velocity for each delay time interval as well as the resultant K values which were obtained. The smallest K value was found when the delay time interval was set at 17 ms. Additionally, Figure 5 shows the result of analyzing the power spectrum of the vibration waveform, from the closest measurement point G, for each frequency using the Fast Fourier transformation. The dominant frequency appears in (n-multiple of) the reciprocal number of time set for each interval. It was conformed that the electronic detonator was extremely precise because the resonance frequency and the dominant frequency match.

2.4 Relationship between delay time interval and noise/low frequency sound

Figure 6 shows the relationship between the distance from the tunnel portal to the residential area and the noise level & the low-frequency sound pressure level for each delay time interval. As clearly seen from the same figure, even when the delay time interval is changed, there is no clear changes in the noise level, but for the low frequency sound pressure level, uniformly setting the time interval at 17 ms proved to be the most effective way to reduce environmental impact on the residential area.

Table 2. Specifications for the experimental blasting.

Number of blasts	Delay time Interval	Number of shot hole	Maximum quantity of instantaneous shot	Total quantity of charge	Duration of blasts
1st time	30ms	72	0.6kg	35.0kg	2.26sec
2nd time		72	0.6kg	35.0kg	2.26sec
3rd time	7ms	73	0.6kg	22.8kg	0.68sec
4th time		63	0.6kg	23.4kg	0.68sec
5th time	17ms	68	0.6kg	24.6kg	1.14sec
6th time		64	0.6kg	14.7kg	1.07sec

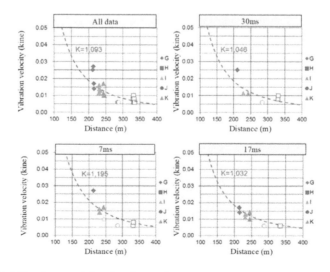

Figure 4.   Estimated K value per delay time interval.

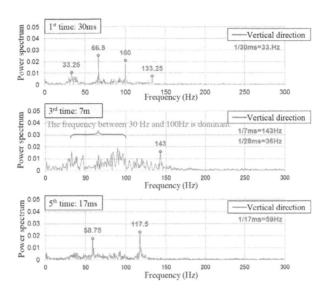

Figure 5.   Power spectral analysis results of vibration waveforms.

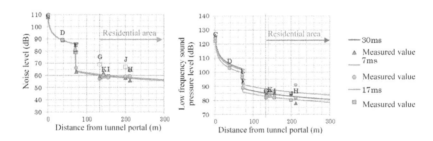

Figure 6.   Distance from tunnel portal and noise, low frequency sound level per delay time interval.

From the results of interviews with neighboring residents, the impressions of blasting with uniform time intervals of 17 ms was optimum, so that the results of measurements and human physical sensation were in accord. Setting a short delay time interval increases the dominant frequency and shortens the blasting duration. Increasing the dominant frequency is valid to attenuate the low frequency sound level and shorting the blasting duration is effective in mitigating the psychological anxiety of neighboring residents.

3   REDUCTION OF IMPACT ON RESIDENTIAL ENVIRONMENT BY THE SETTING OF OPTIMAL DELAY TIME INTERVAL

Hard rock zones that required blasting were discovered at about 210 m from the portal for the inbound lane and about 125 m from the outbound lane. The starting position of blasting was extremely close to the residential area and it was necessary to reduce the environmental impact on the residential area caused by the ground vibration and the noise and low frequency sound accompanying blasting as much as possible.

For this reason, excavation was carried out using controlled blasting with high-accuracy electronic detonators examined by the experimental blasting while attempting to suppress blasting vibrations and noise for residential areas adjacent to the portal. Here, we explain the method to reduce impact on the residential environment caused by low frequency sound.

3.1 Delay time interval and decrease in low frequency sound pressure level

Figure 7 shows the result of spectral analysis of the sound pressure level measured inside the tunnel when blasting with a delay time interval of 17 ms. Based on the results, it can be confirmed that the sound pressure for 59 Hz (≒ 1000 ms/17 ms), which is the inverse of the 17 ms time interval, is dominant.

In order to investigate the sound attenuation effect due to the soundproof equipment, blasting was conducted at the positions shown in Figure 8 and the low frequency sound pressure was measured for each frequency at observation points A to D. There are two soundproof doors between observation points A and B, and a soundproof wall was placed between points C and D.

Based on the measurement difference between observation points A and B, it is possible to calculate sound attenuation for each frequency due to the soundproof door, and sound reduction for each frequency due to the sound proof wall using the measurement difference between observation points C and D. From Figures 9 and 10, it was found that the noise reduction effect was difficult to achieve n the frequency range of 6.3 to 40 Hz for the soundproof door and 20 to 40 Hz for the soundproof wall.

To obtain high noise reduction effect from soundproof doors and soundproof wall, as shown in Figure 11, it is necessary to avoid the inherent frequency bands of the soundproof equipment where it is difficult to achieve noise attenuation. However, by setting an appropriate interval thanks to the high-accuracy electronic detonator, we learned that the above frequency bands can be avoided. The soundproof equipment installed for this project requires a

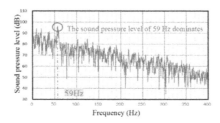

Figure 7.   Spectral analysis results of sound pressure levels (The delay time interval is set to 17ms).

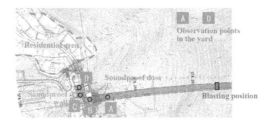

Figure 8.   Blasting positions and measuring points of low frequency sounds.

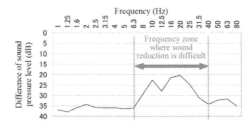

Figure 9.   Sound reduction by frequency due to sound proof door (Observation points A – B).

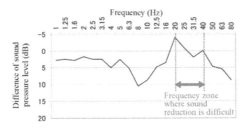

Figure 10. Sound reduction by frequency using a sound proof wall (Observation points C – D).

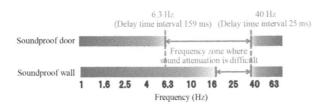

Figure 11. Frequency range where it is difficult to reduce sound, using soundproof equipment.

dominant frequency of 6.3 Hz or lower or 40 Hz or higher, but if the dominant frequency is 6.3 Hz or lower, the blast duration increase and psychological discomfort is amplified. It was deemed desirable to set the time interval at 25 ms or lower so that the dominant frequency would be 40 Hz or higher.

3.2 Considerations related to rattling of fixtures in nearby houses from low frequency sound

At the first stage of this project, excavation was proceeded using DS (Deci second) delay electric detonators, but due to an increase in the quantity of explosives utilized, complaints on household fixture's rattling were received. For this reason, blasting was conducted using high-accuracy electronic detonators. The low frequency sounds generated at the time of blasting were compared, and the factors causing rattling of the fixtures were examined.

3.2.1 Residence where the relationship between the rattling and the blasting conditions was investigated

On the inbound and outbound lanes, blasting with DS delay electric detonators and high-accuracy electronic detonators were conducted, and as shown in Figure 12, rattling of the household fixtures in residences were tested and low frequency sound was measured in the yards of the houses–in question.

3.2.2 Results of investigation of household fixtures rattling

Rattling of household fixtures at the location shown in Figure 12 was investigated with the blasting specification shown in Table 3. The results, as shown in Table 4, demonstrates that only the blasts when the DS delay electric detonators were used on the inbound lane made the fixtures rattle, such rattling never occurs when the blasting was conducted with the high-accuracy electronic detonators. Furthermore, no rattling was observed with the DS delay electric detonators on the outbound lane. This was thought to be due to the fact that the distance between the inbound lanes and the residence in question was greater than the distance between the outbound lanes and the residence and that the portal for the outbound lanes was more recessed than for the inbound lane.

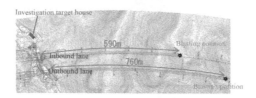

Figure 12.   Blasting positions and houses investigated for rattling.

Table 3.   Blasting specifications (when verifying rattling of fixtures).

Number of blasts	Blasting position	Distance from portal	Detonator used	Total charge
1st time	Inbound	570m	DS delay electric detonator	42.8kg
2nd time		593m	DS delay electric detonator	72.5kg
3rd time		594m	High-accuracy electronic detonator	71.0kg
4th time	Outbound	756m	DS delay electric detonator	75.3kg
5th time		757m	High-accuracy electronic detonator	38.4kg

Table 4.   Investigation result for fixture rattling.

Number of blasts	Blasting position	Detonator used	Investigation result for fixture rattling
1st time	Inbound	DS delay electric detonator	Rattled violently
2nd time		DS delay electric detonator	Rattled a little
3rd time		High-accuracy electronic detonator	No rattling
4th time	Outbound	DS delay electric detonator	No rattling
5th time		High-accuracy electronic detonator	No rattling

3.2.3 *Measurement results of low-frequency sound and causes of fixtures rattling*

The low frequency sound pressure level was measured in the yards of the residential housing. Figure 13 shows the measurements at the time of blasting in the inbound lane and Figure 14 shows the measurements at the time of blasting in the outbound lane. For comparison, these figures also show the reference values, regarding physical complaints with low frequency sound pressure level, presented by the Ministry of the Environment, Government of Japan.

When high-accuracy electronic detonators are compared with DS delay electric detonators in terms of frequency, it is found that in the case of the high-accuracy electronic detonators, the frequency band involving a peak of low frequency sound pressure level is shifted to a high frequency side, and that the numbers of physical complaints are below than the references by the Ministry of Environment.

In addition, when blasting with DS delay electric detonators in the inbound lane, a relatively large low frequency sound exceeding 80 dB was observed in the frequency band of 10 to 20 Hz (Figure 13), but at the time of blasting in the outbound lane, the low frequency sound exceeding 80 dB did not occur (Figure 14), it was presumed that if low frequency sound exceeding 80 dB had occurred in the frequency band of 10 to 20 Hz, the household fixtures would rattle. Furthermore, when the change was made to high-accuracy electronic detonators for excavation of the inbound lane, no rattling of fixtures occurred.

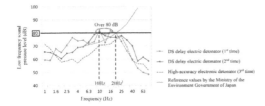
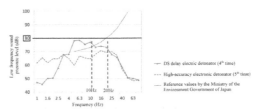

Figure 13. Low frequency sound pressure level for inbound lane blasting.

Figure 14. Low frequency sound pressure level for outbound lane blasting.

3.3 *Difference in low-frequency sound pressure characteristics between DS delay electric detonators and high-accuracy electronic detonators*

3.3.1 *Peak frequency of low-frequency sound pressure*

In order to suppress rattling of fixtures in nearby houses due to low frequency sound from blasting during this project, high-accuracy electronic detonators were used continuously. Furthermore, experimental blasting was conducted using DS delay electric detonators at every 100m progress interval. It was therefore confirmed the difference on the characteristics of the low frequency sound pressure level transmitted to the residential area.

Figures 15 and 16 show the results of measurement of low frequency sound pressure level at each frequency in yards of surveyed houses when blasting with DS delay electric detonators and high-accuracy electronic detonators for the inbound lane.

From the average value in Figure 15, a peak occurring at the sound pressure level of 4 Hz in DS delay electric detonators can be observed. The reason was considered that the delay time on DS delay electric detonators are at 250 ms, the sound pressure level for 1 s/ 250 ms = 4 Hz was maximized. Contrarily, when blasting was pursued with high-accuracy electronic detonators at 17 ms. The region from 4 Hz to 16 Hz shows constant level of low frequency sound pressure displayed in Figure 16, where the well-defined peak does not appear no more.

3.3.2 *Difference between distance attenuation characteristics and low frequency sound level*

Generally, the magnitude of low frequency sound is dominated by the total charge and the outside distance from the tunnel portal to measuring points, and is not assumed to depend on the inside distance from the tunnel portal to the blasting position.

Figure 17 shows the low frequency sound level (flat values) by distance from the portal to the blasting position for each detonator. The average value of the total explosive charge is 80 kg for the DS delay electric detonator and 76 kg for the high-accuracy electronic detonator. They are roughly equivalent.

From the same figure, the expected decline of the sound pressure level with the increase of the tunnel excavation distance cannot be confirmed. Furthermore, since the average value of

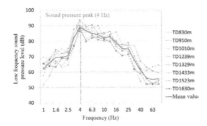
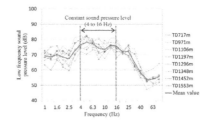

Figure 15. Low frequency sound pressure level when using DS delay electric detonators.

Figure 16. Low frequency sound pressure level when using high-accuracy electronic detonators.

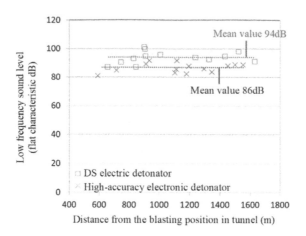

Figure 17. Low frequency sound level by distance from the portal to the blasting position for each detonator.

the low frequency sound level (flat values) is 94 dB for the DS delay electric detonator and 86 dB for the high-accuracy electronic detonator, it was also found that the low frequency sound level can be reduced by about 8 dB using the electronic detonator.

4 CONCLUSION

In construction of the Minoh tunnel, the blasting location and the residential area was in close proximity, bringing the concerns that vibration and low frequency sound from blasts would adversely affect the environment of nearby residential areas. For this reason, high-accuracy electronic detonators were used to successfully reduce blasting vibration impact. At the same time, vibrations and noises were monitored and interviews with nearby residents were conducted to gather information about their physical feeling as well as the conditions in and around the residences. The information gathered was appropriately reflected into the blasting operation. As a result, we were able to suppress the ground vibration in the residential area down to the level it could not be felt. It was also possible to suppress the ground vibration level even further and to reduce low frequency sound pressure by adjusting the appropriate delay time interval thanks the accuracy of the electronic detonator.

REFERENCE

Office of Odor, Noise and Vibration, Office of Environmental Management *Handbook to Deal with Low Frequency Noise (2004)*, pp.9–10, June 2004

Tunnels and Underground Cities: Engineering and Innovation meet Archaeology,
Architecture and Art, Volume 2: Environment sustainability in
underground construction – Peila, Viggiani & Celestino (Eds)
© 2020 Taylor & Francis Group, London, ISBN 978-0-367-46579-7

Environmental reclamation for the Gotthard Base Tunnel effects of spoil management on landscape

P. Lanfranchi & E. Catelli
Pagani + Lanfranchi SA, Bellinzona, Switzerland

T. Bühler
AlpTransit Gotthard Ltd, Bellinzona, Switzerland

ABSTRACT: Nowadays it is impossible to imagine large infrastructure projects without considering the consequences on the landscape and, therefore, the necessary measures to lessen their impact. Tunnel portals, construction sites and final landfills for the excavated material all require large swaths of land and countryside. What are the visible effects on the landscape, three years after the opening of the Gotthard Base Tunnel and what was done with the 24 million tons of excavated material? The strict Swiss regulations specify suitable replacement and compensatory measures, from a large number of small actions in close proximity to the new track, extending to some macro-interventions involving substantial investment. The large deposits of excavated material are an emblematic example of the macro-measures taken. Today the material deposits appear, to the external observer, as green planted hills and naturalistic areas. This helps to ensure public acceptance of large underground projects.

1 THE PROJECT

Already in antiquity and all through the Middle Ages until the present, crossing the Alps has represented a challenge to which constructors have responded with courageous infrastructural works. Italian ports, the critically important economies of Lombardy and Piedmont and the industries of northern Europe all require efficient and sure connections.

Figure 1.   Longer and heavier trains made possible by "flat" railway lines (source: ATG).

However, the Alps are located between the North Sea and the Mediterranean. The trains which transit on tracks constructed over 100 years ago pass through narrow gorges and climb inaccessible valleys with steep gradients, often using systems based on spiral tunnels (Figure 2).

The Gotthard transalpine railway, termed NRLA (New Railway Link through the Alps) and known in common parlance as the AlpTransit project, begins not more than 80 km north of Milan and links Lugano with Zurich. Based on its central position in Europe, Switzerland is a crucial hub for European railway traffic.

The new Gotthard Base Tunnel, at 57 km the longest tunnel in the world, conquers plan and profile obstacles. With the addition of the Ceneri Base Tunnel (15 km long), which will be put into service in 2020, the mountains along this railway line will literally be eliminated (Figure 3).

The project, conceived with the aim of transferring cargo transportation from road to rail, will permit heavy cargo trains, as well as passenger cars, to travel along an efficient flat line.

Figure 2.   Historical and new Gotthard transalpine railway lines (source: ATG).

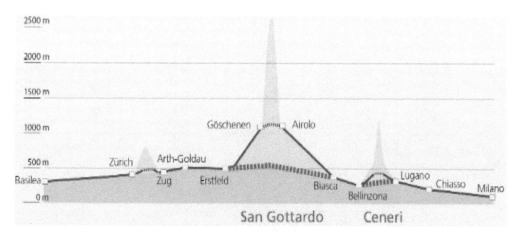

Figure 3.   The new flat railway line through the Alps (source: ATG).

Seventeen years of work and 11 billion euros were necessary to construct the Gotthard Base Tunnel in its present form. Its construction provided complex technical challenges for the designers and builders, some of which are mentioned here:

– Rock overburden greater than 2400 m,
– Ambient temperature in the tunnel up to 40° C,
– Extremely difficult geological conditions in some sections,
– The excavation itself, mainly carried out with powerful TBMs,
– Drilling through the Alps requiring topographical with centimetre precision,
– The enormous amount of excavation material to manage and landfill,
– Guaranteeing high levels of safety during construction and service.

This paper will now focus on the aspects of the project which are visible and impact the territory, such as the architectural insertion in the landscape, the environmental reclamation and in particular the management of the excavation material. These three overarching themes were studied in a unified and coherent manner for every area of territory touched by the project and in collaboration with the federal, regional and local authorities.

The works, in fact, were not only carried out underground. The large worksites, portals, external connection sections and final landfills for the excavation material had a strong territorial impact and indirectly required particular attention to guarantee the attainment of consensus of the population for the "project of the century".

It should not be forgotten that the project was financed essentially through the raising of taxes on heavy road traffic and on mineral oils as well as through an increase in VAT. Indirectly, the population itself, directly or indirectly involved in the financing of the work, enjoys particular attention and anticipates the completion of the works with interest on the returns as well as added value to their own advantage.

2 THE ARCHITECTURAL-LANDSCAPE INSERTION OF THE WORK

The project management by AlpTransit Gotthard Ltd (ATG), owner of the work, following in the wake of all that had been accomplished thirty years previously for the construction of the Gotthard motorway network, firmly believed in conferring a coherent and unified architectural image to the entire line. With this aim, a "consulting architectural group" was formed with the task of designing the image of all of the structural elements and of all of the landscape modifications with high functional and aesthetic quality. The architectural design team studied the following elements [1]:

– The portals of the main tunnels, including the terracing of the surrounding terrain;
– The detailed structural elements such as viaducts and bridges, underpasses as well as service structures (buildings for railway operations, ventilation plants, etc.);
– The artificial tunnels with their cut-and-cover constructions;
– The landfills for excavation materials, which leave a new imprint on the entire surrounding landscape.

With the aim of obtaining quality, coherence and unified results, guidelines were drawn up for each type of structural element for designers and constructors [4]; these were actually repeatable modules which could be adapted and applied as a function of each situation (Figures 4.1 & 4.2).

The example of the entrance zone of the tunnel is indicative. Large curve radii and minimum gradients for the railway conditioned the placement of the portals, which were in danger of disappearing laterally into the mountain flanks in unobtrusive points. Thus it was decided to accent these points, that is the portal zones, creating a crescent shape to mark the countryside, as a sort of "landscape staging".

The surface of these "crescents", covered in stone face work, were designed to be similar to those already existing along the historical Gotthard railway line.

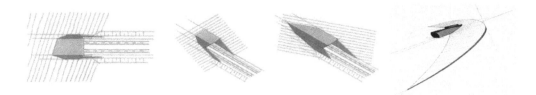

Figure 4.1.   Portals of the base tunnel: architectural guidelines (source: ATG).

Figure 4.2.   Portals of the base tunnel: finished examples (source: ATG).

3   THE ATTENTION TO ENVIRONMENTAL ASPECTS

As stated, the main aim of the new transalpine Gotthard line is that of contributing substantially to the transfer of heavy traffic from road to rail: therefore, a very "green" aim by definition. The construction of a large infrastructure project must include, however, the sacrifice of extensive portions of territory and ecologically fragile spaces.

Swiss laws in this domain are strict: if landscape or vital spaces are modified or sacrificed and if emissions are produced into the environment, adequate measures of substitution or compensation must be foreseen. The commitment of the project management in this direction was absolute both in form and in substance. The measures of compensation undertaken varied from a large number of small interventions distributed over the territory touched by the project to large-scale interventions of greater cost and linked to the creation of added value for the population.

For example, some of the smaller interventions included the creation of new water surfaces near the railway line: in some cases new habitats for amphibians and reptiles and in other cases small lakes for leisure activities (Figure 5). The Claus Lake, located at an altitude of 1300 m above sea level, that is 800 m above the tunnel elevation, was created on a fill made with excavation material from the intermediate access shaft of Sedrun.

A further example which links environmental aspects to landscape insertion, in addition to representing added value for the resident population, is the large-scale intervention of the shifting of the historical railway line, which up until now virtually touched the village of Pollegio (Figure 6). In this zone, the pre-existing line was moved for a distance of more than 2 km along the new line, which was itself located along the motorway. The historic village centre was thus able to be reunited with its original surrounding lands after a separation of over 100 years, also enabling the development of other local projects on the newly liberated territory.

Figure 5.   New Claus Lake for leisure activities (source: ATG).

Figure 6.   Moving of the historical railway line and new habitats for amphibians/reptiles (source: ATG).

4   THE MANAGEMENT OF EXCAVATION MATERIAL

4.1   *Principles*

The 24 million tons of excavation material extracted from the Gotthard Tunnel corresponded to a volume 5 times that of the Cheops Pyramid.

| Greatest possible on-site reuse | Reduction of noise and dust | „Green" transport facilities |

| Landfill solution involving the population | Optimal economic operation |

Figure 7.   Sustainable management principles for excavation materials (source: ATG).

The following ambitious aims were defined for the management of the excavation material (Figure 7):

– Favouring the reuse of the excavation material as much as possible, if possible using it on site for the preparation of aggregates for concrete;
– Reducing the environmental impact of all management processes to a minimum;
– Finding solutions for long-term stockpiling of left-over material, when possible involving the population and providing added-value solutions for them;
– Guaranteeing an optimal economic operation, in any case.

The AlpTransit work sites were an excellent example, with maximum attention given to the recycling of excavation material, transport by rail and favourable landscape solutions for landfills [5], [6], [7].

4.2   *Work sites with plants for on-site material processing and loading of trains*

The production of aggregates for concrete through the reuse of excavation material permitted:

– Avoiding the use of large amounts of alluvial resources (sand and gravel) from existing sources;
– Reducing two-way transport and its impact on the population;
– Limiting landfill volumes and thus limiting territorial impact;
– And with this, also permitting notable savings in costs.

The large treatment plants were intentionally planned in the immediate vicinity of the exit zones for materials from the tunnel.

A key point was that of well-organized and clearly delimited work site areas. The efficient design of the plants contributed to the limitation of territorial use.

From the intermediate excavation point in Amsteg, located north of the tunnel, the non-recyclable material was loaded onto trains. Using a special siding, the trains thus loaded with material continued to the port of Flüelen on Lake Lucerne. A temporary bridge was constructed along the siding to cross the Reuss River using an old military bridge now considered to have historical value. At the end of the works, the siding was dismantled but not the bridge, which was taken over by the Municipality of Erstfeld and reconverted into a bicycle path. The

Figure 8.   Clearly delimited worksite with plants for loading trains and processing materials (source ATG).

bicycle path, now much appreciated, is other example of a "win-win" project between the project owner and the municipality in question.

4.3   *Transport by barges and deposit in Lake Lucerne*

At the port of Flüelen on Lake Lucerne, the non-recyclable material left over from the excavation of worksites north of the tunnel was transferred from train to barge. The loaded barges were then dumped in a controlled way into the lake, with the appropriate monitoring through bathymetric and GPS navigation systems.

Strict environmental control systems were also used in this case, both at the beginning concerning the degree of non-polluted material and for the actual dumping operations in the lake.

The dumping was carried out in areas of shallow water using conveyor belts directly from the barges and in deeper water through openings in the bottom of the boat with special devices to dump the material on the lake bottom to avoid water turbidity (Figure 9).

As shown in Figure 10, the bottom areas, exhausted by the intense extraction of gravel and sand over the last decades, have now become viable zones for fish, natural islands and pleasant leisure areas.

This is an example of a large-scale intervention which combines the intelligent reuse of materials, the creation of new natural vital spaces and, not least, a valid contribution to the local recreation needs [3].

4.4   *Landfill for material in the Buzza di Biasca*

The landfill of excavation material in the so-called "Buzza" (in local dialect meaning alluvium, overflowing) is an example of a another large-scale intervention symbolic of the shrewd management of the excavation material, the architectural and landscape aspects and the involvement of the local population.

Figure 9.   Dumping from barges in shallow and deeper water (source: ATG).

Figure 10.   Revitalizing the Reuss River delta in Lake Lucerne (source: ATG).

Various environmental reclamation aspects may be seen in the carrying out of this project:

– The undeniable necessity to landfill 3 million m3 of excavation material in an adequate way;
– The necessity of guaranteeing the "ecological" transport exclusively by conveyor belt and through a special tunnel to avoid the residential area at the southern portal of the tunnel;
– The reclamation of the surface areas through the creation of areas planted with local species and new agricultural areas;
– And not last, the integration of regional projects ("win-win" solutions).

The area chosen for the landfill has an "historical" past. The landslide in 1513 of Monte Crenone overhead caused the formation of a dam in the Brenno River, which then failed and destroyed the village of Biasca, flooding the city of Bellinzona and devasting the Magadino Plain down to Lake Maggiore. The number of victims is estimated to have been about 700.

In the boom construction years of the last century (70s and 80s), the site at the foot of the landslide became a profitable quarry of sand and gravel which supplied the concrete aggregates for the construction of the Gotthard motorway. The foot of the landslide cone, removed during the years of gravel quarrying, was thus reconstructed. The material placed there

Figure 11.   Buzza di Biasca landfill during filling operations (source: ATG).

Figure 12.   Final landscaping step with containment basin (source: ATG).

contained components of the excavation material which were petrographically unsuitable and therefore worthless for other uses. (Figure 11).

Following these territorial interventions (Figure 12), to the casual observer, the landfill material appears as a green hill planted with local species such as chestnut, oak, etc. The replanting was carried out in steps, compensating the loss of pasture lands.

A containment basin was also integrated into the formation of the landfill in case of any further debris flow from Monte Crenone. This is a work which will ensure territorial safety, which had been a subject of concern to the local authorities for some time. This interesting "win-win" solution is advantageous to all concerned [2].

5   FINAL CONSIDERATIONS

The shrewd management of the enormous volumes of excavation material, the care with which aspects of impact and environmental reclamation were handled and the particular

Figure 13.   Final landscaping project of the north portal of the tunnel (source: ATG).

attention given to the architectural and territorial insertion of the works compared with added value solutions for the local population are results of the key factors of success for the construction of the longest railway tunnel in the world.

The various interventions were carried out by involving, from the very first steps of the project, all of the institutional partners, environmental associations as well as the population touched by the worksites.

Only in this way was it possible to establish the basis for a profitable and long-lasting collaboration and to guarantee the acceptance of the project with "win-win" solutions for the benefit of all of the parties involved, also taking into consideration the main objectives of the work such as costs, construction time and the quality of the resulting structures.

REFERENCES

Gruppo BGG, 2002-2013. Progettazione architettonica della nuova linea. *In La Galleria di base del San Gottardo: Inizia il futuro (volume 1) -. Stämpfli Verlag AG.* S 85.
Svaluto-Ferro, R., 2011. Il settore San Gottardo Sud. *In La Galleria di base del San Gottardo: L'opera del secolo diventa realtà (volume 2) -. Stämpfli Verlag AG.* S 89-93.
Regli, A., Wildbolz, A., Zistler, H., 2017. Sostenibilità ambientale. *In La Galleria di base del San Gottardo: Via libera alla Galleria (volume 3) -. Stämpfli Verlag AG.* S 245-249.
Gantenbein, K. & Petersen, P. 2016. La costruzione della velocità: 100'000 tonnellate di volontà progettuale, Hochparterre (ed) in collaborazione con AlpTransit Gotthard Ltd, Atelier Feddersen & Klostermann e Uli Huber. *In Quaderno tematico di Hochparterre, aprile 2016*: 8-13.
Hitz, A., Kruse, M. 2016. "The mountain from the mountain". Management of the material excavated from the GBT. *In Tunneling the Gotthard. STS Swiss Tunneling Society.* S 90-95.
Kruse, M., Ehrbar, H., Thalmann, C., 2016. Materials management in Erstfeld/Amsteg. *In Tunneling the Gotthard. STS Swiss Tunneling Society.* S 456-461.
Lanfranchi, P., Röthlisberger, B., 2016. Materials management on the Faido and Bodio sections. *In Tunneling the Gotthard. STS Swiss Tunneling Society.* S 468-473.

Tunnels and Underground Cities: Engineering and Innovation meet Archaeology,
Architecture and Art, Volume 2: Environment sustainability in
underground construction – Peila, Viggiani & Celestino (Eds)
© 2020 Taylor & Francis Group, London, ISBN 978-0-367-46579-7

Habitats and protected species compensations for Alpine underground works: A pilot experience between France and Italy

E. Luchetti, S. Viat & H. Besançon
Tunnel Euralpin Lyon-Turin, Bourget du Lac, France

P. Grieco & S. Bellingeri
Tunnel Euralpin Lyon-Turin, Torino, Italia

ABSTRACT: The works for the construction of the 57 km of Transalpine Base Tunnel of the Turin-Lyon railway link involve the installation of outdoor building sites in the Alpine ecological context. The areas affected by these installations have been minimized to reduce the use of soil as much as possible and areas with low ecological value have been chosen as far as possible. The residual impact of the works on habitats, protected species and ecological corridors has been assessed on the basis of detailed inventories. In line with the European doctrine of "avoiding – reducing – compensating", an extensive program of compensatory environmental measures, covering an area of more than 170 hectares, is being implemented to guarantee a balance sheet without net biodiversity losses. This article describe the compensation program implemented and analyses the similarities and differences in the approach to environmental compensation in the two countries – France and Italy.

1 THE BASE TUNNEL OF THE NEW TURIN-LYON LINK

1.1 *The project*

The new rail link from Turin (Italy) to Lyon (France) will complete the European rail network. It will constitute the key element of the east-west axis of the Mediterranean corridor, and will be one of the three main rail routes south of the Alps planned by the European Community.

This new rail link will have its profile at the base of the Alpine massif, at an altitude of around 600 m, and will have a maximum gradient of 1.2%, allowing the development of combined transport and authorizing the introduction of high-gauge and high-performance "rail-motorway" services. The project is designed for mixed traffic, it will allow freight trains as well as passenger trains to circulate.

The bi-national cross-border section between Italy and France includes a 57.5 km-long 8.70 m-diameter single-way twin-tube base tunnel, one of the longest in the world, which crosses the Alps roughly 45 km in France and 12.5 km in Italy (see figure 1) between Saint-Jean-de-Maurienne and Susa. The base tunnel incorporates many ancillary works: communications between tubs, sidings, exploratory adits and emergency access tubs, wells and ventilation tunnels, technical rooms for a total of 164 km of underground works. Design of the base tunnel includes four exploratory adits and geognostical tunnels.

Three French exploratory adits were completed between 2007 and 2010: Saint-Martin-La-Porte (2.4 km), La Praz (2.5 km) and Modane (4.0 km). The Italian exploratory tunnel of La Maddalena (7.1 km) was completed in February 2017 and the French exploratory tunnel of Saint-Martin-La Porte (9 km) has been under construction since 2014, more than 5 km of which have already been excavated.

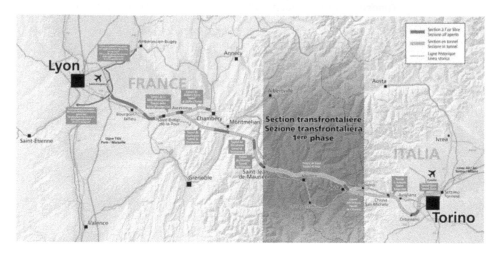

Figure 1. Location of the cross-border section of the new Lyon – Turin rail link.

1.2 *The Alpine natural environment and the areas occupied by the project*

As 90% of the cross-border section of the new Lyon-Turin rail link consists of tunnels, the ecological impacts and land use are significantly lower compared to a rail line of the same length located outdoors. The ecological impacts are therefore those of the outdoors construction sites. The problems of ecological continuity, which constitute one of the main issues of linear transport infrastructures when they are located outdoors, are rather limited.

The ecological dimension has been taken into account in the choice and definition of work sites. For example, in France, material transit sites are ruderal sites on the edge of the Arc river with few ecological stakes, while Italians final disposal sites are old quarries already in a compromised state from an ecological point of view. Moreover, it should be noted that the vast majority of the surfaces used during the work will be rehabilitated and restored to their natural state at the end of the work.

The precise knowledge of natural habitats, fauna and flora species, ecological functionalities, etc., enabling us to understand the ecological issues present on construction sites and around, is based on bibliographic studies but, above all, on numerous hours and days of ecological inventories in the field. On the French side, the over-sides investigated represent almost double the impacted areas presented below. This represents a cumulative inventory pressure of more than 200 man-days. On the Italian side, the areas involved are smaller but the planned investigation activities are proportional.

On the French side, the work sites are located in the central part of the Maurienne. This area, around the Arc valley, benefits of a climate with some precipitation and warm weather in summer, especially on the well exposed slopes. These conditions are suitable to the extension of natural environments of southern affinity. There is a great diversity of plant groups, related to the variety of local substrates. The sector allows the observation of some species of remarkable messicolous plants (plant species whose life cycle are adapted to the harvest cycle, in particular of cereal grains and are associated with traditional cultures). The fauna is also very varied, with alpine species occasionally reaching the lower slopes, forest species or species associated to more open environments. The lower valley, heavily urbanized near the river and the roads of communication, quickly finds all its naturalistic interest when one progresses towards the tops.

In terms of habitats and natural environments, hay meadows, cliff and scree areas, coniferous forests and wetlands are particularly noteworthy habitats in the Maurienne, home to a diversified fauna and flora; but there are also uplands and high moors, including steppe grasslands. The Maurienne steppe grasslands are very localized environments on a national scale: they are present only in the internal valleys of the Savoy and Queyras Alps. These grasslands have a major ecological interest, both because of their rarity and because of the specific flora they host, including *Festuca valesiaca* or *Thesium linophyllon*. *Cypripedium calceolus*, in forest gaps, and *Erica carnea*, in dry pine forests very poorly represented in France, are present in forest environments. Remarkable species of flora found in open environments, sometimes formerly agricultural, are also contacted, including several species of tulips, some of which are endemic to Savoy. Wildlife representatives include *Parnassius apollo* for insects, *Bufo calamita* for amphibians, *Upupa epops* and *Otus scops* for birds, and *Rhinolophus ferrumequinum* for chiropterans. Some of these species are representative of the Sites of Community Importance (SCI) of the "Perron des Encombres" and of the dry forest and herbaceous formations of the internal Alps listed in this part of the valley.

On the Italian side, the areas affected by the project are mainly in the lower Susa Valley, crossed by the Dora Riparia river. Characterized by a typically alpine climate, as in the valley of the Arc in France, the Susa Valley is home to a vast natural heritage, with a rich variety of fauna and flora. In fact, the strongly anthropized areas of the central part of the valley contrast with a well-preserved context, especially as far as the slopes are concerned, characterized by important naturalistic values, both in terms of habitat and of single species.

In flat areas, there are agricultural surfaces with stable meadows, grasslands or alfalfa fields and wooded areas, partly dominated by oaks and partly by *Robinia*. The slopes are characterized in part by xeric grasslands because of the particularly mild microclimate that favours Mediterranean gravitation species, many of which are extremely rare and localized, while in the lower areas a sub-Mediterranean thermoxerophilic and steppe vegetation prevail, as well as numerous grassy formations and shrubs.

There are numerous habitats of conservation interest and some rare plant species of considerable floristic importance, such as the *Carex Alba* and the *Typha minima*. Regarding the fauna, the area is characterized by a great biodiversity, with presence of numerous species of interest such as the lepidopteran *Zerynthia Polyxena*, species of Community interest in the "Habitats Directive," many species of chiropterans of conservation interest including the bat *Myotis bechsteinii*, in addition to a significant presence of wild ungulates, especially deer, and wolves. The great naturalistic value of the areas is confirmed by the presence of two Sites of Community Interest (SCIs) in the immediate proximity: the SCI of the "Oasi Xerotermiche" and the SCI of the "Gran Bosco di Salbertrand."

The table below summarizes the surfaces affected by the project in France and Italy, by major type of environment, and the areas rehabilitated in general (it being understood that "rehabilitation" tends to recreate woodland or open and shrubby environments).

The affected areas are significantly higher in France than in Italy (71% of the total), but this remains proportional to the location of the route of the cross-border section, mostly located in France (78% of the total). In France and in Italy, in particular, it can be observed that more than 50% of these impacted areas are not natural environments in their own right but artificial and ruderal zones, which confirms that the ecological dimension is taken into account in the choice and definition of work sites. Moreover, on each side of the border, more than 55% of the impacted areas are subject to ecological rehabilitation, which corresponds to an area larger than the truly natural environment impacted.

Table 1. The number of hectares affected and restored by the project.

Major types of environment	France		Italy		Total	
	Affected Surface	Restored Surface	Affected Surface	Restored Surface	Affected Surface	Restored Surface
Woodland	49	94	23	38	72	132
Open/shrubby environments	33		5		38	
Artificialized and ruderal environments	84		37		121	
Total	167	94	65	38	232	132

2 THE REGULATORY AND PROCEDURAL CONTEXT FOR IMPACTS ON PROTECTED SPECIES

2.1 *International conventions and European directives*

Certain species are protected "at all times, in all places and over the entire territory" on a European scale, or more widely, within the framework of international conventions. At European level, the texts and conventions relating to the protection of species are:

- The Berne Convention, in force since June 1982 and regarding the conservation of Europe's wildlife and natural environment, commonly aims to ensure the conservation of wild flora and fauna and their natural habitats, in particular species and habitats whose conservation requires the cooperation of several States.
- The Bonn Convention, in force since November 1983 and relating to the conservation of migratory species of wild animals, has as its objective the protection and management of all migratory wild species (terrestrial, marine and aerial). A significant portion of those populations traverses one or more parts of the national territory in a predictable cyclical pattern.
- Directive 2008/99 of 19 November 2008 on the protection of the environment through criminal law defines a range of serious environmental crime at Community level and requires Member States to provide for criminal penalties.

Projects likely to have impacts on protected species or their habitats are governed by European Directive 85/337 of 27 June 1985 on the assessment of the effects of certain public and private projects on the environment, better known as the "impact assessment" directive.

This assessment shall include in particular the data necessary to identify and assess the main effects that the project is likely to have on the environment and a description of the measures proposed to avoid, reduce and, if possible, remedy significant adverse effects.

2.2 *The concrete application of protected species procedures in France and Italy*

TELT, the contracting authority of the project, has obtained the environmental authorizations, on both the French and Italian sides, which enable it to carry out the work of the cross-border section, on the basis of dossiers resulting from the impact study drawn up in accordance with the European directive. The particular procedures concerning protected species comply, both in France and in Italy, with the conventions and directives previously presented.

In particular, in France, in order to be able to derogate from the prohibitions on the destruction of protected species or their habitats (Nature Protection Law of 1976 – Art. L. 411-2 of the French Environmental Code), it is necessary to submit a specific request for authorization, in accordance with the conditions indicated in Art. R. 411-6 et seq. of the Environmental Code.

The Environmental Code sets three conditions for obtaining the exemption:

- The project for which a derogation is requested must be based on an overriding reason in the public interest;
- There is no other more satisfactory solution;
- The derogation does not prejudice the favourable conservation status of the species in its natural range.

The applicant for a derogation must therefore be able to demonstrate that these conditions are met by describing in an impact study the size of the project, the justification for its major public interest, the methodology used to design a project that minimizes its impact on the natural system (fauna, flora, ecosystems) and the measures it undertakes to put in place in accordance with the European doctrine of "Avoid-Reduce-Compensate" for residual impacts.

The derogation, which specifies the conditions of execution of the authorized operations, is granted by prefectural decree, based on the opinion of the National Council for the Protection of the Nature (CNPN) and after collecting the observations of the public.

TELT was granted a derogation by a prefectural decree in 2016.

In Italy, Legislative Decree no. 152 of April 3, 2006, as amended, implemented the mandate conferred on the Government by Law no. 308 of 2004 to reorganize, coordinate and integrate environmental legislation. This legislation requires an Environmental Impact Assessment (EIA) procedure to be activated in order to obtain the necessary authorizations. The procedure, introduced in Europe by Directive 85/337/EEC of 1985 on the assessment of the effects of certain public and private projects on the environment, is based on the principle of prevention, i.e. the identification and assessment during the design phase of potential impacts produced by human intervention on the environment, meaning by Environment a system consisting of man, flora and fauna, soil, water, air, climate, landscape, material assets and cultural heritage. Specifically, the environmental assessment of projects aims to verify the effects on the components, providing for the maintenance of species and the preservation of the reproductive capacity of the ecosystem, as an important resource for life. To this end, the environmental impact study identifies, describes and evaluates the direct and indirect impacts of the project on fauna and flora, also studying all the measures to mitigate the effects generated by the work and the necessary environmental compensations.

On the basis of the documentation submitted by the applicant throughout the procedure, and taking into account any comments and opinions received during the public consultation phase, the Technical Commission for Environmental Impact Assessment ("CTVA" in Italy) of the Ministry of the Environment, carries out the technical investigation to verify whether the project has potential significant environmental impacts. On the basis of this preliminary investigation, the Minister for the Environment, the Protection of Natural Resources and Sea adopts an environmental compatibility assessment, which is then sent to the Ministry of Infrastructure and adopted by the Inter-ministerial Committee for Economic Planning (CIPE) at the same time as the project is approved.

TELT obtained the approval of the project with regard to environmental compatibility with the CIPE Resolution 19 in 2015 and environmental compatibility of the variant project with the CIPE Resolutions 30 and 39 in 2018.

In conclusion, it may be noted that the French part of the project was authorized by a particular concept of "derogation" from the prohibitions on protected species, constituting a specific authorization procedure distinct from other environmental procedures and which doesn't exist in Italy. In Italy, the project was authorized by a single authorization procedure which groups together all environmental subjects: the assessment of impacts, which therefore groups together different environmental themes, is more general.

3   ENVIRONMENTAL COMPENSATION FOR PROTECTED SPECIES

3.1   *Definition of "compensatory measure"*

The purpose of compensatory measures is to compensate or offset the expected or foreseeable damage to biodiversity caused by the realization of a work project, through the implementation of field actions favourable to the species, habitats and functionalities impacted. They are basically to be distinguished from avoidance and reduction measures – the purpose of which is to eliminate or mitigate the direct impacts of works – and can in no way replace them.

Aiming at an objective of zero net loss or even gain of biodiversity, a compensatory measure must theoretically lead to a state of the environment considered functionally normal or ideal. It must result in an obligation of results and be effective throughout the duration of the effects. The compensatory measures relate exclusively to sites distinct from the work areas and their implementation must begin before the residual impacts that they must offset occur. These are ecological and non-financial or social measures, which may consist, for example, of actions to restore and manage environments and species, and also include operations to protect or raise awareness, always with the aim of maintaining the good conservation status of species and habitats.

Beyond the purely surface aspect, for compensatory measures to be effective, they must satisfy several rules:

- Targets: all protected habitats and species with residual impacts must be represented, in proportion to these impacts, within the over-sides of the compensatory program;
- Geographical proximity: compensatory measures must be located in the same geographical area as the works;
- Feasibility: compensatory measures must be technically and ecologically feasible; this includes not implementing actions with uncertain success and ensuring that they can actually be put in place: land control, partnerships to be set up, possible administrative procedures, etc.
- Anticipation: compensatory measures must be anticipated as far upstream as possible by the contracting authority so as to disturb the conservation status of the target species as little as possible;
- Additionality: compensatory measures must generate environmental added value that would not have been achieved in their absence.
- Objectives: compensatory measures must be accompanied by performance objectives and arrangements for monitoring their effectiveness and effects.
- Duration: compensatory measures must be long-term (30 to 50 years); sustainability can be ensured through the acquisition of land or long-term contractual arrangements with landowners.

3.2   *Compensation projects in progress for the France side project*

3.2.1 *Definition of compensatory need*
As indicated in paragraph 2.2 above, the exemption from the prohibitions on protected species was obtained on the basis of the impact study carried out, in agreement with the administration, in accordance with current practices in France, namely according to the doctrine known as Avoid-Reduce-Compensate.

As part of this study, an articulated set of measures to avoid and reduce impacts, both during the construction phase and during the operation phase, was identified and integrated into the design of the infrastructures themselves.

Despite this, residual impacts have been identified; the residual impact represents the foreseeable gross impact of the work less the effects of the implementation of avoidance and reduction measures.

Once the residual impacts have been calculated habitat by habitat and species by species, the applied method foresees that the compensatory need is assessed by assigning, according to the level of conservation stake of the species/habitat concerned by the impacts, a specific coefficient that multiplies the residual areas impacted. The coefficients for calculating the compensatory need are logically progressive from the lowest to the highest stake levels and vary from 1.2 to 2.

According to the concept of mutualisation, an area compensates for several species which frequent the same type of habitats. By applying this method, it was calculated that the overall compensatory need amounts to 82 ha for wooded areas and 86 ha for open and shrubby areas. In addition, compensation is provided in pioneering environments, mainly for the *Bufo calamita*, which is the subject of functional compensation through the creation of permanent breeding sites.

Following the calculation of the compensatory need, the compensatory measures had to be identified and defined in accordance with the principles set out in paragraph 3.1 above. The knowledge of the general and specific ecological and territorial contexts made it possible in the first place to target a certain number of compensatory measures areas. Afterwards, precise investigations have been carried out on the surfaces where the compensatory measures areas were located, in order to define the actual perimeter of the compensatory measure according to the environmental issues found and particular feasibility problems.

3.2.2 *Summary presentation of measures*
A total of 23 compensatory measures are being implemented on the French side of the project. A summary description is given in the table below.

Table 2. The number of hectares of the surfaces involved and the duration of activities in years.

"Biodiversity" environmental compensatory measures in France

Environment	Measures	Surface	Actions to be implemented	Duration
Forester	Senescence Islands	20	Lack of management for biodiversity enhancement, targeting in particular avifauna and bats living in caves	30 years
	Preserved shoreline woodland	0.5	Improvement of bird and bat reception capacities: installation of nesting boxes, removal of exotic species, etc.	30
	Forest holes	5	Forestry management favourable to biodiversity, targeting in particular the *Cypripedium calceolus*	30
	Pine forest for heather	3	Lack of management for the preservation of plant species, targeting in particular the *Erica carnea*	50
	Shoreline forest	1	Biodiversity-enhancing forestry management, targeting in particular bryophytes	50
	Extended forest domain	28	Biodiversity-enhancing forestry management, targeting in particular birds and bats	50
	Sensitive wooded plots	25	Acquisition and management of wooded plots of high environmental value, various targets	50
Open and shrubby	Maurienne steppe grasslands	50	Inventories, diagnosis, animation and management of Middle Maurienne steppe grasslands	5–8
	Tulip Talus	0.1	Opening and management favourable to the expansion of tulips	30
	Mosaic of shrub and open environments	12	Reopening and management of dry mosaic grasslands and meadows, various targets	30
	Mow meadows	4	Extensive grassland management and hedgerow planting, targeting grove flora and fauna	30
	Shrub Talus	0.5	Reopening and management of tulip and *Ornithogalum nutans* environments	30
	Fallow meadow and terraces	1	Reopening and management favourable to the expansion of tulips	30
	Dry lawn with shrubs	0.5	Conservatory management for flora and fauna	30
	Steppe lawns of Châtel	7.5	Conservatory management of steppe mosaic lawns, various targets	30
	Upper Slope Grasslands	10	Various agro-environmental management for conservation purposes, various targets	30
	Dry lawns	20	Various agro-environmental management for conservation purposes, various targets	30
	Forgotten grasslands and hedgerows	1.5	Grassland management and maintenance for tulip and garlic expansion	30
Pioneer	Babylon ponds	0.1	Creation of 2 ponds and habitats in favour of amphibians and reptiles	5
	Functional pond network	1	Creation of a network of 6 temporary ponds and associated habitats, targeting *Bufo calamita* and pioneer flora in particular	30
	Ponds of the water body	1	Management of water environment in favour of amphibians	30
	Pond of Ferropem	0.2	Creation of a reproduction site for *Bufo calamita*	30

All these measures are the subject of detailed management plans describing the precise and complete initial ecological state of the site, the details of the actions and their implementation methods, and including protocols for scientific monitoring of the effectiveness of the measure, which will be carried out on a regular basis throughout the duration of the compensatory measures. All these measures were defined before obtaining of the administrative authorization.

3.3   *The flora-fauna habitat and ecosystem protection measures for the Italian side project*

As indicated in paragraph 2.2 above, the Italian side project was approved by the Ministry of the Environment on the basis of the results of the Impact Study, analysed during the technical investigation phase by the Technical Commission for Environmental Impact Assessment.

As already mentioned, one of the main objectives to be pursued with an analysis of the impacts carried out in parallel with the design of the work is to avoid or minimize the negative impacts and to enhance the positive ones. As part of the Impact Study, measures to avoid and reduce the impacts (mitigation) of the project were identified and sized, both during construction and during operation, in line with the natural context in which the project is located.

In addition, a highly developed monitoring system was defined in order to verify, during the entire construction period and for a year after works or beyond according to the stake of the species followed, that local disturbances linked to the implementation of the project do not cause permanent damage to the ecosystem and that the ecological balance is not compromised, as provided for in the impact assessment.

The conclusion of the impact assessment of the project on the Italian side is that, taking into account the context and the reduction and mitigation measures put in place, compensation works, as indicated in paragraph 3.1, are not needed, excepted for the ecological restoration of the areas around two specific construction sites (La Maddalena and Salbertrand) where species of great conservation interest have been found, such as *Aristolochia*, fundamental for the existence of *Zerynthia polyxena*, *Carex alba*, *Typha minima* and *Epipactis palustris*.

The "Plan" related to ecological restoration measures includes principles and guidelines aimed at promoting the use of local ecotypes as part of site restoration activities. The operational details of the implementation of this plan, including the physical extension of the measure (of approximately 20 ha), are being defined.

4   CONCLUSIONS

On the French and Italian sides, the project for the cross-border section of the new Turin-Lyon link pursues the same objective of minimizing impacts and preserving ecosystems, habitats and biodiversity. On both sides of the border, the technical authorities of the respective administrations have expressed their favourable opinion on the contents and results of the respective Impact Assessments which have shown how this objective is concretely pursued.

However, it should be noted that the volume of environmental compensatory measures for the fauna-flora, habitat and ecosystem components to be implemented is clearly unbalanced on the French side: more than 20 projects covering a total area of more than 170 ha on the French side, compared with a single equivalent project of about 20 ha on the Italian side.

This difference is explained, in part, by the fact that the project is less extensive on the Italian side and that the species directly involved in the project have overall a lower critical level than those on the French side of the project.

Nevertheless, it is also possible to highlight a certain difference in the scientific-cultural approach to the issue of species conservation and resilience. In fact, given equal mitigation measures, the competent authorities required an application "to the letter" of the doctrine of the sequence "Avoid-Reduce-Compensate" on the french side and aim at the absence of net loss of biodiversity, by applying multiplying coefficients to the areas affected by the project.

Necessarily, this approach requires the implementation of compensatory measures. Even when faced with a temporary and reversible impact like that of a construction site, the fact that certain species cannot absorb the disturbance generated by that impact is taken into account, unless favourable habitats where these species can find refuge are made available, even before causing the impact.

On the Italian side, the impact study showed that impacts are temporary and reversible and monitoring is constantly carried out to make sure that this statement is justified. In fact, the ecosystem is considered to be capable of restoring its equilibrium on its own and monitoring activities are carried out to ensure that this is the case.

The fact that in France there is a specific procedure to ask for derogation in order to destroy protected species accentuates the approach differences.

It should also be noted that, although the implementation of the compensatory program on the French side is well under way, the implementation of environmental compensatory measures in favour of biodiversity is facing many difficulties. These can be inherent to the technical aspects of management operations: they must be as non-intrusive as possible. In addition, since this is not an exact science, differences between the expected responses of nature and those actually observed via ecological monitoring often exist. Moreover, supporting the development of one species is sometimes to the detriment of another one. However, the main difficulties are the surpassing of the "feasibility" rule mentioned above with regard to land control and the partnerships to be set up. There are only a few pragmatic legal tools to obtain and control the land needed to implement environmental compensatory measures. In a context of severe land fragmentation, the amicable acquisition of private land is a difficult task. The existence of pastoral land groupings represents an opportunity, but the dialogue between the biodiversity world and the agricultural world, which theoretically work in symbiosis, is sometime difficult. As for municipalities, which are essential partners, despite the rhetoric of well-meaning ideas, the implementation of environmental compensatory measures on their territory is almost unanimously perceived as the appearance of constraints hindering economic developments.

From this experience, it can be concluded that a middle ground between the French and the Italian approaches should be favoured.

REFERENCES

Grizard, S. & Marie, A. 2013. *Liaison Lyon-Turin – Travaux de reconnaissance à partir de la descenderie de Saint-Martin-la-Porte – Dossier de demande de dérogation aux interdictions portant sur les espèces protégées*. Le Bourget du Lac: TELT

Uster, D. 2015. *Liaison Lyon-Turin – Projet de Référence Final – Dossier de demande de dérogation aux interdictions portant sur les espèces protégées, travaux liés au creusement du tunnel de base*. Le Bourget du Lac: TELT

TCC & Lombardi 2013. *Nuova Linea Torino Lione – Quadro di riferimento ambientale – Analisi dello stato attuale*. Torino: TELT

Tunnels and Underground Cities: Engineering and Innovation meet Archaeology,
Architecture and Art, Volume 2: Environment sustainability in
underground construction – Peila, Viggiani & Celestino (Eds)
© 2020 Taylor & Francis Group, London, ISBN 978-0-367-46579-7

Water monitoring for Alpine underground works: Differences and similarities between France and Italy

E. Luchetti, S. Viat & H. Besançon
Tunnel Euralpin Lyon-Turin, Bourget du Lac, France

P. Grieco & S. Bellingeri
Tunnel Euralpin Lyon-Turin, Torino, Italia

ABSTRACT: The works for the construction of the 57 km Transalpine Base Tunnel of the Turin-Lyon railway link are subject to extensive monitoring of ground and surface water. The purpose of the monitoring is to acquire the necessary knowledge of the natural context to guide the design choices towards less impactful solutions, to enable the facilities for the management of tunnel and platform water to be correctly sized and to anticipate the implementation of the measures to avoid or reduce the disturbance of water environments. The monitoring carried out for the Base Tunnel complies with the regulatory framework imposed by the country in which the work takes place: France and Italy respectively. This article analyses the similarities and differences between the two systems of water monitoring of the same work in two neighbouring countries.

1 THE BASE TUNNEL OF THE NEW TURIN-LYON LINK

1.1 *The project*

The new rail link from Turin (Italy) to Lyon (France) will complete the European rail network. It will constitute the key element of the east-west axis of the Mediterranean corridor, and will be one of the three main rail routes south of the Alps planned by the European Community.

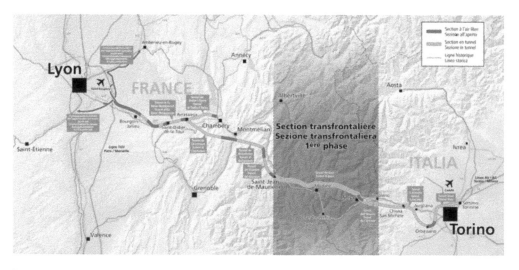

Figure 1.   Location of the cross-border section of the new Lyon – Turin rail link.

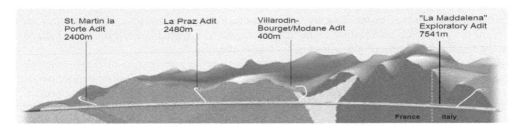

Figure 2. Simplified representation of the base tunnel and its ancillary works.

This new rail link will have its profile at the base of the Alpine massif, at an altitude of around 600 m, and will have a maximum gradient of 12%, allowing the development of combined transport and authorizing the introduction of high-gauge and high-performance "rail-motorway" services. The project is designed for mixed traffic, it will allow freight trains as well as passenger trains to circulate.

The bi-national cross-border section between Italy and France includes a 57.5 km-long 8.70 m-diameter single-way twin-tube base tunnel, one of the longest in the world, which crosses the Alps roughly 45 km in France and 12.5 km in Italy (see figure 2) between Saint-Jean-de-Maurienne and Susa. The base tunnel incorporates many ancillary works: communications between tubs, sidings, exploratory adits and emergency access tubs, wells and ventilation tunnels, technical rooms for a total of 164 km of underground works. Design of the base tunnel includes four exploratory adits and geognostical tunnels.

Three French exploratory adits were completed between 2007 and 2010: Saint-Martin-La-Porte (2.4 km), La Praz (2.5 km) and Modane (4.0 km). The Italian exploratory tunnel of La Maddalena (7.1 km) was completed in February 2017 and the French exploratory tunnel of Saint-Martin-La Porte (9 km) has been under construction since 2014, more than 5 km of which have already been excavated.

It is an ambitious infrastructure, including as regards the management of hydro-geological components related to boring the Alpine massif.

1.2 *The hydrogeological and hydrological context of the base tunnel*

Mastery of the hydro-geological situation traversed by underground structures makes it possible to understand the potential impact that base tunnel works can have. Broadly, two things have to be considered in particular: the modification of the hydro-geological system caused by the drainage carried out through the tunnel that could affect the surface water network on one hand, and the interception of underground water with physical and chemical characteristics different from those of the surface water network in which they will be emptied and whose quality they could change.

The base tunnel cuts, from west to east, across the Alpine massifs of Vanoise, Ambin (Mont Cenis) and the Graie Alps marginally.

Generally speaking, the hydro-geological description is divided into sections found in a somewhat homogeneous and consolidated rock, separated by shorter sections characterised by the presence of a tectonic contact, and thus potentially a fault, where permeability is lowest. All these sections have secondary porosity (pores, karst, micro and macro fissures) as opposed to the primary porosity made up of intergranular spaces and observed in sediments deposits. Most areas are characterized by a great depth and low permeability (close to 90% of the tunnel length is found at a depth of more than 300 m). The main impact risks with regard to considerable modifications of hydro-geological systems are thus found at the level of tectonic contacts with fault zones, which are sometimes found vertically to surface streams, as well as in some sections with special features. For example, fracturing strips and faults considered directly connected to the surface were found among the crystalline rocks (mica schist and gneiss) of the Ambin massif: the water intercepted during the excavation in this case should be a mixture of water from the surface more than groundwater. All these areas are in fact those over which uncertainty looms the most, and where real hydro-geological behaviour can be reliably understood only after a gallery has been dug.

In the other sections, the prevailing environment issue is the interception of highly-mineralised water: "fossil" water that can be found in the karst and fracturing systems not connected to the surface. Since hydrogeological data allow to forecast the general mineral characteristics of underground water intercepted by the works, the content in sulphates found in the geological context is the most representative indicator of water aggressiveness. In some sectors it exceeds regulatory standards for potable water. This is especially the case with strips of anhydrites.

The hydro-geological environment and the hydrological environment meet in springs and surface water feed and loss areas, and in the discharge of mine water in attack sites (adit platform, west and east portals of the tunnel). The latter are found in the two valleys of the internal Alps: the Maurienne valley and the Susa valley. These valleys are respectively bathed by rivers Arc and Dora Riparia, which are quite similar, with a length of between 120 and 130 km. They both take their rise at the foot of glaciers at an altitude of over 2500 m and have, in the relevant sections affected by works, a nival regime and high flow rates of several m3/s including during low-flow periods.

The tunnel's underground route is located westwards to the north of the Arc river over thirty kilometres, following the valley's curve, to the right of the river's right bank tributaries, which constitute the torrential watercourses flowing down in ravines, subject, in the central part of the valley, to torrential melts that from time to time (and regularly) fill the Arc with huge quantities of fines. All attack sites on the French side are directly connected to the Arc, located in a valley floor marked by the presence of metallurgical industries, hydroelectric structures, and high anthropization of shores.

All these elements account for the general quality characteristics of the Arc: the physical and chemical quality, in which the influence of the area's special hydro-geological context is felt for example through its high sulphate content, generally ranges from good to very good, despite the existence of potential one-off alterations in heavy metal concentrations. The biological quality is generally mediocre, in terms of habitability for macroinvertebrates as well as for fish stock while aquatic and wetland habitats suffer from the artificialisation of banks over numerous sections.

In the east, after going past the valley of the Arc and the Ambin massif, the tunnel's underground route is set over ten kilometres on the left shore of the Dora Riparia, shifting to the right of the Clarea Valley and the Cenischia Valley. The attack site on the Italian side is located perpendicularly and upstream of the confluence between the Clarea torrent and the Dora Riparia. Due to the low regime of the former's flow rate, water from tunnel boring operations are discharged into the Dora Riparia, which has dug at this location a steep-sided valley with no human impact.

The Dora Riparia runs along the entire Susa Valley until it reaches the plain surrounding Turin. It is an important tributary of the Po river. The Dora has a very articulated and wide drainage basin with both left and right tributaries such as Clarea, Cenischia and other minor streams.

The Dora Riparia basin is also characterized by the presence of numerous withdrawals for hydroelectric purposes close to the head of the basin and the tributaries (in particular Clarea and Cenischia for the Chiomonte and Susa plants) as well as downstream of Susa.

Therefore the main river channels are affected by numerous works in the riverbed, while at the head of the valleys there are some reservoirs with considerable impact on the hydrological regime.

2 MONITORING OF WATER DURING EXCAVATION PHASES

2.1 *The potential impacts of the work on the water environment and mitigation measures*

The same type of potential environmental impacts on water and aquatic environments related to the base tunnel building activities are found on both the French and Italian flanks of the infrastructure. This refers on one hand to all impacts that can be caused by open-air construction site activities and that mainly concern the Arc, the Dora Riparia and their associated groundwater: discharge on the surface environment of effluents from platform activities, run-off water from the construction site platforms and from provisional and final stocks of material, accidental infiltration of polluted water from construction site areas, other contaminations of groundwater.

On the other hand, there are potential impacts caused by the underground structure itself ("hole in the mountain") and that mainly concern springs and tributary streams of the Arc and the Dora Riparia located to the right of underground structures: changes in the spring flow rates and in the hydrometric systems of streams. The discharge of mine water in the surface water environment, made of naturally-drained underground water potentially altered during the construction phase by industrial activities in galleries, takes place where these two impact categories intersect.

In any case, these impacts can produce indirect effects, causing the change in the hydro-biological quality of aquatic environments.

These incidences were assessed in order to improve the project in the logic of avoiding impacts and to identify and measure the necessary mitigation measures.

Focusing on the construction phase, based on the activities related to the construction of the work, the families of mitigation measures are listed in the table below.

Table 1. The mitigation measures in the construction phase.

Nature of the impact	Mitigation measure family at construction site
Quantitative (disturbance/change of flow)	- Reduction of water flow rates in the tunnel (waterproofing, injections) - Reasoned management of the water pumped during excavation activities (limitation of the water withdrawn/drained with respect to the amount of water that can be renewed by precipitation, recycling of treated industrial water)
Qualitative (physical-chemical and bacteriological characteristics)	- Collection and treatment of construction site water before discharge into natural water bodies (filters and tanks for dealing essentially with suspended materials and correction of all parameters not compliant with the discharge) - Collection and treatment of the drainage water of the massif to regulate any physical-chemical differences (temperature and mineralization) with respect to the receiving water body - Waterproofing of work areas, product storage, parking areas and machine maintenance areas and other containment measures in case of accidental events.

The study of potential impacts of the work and of the mitigation and compensatory measures was carried out in the context of the Environmental Impact Assessments made on both the French and Italian sides according to the respective national regulations. There are no technical differences of approach in the conduct of Impact Assessments on the two sides of the project; the only difference lies in the fact that the Italian side of the project has been the subject of a "single environmental authorization" procedure, while in France the different environmental authorizations have been obtained in a disjointed way.

However, on the basis of the Environmental Impact Assessments, the project has been approved by both States, obtaining environmental compatibility on the Italian side (CIPE Resolutions 19/2015 and 30 and 39/2018) and the authorization concerning the "Law on water" on the French side (Prefectural Decree of 12/02/2007).

2.2 The objectives of water monitoring

By law, both on the Italian side and on the French side, the correctness of the assessments carried out during the Impact Assessment phase and the compliance with the protection requirements of the water environment must be confirmed by monitoring various parameters during both the construction and operation phases of the project.

In particular, during the excavation phase, the monitoring objectives are as follows:

- To verify the relevance and effectiveness of mitigation measures designed according to the expected impacts;
- To anticipate potential critical situations on sensitive water points to immediately activate the mitigation and compensatory measures necessary in close contact with local administrations and the population;
- To promptly identify any critical factors attributable to site activities to implement the intervention procedures envisaged in order to resolve anomalies in the shortest possible time;
- To provide the control authorities with the elements of verification of the execution of the works in compliance with environmental requirements.

These purposes are conceptually identical on both sides of the border and there are no differences in approach between the two countries in these terms.

2.3 The contents of the water monitoring plans during the works

The contents of the monitoring to be carried out to reach the objectives described in the previous paragraph, in terms of parameters to be measured, frequency of measurements, number of measurement points, measurement methodologies, etc., are established according to the expected impacts and submitted to the approval of the competent control authorities of the two countries.

In particular, the base tunnel water monitoring protocols have been approved:

- in Italy, as part of the "Environmental monitoring plan" of the project and therefore through the aforementioned CIPE Resolutions 19/2015 and 30 and 39/2018;
- in France, as part of the "initial state and monitoring protocols related to the water law", through the Prefectural Decree of 4/03/2011.

These protocols are compared by subject in the tables of the following paragraphs.

2.3.1 Physical and chemical quality of surface water
This comparison highlights the fact that:

- In general, regarding platform and mine water, clearly more parameters are analysed in Italy: chlorine and aromatic components are not analysed in France, and the list of general, metal and hydrocarbon parameters is less complete. For example, manganese, whose eventual effects on health have not been proven, is not in the list of parameters analysed in France for this reason.
- In Italy, except in special cases, water management is the same for mine water and platform water.
- Measurements on sediment samples at the bottom of the bed are not carried out in Italy.
- The physical and chemical quality of watercourses to the right of the tunnel's underground route (which thus do not receive any effluent from the construction site) can be analysed on the French side by monitoring diatoms (see the paragraph on the lower hydro-biological quality).

Table 2. Comparison for physical and chemical quality of surface water.

Secondary subject areas	France			Italy		
	Measurements carried out	Minimum frequencies	Spatial extent	Measurements carried out	Minimum frequencies	Spatial extent
Platform water	In situ measurements and laboratory measurements on water samples collected	Daily or continuous for 3 in situ parameters, monthly for others	A measuring point for each platform	In situ measurements and laboratory measurements on water samples collected	Every fifteen days for all parameters	A measuring point for each discharge (except in exceptional cases, there is only one discharge for mine and platform water)
Mine water	In situ measurements and laboratory measurements on water samples collected	Weekly for 8 parameters, monthly for others	A measuring point for each discharge			
Receiving environment	In situ measurements and measurements on water samples collected, measurements on sediments	Weekly for 8 parameters, monthly for others on raw water, bi-annual for sediments	Upstream and downstream measurements with regards to discharges (2 points for each discharge)	In situ measurements and laboratory measurements on water samples collected	Quarterly, monthly for in situ parameters to the right of the attack site	Upstream and downstream measurements with regards to discharges (2 points for each discharge)
Other watercourses (to the right of the tunnel's underground route)	No measurement, excluding an exceptional situation			In situ measurements and eventual laboratory measurements and for bacteriology on water samples collected	quarterly	2 measuring points for each relevant watercourse

2.3.2 Bacteriological and toxicological quality of surface water

On the French side, the site facilities are systematically connected to municipal waste water systems. Therefore, the very low impact risk of the construction site for this subject area can be the reason for lack of monitoring.

Table 3. Comparison for bacteriological and toxicological quality of surface water.

France			Italy		
Measurements carried out	Minimum frequencies	Spatial extent	Measurements carried out	Minimum frequencies	Spatial extent
No measurement, excluding an exceptional situation			Laboratory measurements on water samples collected	Quarterly for bacteriology, bi-annual for toxicology	Upstream and downstream measurements in the receiving environment with regards to discharges (2 points for each discharge)

2.3.3 *Quantity of surface water and hydrometry*

- Given that the receiving environment in France is almost exclusively the Arc, which is a river with a considerable flow rate, monitoring the speed and flow rate under the current metre would not be significant when compared to the construction site's potential impact.
- As far as the other watercourses are concerned, different approaches exist to measure the flow rate, but the objective sought is the same.

Table 4. Comparison for quantity of surface water and hydrometry.

	France			Italy		
Subject areas	Measurements carried out	Minimum frequencies	Spatial extent	Measurements carried out	Minimum frequencies	Spatial extent
Platform water	Flow rate of the discharge where necessary	Weekly	A measuring point for each platform	Flow rate of the discharge	Daily or continuous	A measuring point for each discharge (except in special cases, there is only one discharge for mine and platform water)
Mine water	Flow rate of the discharge	Daily or continuous	A measuring point for each discharge			
Receiving environment	No measurement, except in exceptional situations			Speed and flow rate under the current metre	quarterly	Upstream and downstream measurements with regards to discharges (2 points for each discharge)
Other watercourses (to the right of the tunnel's underground route)	Flow rate for each gauging station monitoring operation and mark-out operations	Weekly to monthly for the gauging station, and quarterly to none for mark-out operations, depending on the nearness of underground works	Approximately one gauging station reading and four mark-out operations for each watercourse	Speed and flow rate under the current metre	quarterly	2 measuring points for each relevant watercourse

2.3.4 *Quality of underground water*

- In general, regarding the physical and chemical quality of groundwater, it is clear that more parameters are analysed in Italy: chlorine and aromatic components are not analysed in France, and the list of general, metal and hydrocarbon parameters is less complete.
- Regarding the physical and chemical quality of springs, in France only basic parameters are measured *in situ*, making it possible to detect a change in hydro-geological functioning, considering that underground works do not represent pollution risks for these springs beforehand.
- Regarding the bacteriological quality of springs, in Italy over 100 parameters are analysed, and additional analyses are carried out for potable water.

Table 5. Comparison for quality of underground water.

Subject areas	France			Italy		
	Measurements carried out	Minimum frequencies	Spatial extent	Measurements carried out	Minimum frequencies	Spatial extent
Physical and chemical quality of groundwater	Laboratory measurements on water samples collected in piezometers	quarterly	Piezometers located to the right, upstream and downstream of the open air construction sites	In situ and laboratory measurements on water samples collected in piezometers	Monthly for in situ measurements, quarterly for laboratory analyses	Piezometers located to the right, upstream and downstream of the open air construction sites
Physical and chemical quality of springs (to the right of the tunnel's underground route)	In situ measurements of temperature and conductivity	Weekly to monthly depending on nearness of underground works	Network of springs and catchments	In situ and laboratory measurements on water samples collected	Monthly for potable water and in situ measurements of non-potable water, quarterly for laboratory analyses of non-potable water	Network of springs and catchments
Bacteriological quality of springs (to the right of the tunnel's underground route)	No measurement, excluding an exceptional situation			Laboratory measurements on water samples collected	Monthly	Network of potable water catchments among the network of springs and catchments

2.3.5 *Quantity of underground water*

In France, the frequency is systematically weekly when underground works take place at less than 1 to 2 km from springs. In Italy, the frequency is monthly in standard situations, but more frequent measurements may nevertheless be carried out if an anomaly is noticed.

Table 6. Comparison for quantity of underground water.

France			Italy		
Measurements carried out	Minimum frequencies	Spatial extent	Measurements carried out	Minimum frequencies	Spatial extent
Measurements of the flow rate of springs, the piezometric levels of groundwater and the piezometric level of deep drilling	Weekly to monthly depending on nearness of underground works	Network of springs, catchments, network of piezometers, network of deep drilling	Measurements of flow rate of springs and piezometric level of groundwater	Monthly	Network of springs and catchments, network of piezometers

2.3.6 *Sediment transport*

Although in France numerous hydraulic safeguards to the right of construction site platforms and the intense alluvial activity of the Arc explain this monitoring, the situation and stakes, which are quite different in the Dora valley, seem to suggest that this monitoring is not relevant in Italy.

Table 7. Comparison for sediment transport.

France			Italy		
Measurements carried out	Minimum frequencies	Spatial extent	Measurements carried out	Minimum frequencies	Spatial extent
Measurement of transversal profiles and longitudinal reconstitution of profiles	Biennial	Mainly on the Arc (approximately 60 profiles, area of about 17 km)	No measurement, excluding an exceptional situation		

2.3.7 *Hydrobiological quality*

- Regarding macroinvertebrates, in France the characteristics of the Arc, marked by very low habitability, make the results obtained from the monitoring operation carried out on this watercourse unusable, explaining the lack of monitoring of the receiving environment.

Table 8. Comparison for hydrobiological quality.

Subject areas	France			Italy		
	Measurements carried out	Minimum frequencies	Spatial extent	Measurements carried out	Minimum frequencies	Spatial extent
Macro invertebrates	Analysis of macroinvertebrates using the IBGN (Standardized Global Biological Index) method	Quarterly to annual	Other watercourses to the right of the tunnel's underground route	Analysis of		
	macroinvertebrates using the IBE (Extended Biotic Index) or Macrobenthos multihabitat method	quarterly	On the receiving environment of discharges and watercourses to the right of the tunnel's underground route			
Diatoms	Analysis of diatoms using the IBD (Biological Diatom Index) method	Quarterly to annual	On the receiving environment of discharges and watercourses to the right of the tunnel's underground route (20 points)	No measurement, excluding an exceptional situation		

(Continued)

Table 8. (*Continued*)

Subject areas	France			Italy		
	Measurements carried out	Minimum frequencies	Spatial extent	Measurements carried out	Minimum frequencies	Spatial extent
Aquatic and wetland habitats	Monitoring of watercourses using the method based on physical, morphological and ecological criteria	Annual	On the receiving environment and watercourses found to the right of the tunnel's underground route	Monitoring of watercourses using the river functionality index	Annual	On the receiving environment and watercourses near open-air construction site areas
Fish stock inventories	Electrofishing for inventory purposes	Annual	On watercourses to the right of the underground route (12 points), exceptionally on the Arc	No measurement, excluding an exceptional situation		

- Diatoms: this indicator is correlated to the physical and chemical quality of water. Although it is not monitored in Italy, the physical and chemical quality is on its part directly studied (see the paragraph on the physical and chemical quality of surface water above).
- The methodologies used for aquatic and wetland habitats are based on similar criteria although they seem to be slightly more specific on the Italian side. The area concerned by this monitoring operation is clearly weaker in Italy; the risks of impact on this subject area do not seem to justify wider monitoring on the French side.
- Fish stock inventories: this monitoring may suffer considerable bias caused by artificial fish populations, limiting its relevance.

3 CONCLUSIONS

The monitoring has the same objectives in both countries, it is based on the same principles and is organized in a very similar way.

The impact of specific regulations and standards in each country does not seem to be the reason for the differences between monitoring operations conducted in France and in Italy for the base tunnel construction work.

In fact, the several discrepancies noticed between the two monitoring protocols seem to hinge on the different methodologies in the countries (in particular regarding hydrobiological quality), the approach for the choice of parameters (complete, or more focused on those that seem to represent *a priori* the main risks), and even the distribution of tasks between the contracting authority and the company providing water monitoring services (regarding the frequency of monitoring operations for the physical and chemical quality of surface water, and the manner in which a link is established between the monitoring of discharges and the monitoring of the receiving environment).

Furthermore, the specificities of the hydrogeological and hydrological contexts encountered in France and in Italy are also the reason for the different considerations underlying some subject areas: for example, the transport of sediments on the Arc, which is not realized in Italy, or the monitoring of macroinvertebrates, which does not give convincing results on the

Arc. The location of construction site areas hosting site facilities has also an effect on the method for managing waste water and the relevance or non-relevance of bacteriological monitoring operations.

Moreover, the spatial extent of monitoring operations and the number of measuring points are linked to the number of attack and construction sites, which are higher but more grouped in France, as well as the tunnel area, which is less extensive in Italy. In any case, proportionality seems to be generally well respected.

Although there are slight heterogeneities regarding the frequencies of measurements, in France and in Italy these frequencies are adapted to the type and intensity of construction site activities.

Overall, the fact that the nature of potential impacts relating to TELT activities on water and the aquatic environment – and consequently monitoring objectives – are similar in France and Italy does not produce any significant and substantial heterogeneity in the monitoring operations carried out. Differences are found mainly in operational details.

REFERENCES

Bernagaud, C. & Others 2011. *Dossier pilote des tunnels, document n°5: environnement*. Bron: CETU.

RBA, BCO & EME (Setec) 2010. *Liaison Lyon-Turin – Protocole Loi sur l'eau de suivis*. Le Bourget du Lac: TELT

TCC & Lombardi 2017. *Nuova Linea Torino Lione – Piano di monitoraggio ambientale*. Torino: TELT

Viat, S. 2014. *Liaison Lyon-Turin – Valorisation des eaux chaudes collectées par le tunnel de base*. Le Bourget du Lac: TELT

Tunnels and Underground Cities: Engineering and Innovation meet Archaeology, Architecture and Art, Volume 2: Environment sustainability in underground construction – Peila, Viggiani & Celestino (Eds)
© 2020 Taylor & Francis Group, London, ISBN 978-0-367-46579-7

The role of Cairo metro in saving the social costs of air pollution

M.A. Madany
National Authority for Tunnels, Cairo, Egypt

ABSTRACT: Air pollution is regarded as one of the highest priorities in environmental protection in both developed and developing countries. Air pollution is a serious and growing problem, particularly in rapidly expanding cities like Cairo, where pollution levels exceed allowable thresholds. High levels of air pollution have adverse effects on human health which might cause premature death. If any cost arises as a result of pollution, the cost is usually borne by individuals that were harmed, rather than the emitting facility. Underground constructions, especially Metro lines, significantly contribute in saving the adverse effects of air pollutants generated from other surface transport. This study presents the estimate of social costs of adverse human health effects resulting from ambient air pollution. The study describes the measuring and valuing the impacts of air pollution on health. The study shows the monetary values of health damages that have been avoided as a result of construction of Cairo Metro.

1 SOCIAL COSTS

Social cost incorporate the total of all costs associated with an economic activity. It includes both costs borne by the economic agent and also all costs borne by society at large. Environmental pollution is an example of a social cost that is seldom bore completely by the polluter thereby creating a negative externality.

2 IMPACTS OF AIR POLLUTION ON HEALTH

The effect of air pollution on human health takes several forms including; chest discomfort, chronic bronchitis and asthma attacks, and premature death. The most accurate method of measuring air pollution on health in a given area is to conduct epidemiologic studies for that area that establish dose-response relationships linking environmental variables with observable health effects. However, given the time and cost involved in such studies, as well as likely problems with data availability, it may often be the case that dose-response relationships established in other locations will have to be used instead. Therefore, in this study, dose-response relationships have been identified and adapted from published epidemiologic literature.

The main air pollutants are; particulate matter (Pm), sulfur dioxide (SO_2) nitrogen dioxide (NO_2), carbon monoxide (CO), carbon dioxide (CO_2) and Ozone. Unfortunately, there is little quantitative dose-response information linking CO exposure to a meaningful health endpoint. CO dissipates rapidly in the environment, because of these shortcomings, there are no quantitative estimates of the effects of CO are provided. The precise association between the production of CO_2 and global warming is unknown and the relationship between global warming and subsequent health outcomes are unclear. However, there is little information about the impact these changes on health outcomes.

Particulate matter has several adverse health outcomes including: mortality, respiratory hospital admissions, emergency room visits, restricted activity days for adults, lower respiratory illness for children, asthma attacks, and chronic disease (Ostro 1994). Premature mortality is a major problem associated with high levels of particulates. Table 1 lists the summary of mortality and morbidity effects of 10 µg/m³ change in PM10.

Based on the data given in Table 1, it is found that the mean effect of a 10 µg/m³ change in PM10 implied by these studies varies between 0.31% and 1.49%, with a mean of 0.96%. The low effect varies between 0.29% and 0.92% with a mean of 0.62%. The high effect varies between 0.33% and 2.06% with a mean of 1.3%. The central estimate of the number of cases of premature mortality can be expressed as per Equation 1.

Change in mortality = 0.96 * change in PM10 * 1/100 * crude mortality rate * population exposed.

$$\text{Increase in mortality} = \frac{0.96}{100} * \frac{\text{Change in PM10}}{10} * \text{Crude mortality rate} * \text{Population exposed} \qquad (1)$$

Considering an average of crude mortality in Egypt to be 0.0051 (CIA 2007), the low, central and high estimates of premature mortality associated with 10 µg/m³ change in PM10 are; 3.162×10^{-5}, 4.896×10^{-5} and 6.63×10^{-5}, respectively.

On the other hand, Sulfur dioxide effects on the respiratory system. It may affect lung function, the incidence of respiratory symptoms and diseases, and risks of mortality. With respect to mortality, it should be noted that 10 µg/m³ changes in SO_2 is associated with a daily increase of 0.346 deaths with central and lower bounds 0.1367 and 0.057, respectively (Hatzakis et al 1986). Based on crude mortality in Egypt of 0.0051, the estimated upper, central and lower bounds of mortality, associated with 10 µg/m³ change in SO_2, are 0.00176, 0.00069 and 0.00029, respectively. Schwartz et al (1988 and 1991) estimate the low, central and high bounds of respiratory symptoms per thousand Childs per year and chest discomfort per adult per year as a result of 10 µg/m³ change in SO_2.

The effect of nitrogen dioxide on respiratory symptoms is more uncertain than the effects of the other pollutants. This attributed to: 1) large errors related to measuring outdoor NO_2; 2) the occurrence of effects only at the high levels of NO_2; and 3) the possibility that chronic, not acute, effects of NO_2 are important. However, an epidemiologic study has found health effects related to ambient NO_2 and an association between NO_2 and the increased likelihood of phlegm production (Schwartz and Zeger 1990). The change in respiratory symptoms is calculated based on Equation 2.

Table 1. Mortality and Morbidity Effects of 10 µg/m³ Change in PM10

Health Effects	Studies	Low Estimate	Central Estimate	High Estimate
Mortality	London, U.K (Mazurmdar et al. 1982)	0.29	0.31	0.33
	Ontario, Canada (Plagiannakos and Parker 1988)	0.49	0.98	1.47
	Steubenville, Ohio (Schwartz and Dockery 1992)	0.44	0.64	0.94
	Philadelphia, PA. (Schwartz and Dockery 1992)	0.96	1.2	1.44
	Santa Clara Country, CA (Fairley 1990)	0.73	1.12	1.51
	US Metropolitan Area. (Ozkaynak and Thurstonr 1987)	0.92	1.49	2.06
Morbidity	Respiratory hospital admission/100,000 (Pope 1991)	6.57	12	15.6
	Emergency room visits/100,000 (Samet et al. 1981)	128.3	235.4	342.5
	Restricted activity days/person (Ostro 1983)	0.404	0.575	0.903
	Lower respiratory illness/child/per asthmatic (Dockery et al. 1989)	0.008	0.0169	0.0238
	Asthma attacks/per asthmatic (Whiemore and Korn 1980)	0.163	0.326	2.73
	Respiratory Symptoms/person (Krupnick et al. 1990)	0.91	1.83	2.74
	Chronic Bronchitis/l00,000 (Detels et al. 1991)	30.6	61.2	91.8

Change in respiratory symptoms per year = F $*$ Change in NO$_2$(ppm) (2)

Where; F is a factor that equals 14.42, 10.22 and 6.02 for Upper, Central and Lower limits, respectively.

3 MONETARY VALUATION OF HELATH IMPACTS

Once the physical health impacts are estimated, the next step involves imputing monetary values to those health impacts as shown in below sub-sections.

3.1 *Economic Valuation of Mortality*

The value of premature death caused by air pollution is based on establishing the value of a statistical life (VSL). Two broad alternative approaches are used to estimate VSL; the human capital approach (which values an individual's life according to the net present value of his/her productivity) and the Willingness-To-Pay (WTP) approach (which measures the value society places on individual distinct from an individual's wage-earning capacity). The first approach tends to give lower values than the second one.

To get the value of a premature death, the life lost years are multiplied by the average wage rate in the country. Table 2 lists the calculated total lost years based on the death distribution by age, and based on the life expectancy of 68 years (WHO 2000).

According to UNICEF Egypt Statistics (UNICEF 2005), the average annual wage rate and annual rate of inflation are $1,250 and 7%, respectively. Therefore, the value of premature death for Egypt can be calculated as per Equations 3 and 4.

$$\text{Value of a premature death} = \text{average annual wage} \times \text{Total lost years}$$
$$= \$1,250 \times 14.872 = \$18,590 \tag{3}$$

$$\text{Value of a premature death taking into account the compound interest}$$
$$= \$18,590 + \left(18,590 \times \left((1+0.07)^{14.872} - 1\right)\right) = \$50,848 \tag{4}$$

Ideally the Value of a Statistical Life (VSL) should be based on a Willingness-To-Pay (WTP) value in the study area. There is no study in Egypt has been carried out in this regard since the method requires a large survey sample to ensure its reliability. Therefore, a benefit transfer (Brouwer 2000), is used to calculate the value of premature death. Benefit transfer approach involves the use of estimates of environmental loss of a project to estimate the economic value of environmental impact of a similar project on the assumption that the latter has

Table 2. Distribution of Lost Years by Age.

Age	Lost years	Deaths percentage	Total lost years
1<	68	0.11	7.48
1 – 4	66	0.029	1.914
5 – 14	58	0.019	1.102
15 – 24	48	0.022	1.056
25 – 34	38	0.024	0.912
35 – 44	28	0.048	1.344
45 – 54	18	0.108	1.944
55 – 64	8	0.161	1.288
65 – 74	-2	0.23	-0.46
75	-7	0.244	-1.708
Total lost years			14.872

the similar impact. In terms of mortality, this approach is to scale down the WTP by the ratio of per capita income of Egypt to per capita income of the country where the value is adapted from. The ratio could be derived directly from income difference or from relative incomes by using Purchase Power Parities (PPPs) as a conversion factor.

Since most of the VSL studies are conducted in the US and Europe, it is required to compare the values in order to obtain a reasonable estimate for Egypt. European Commission DG Environment (DG Environment 2005) provided a best estimate of VSL of one million Euro, for Europe-based mortality valuation for the year 2000, with a lower bound of 0.65 million and an upper bound 2.5 million. If the central estimate is transferred to Egypt using PPP-based national output ratio between the European Union and Egypt, considering a gross national income per capita (PPP international) for Egypt and European Union as $ 4,200 and $26,900, respectively (World Bank estimates 2004). As such, the VSL can be calculated as per Equation 5, considering exchange rate of €1 equals $1.3 US. Table 3 lists the calculated VSL for Egypt based on different studies carried out in United States, France, Canada, United Kingdom and Chile. Based on benefit transfer technique, it is found that the lower bound of VSL for Egypt is $78,504 and the upper bound is $506,666.

$$\text{VSL for Egypt} = €1,000,000 \times 4,200/26,900 = €156,000 = \$120,000 \qquad (5)$$

3.2 Economic Valuation of Morbidity

The valuation of morbidity can be estimated using human capital approach and WTP Approach. In the former approach, the direct costs of medical treatment and duration of illness are obtained from interviews carried out with specialist physicians and professors. Table 4 shows the average duration of illness and the average cost of medical treatment as a result of interviews. In addition, the table shows the total cost of illness taking into consideration average wage rate of lost day is 50 L.E.

Table 3. Estimated VSL for Egypt based on other Countries

Country	VSL ($)	PPP ($)	VSL for Egypt ($)
United States (Viscusi et al. 1991)	3,300,000	40,100	345,635
France (Le Net 1994)	700,000	28,700	102,439
Canada (Krupnick et al. 2000)	1,200,000 - 3,800,000	31,500	160,000 - 506,666
United Kingdom (Jones-Lee et al. 1998)	1,600,000 - 2,600,000	29,600	227,027 - 368,918
Chile (Cifuentes et al. 2000)	200,000 - 1,000,000	10,700	78,504 - 392,523

Table 4. Average Duration and Cost of Illness in Egypt.

Morbidity	Duration (Days)	Cost (L.E)	Lost days cost (L.E)	Total Cost of illness (L.E)
Respiratory hospital admission	7	2500	350	2850
Emergency room visits	1	300	50	350
Cardiovascular hospital admission	10	3000	500	3500
Lower respiratory illness in children	10	1000	500[*]	1500
Asthma attacks	5	1500	250	1750
Respiratory Symptoms	-	100	20[**]	120
Chronic Bronchitis	10	5000	500	5500

Notes:
* Lost days for parent
** 40% of lost day

Table 5. Estimated Cost of Morbidity for Egypt - US Studies

Morbidity	Duration (Days)	Value ($)	Value for Egypt $	Value for Egypt L.E
Respiratory hospital admission (Cropper and Krupnick 1989)	9.5	7,248	759	4326
Emergency room visits (Rowe et al. 1986)	1	242	25	142
Lower respiratory illness in children (Ostro 1992)	14	326	34	194
Asthma attacks (Rowe and Chestnut 1985)	9.5	578	60	342

Table 6. The number of officially reported plague cases in the world.

Morbidity	Value ($)	Value for Egypt $	Value for Egypt L.E
Respiratory hospital admission	786.0	589.5	3,360
Cardiovascular hospital admission	1,302.0	976.5	5,566
Asthma attacks	11.5	8.7	50
Chronic Bronchitis	10,370	7,777.5	44,330

The WTP values are estimated from other studies and transferred to Egypt taking into account the income differences. Table 5 lists the estimated cost of morbidity for Egypt based on studies performed in USA. Table 6 lists the estimated cost of morbidity for Egypt based on study (Zhou 2005) performed in China (PPP for China = $5,600).

4 MONETARY VALUE OF POLLUTION

The monetary value of damage due to the air pollution is calculated by multiplying marginal physical impact with the monetary unit values. Table 7 summarizes the value of 10 µg/m^3 changes in PM10, SO$_2$ and the monetary value of 1 ppm change in NO$_2$. It is noticed that the value of 10 µg/m^3 change in PM10 ranges from 120 to 5,422 L.E per person. The values of 10 µg/m^3 change in SO$_2$ range from 89 to 5,122 L.E per person. The values of 1 ppm change in NO$_2$ range from 6.0 to 17.3 L.E. per person.

5 CASE EXAMPLE AND DISCUSSION

According to JICA Study (JICA 2002) about 24.9 million trips are made daily in Greater Cairo, 18.3 million are made via motorized modes, about 68% are made by public transport, followed by passenger cars (16.4%). some 6.5 million trips are made by "Shared Taxi" which occupies a 52% share of the public transport trips, followed by "Public Bus" (3.1 million trips; 25%) and "Metro" (2.1 million trips; 17%).

These three major modes share 83% of the total trips made by public transport. Trips by "Public Minibus" are 0.4 million, sharing 3.4% only. Alike, "Light Rails" such as the Heliopolis Metro and CAT tram serve 0.18 million trips a day, sharing a minor portion, 1.4%. "Nile Ferry" trips are marginal (0.1%) in the whole trips in Greater Cairo (see table 8).

Table 9 shows the estimated emission factors in gm per kilometer

Assuming that in case of no metro the share of metro goes to the other public transport modes with the same shared percentages. In that case, based on the above tables the total emission per year for each type of vehicle can be calculated as shown in table 10.

Table 7. Monetary Value of Pollution.

| Health Effects | Unit Value(L.E) | | Dose-Response Estimate | | | Value of pollution | | | | | |
	Min.	Max.	Low	Central	High	Min. Low	Central	High	Max. Low	Central	High
a) Value of 10 μg/m³ Change in PM10											
Premature death	290,000	2,900,000	3.16E-5	4.89E-5	6.63E-5	9.164	14.18	19.22	91.64	141.8	192.2
Respiratory hospital admission	2,850	4,326	6.57E-5	12 E-5	15.6E-5	0.187	0.342	0.444	0.284	0.519	0.674
Emergency room visits	142	350	128.3E-5	235.4E-5	342.5 E-5	0.18	0.33	0.48	0.45	0.82	1.19
Restricted activity days/person	20	50	0.404	0.575	0.903	8.08	11.5	18.06	20.2	28.75	45.15
Lower respiratory illness/child/per asthmatic	194	1500	0.008	0.0169	0.0238	1.552	3.27	4.61	12	25.35	35.7
Asthma attacks/per asthmatic	50	1750	0.163	0.326	2.73	8.15	16.3	136.5	285.25	570.5	4,777.5
Respiratory Symptoms/person	100	120	0.91	1.83	2.74	91	183	274	109.2	219.6	328.8
Chronic Bronchitis	5500	44,330	30.6E-5	61.2E-5	91.8E-5	1.683	3.366	5.049	13.56	27.13	40.69
Total Value (L.E)/person/year						120.0	232.3	458.3	532.5	1,014.4	5,422.0
b) Value of 10 μg/m³ Change in SO₂											
Premature death	290,000	2,900,000	0.00029	0.00069	0.00176	84.1	200.1	510.4	841	2,001	5,104
Lower respiratory illness/child	194	1500	0.0001	0.00018	0.00026	0.019	0.034	0.05	0.15	0.27	0.39
Respiratory Symptoms/person	100	120	0.05	0.1	0.15	5	10	15	6	12	18
Total Value (L.E)/person/year						89.0	210.1	525.45	847.15	2,013.27	5,122.0
c) Value of 1 ppm Change in NO₂											
Premature death	100	120	0.0602	0.1022	0.1442	6.0	10.2	14.4	7.2	12.2	17.3
Total Value (L.E)/person/year						6.0	10.2	14.4	7.2	12.2	17.3

Table 8. Daily Trips at Greater Cairo

	S.Taxi	P. Bus	Metro	P. Minibus	Light rail	T. Coops	S. Rail	N. Ferry	Taxi	P. Car	Other	Motorcycle	E. bus
Public Transport													
Trip total	6.5	3.1	2.1	0.4	0.18	0.13	0.08	0.01	1.26	3.0	0.24	0.13	1.24
						18.3 million							
%				67.9					6.9	16.4	1.3	0.7	6.8

Table 9. Estimated emission factors in gm/Km (JICA 2009)

Type of vehicle	Type of pollutant			
	NO_x	SO_x[1]	PM_{10}	Annual Km[5]
Car	1.600[4]	0.010	0.110[2]	20000
Taxi	1.680[4]	0.012	0.116[2]	80000
Shared Taxi	4.330[3]	0.027	1.408[2]	100000
Minibus	4.330[3]	0.027	1.408[4]	75000
Public Bus	4.763[3]	0.030	2.872[3]	75000

Notes:
1. Sulphur content of gasoline is considered to be 50 wt.ppm of diesel 150 wt.ppm.
2. estimation based on emission factors for Iran, Malaysia, and USEPA
3. Based on Chassis Dynamometer test, conducted in Malaysia
4. estimation based on emission factors of Malaysia
5. SDMP study Report

Table 10. Total emissions.

Vehicle type	Emission gm/km			Annual KM	Emission per car gm/year			No. of cars	Total Emissions ton/year		
	NOx	SOx	PM10		NOx	SOx	PM10		NOx	SOx	PM10
Car	1.6	0.01	0.110	20000	32000	200	2200	0.34	10880	68	748
Taxi	1.68	.012	.116	80000	134400	960	9280	0.14	18816	134.4	1299.2
Shared Taxi	4.33	.027	1.408	100000	433000	2700	140800	0.85	368050	2295	119680
Public Bus	4.76	.030	2.872	75000	357225	2250	215400	0.4	142890	900	86160

Based on Gaussian dispersion model, the maximum ground level concentration of pollutants generated from vehicles can be calculated using Equation 6.

$$Cx = \frac{Q}{\pi \times \delta y \times \delta x \times U} e^{-1/2 \left[\frac{H}{\delta z}\right]^2} e^{-1/2 \left[\frac{y}{\delta y}\right]^2} \tag{6}$$

Where:
Cx: ground level concentration at some distance x downwind (g/m^3)
Q: average emission rate (g/sec)
U: mean wind speed (m/sec)
H: source height (m)
σy: standard deviation of wind direction in the horizontal (m)
σz: standard deviation of wind direction in the vertical (m)
Y: off-centerline distance (m)
For Cairo, the average wind speed is 3 Beaufort (weather 2009), considering Equation 6.

$$v = 0.836\ B^{3/2}\ \text{ms} \tag{7}$$

Where v is the equivalent wind speed at 10 meters above the sea surface and B is Beaufort scale number. Consequently, the average wind speed equal to 4.34 m/sec.

Applying equation 6 using U = 4.0 m/s, H=1.0 m and Y= 0.0. σy = 270 σz = 45 we get the following (see table 11):

Table 11. Pollutants concentration.

	NOx gm/m^3	SOx gm/m^3	PM10 gm/m^3
	0.002260212	1.41263E-05	0.00015539
	0.003908837	2.79203E-05	0.000269896
	0.076458727	0.000476763	0.02486233
	0.029683976	0.000186966	0.017898883
Total	0.112312	0.000706	0.043186

To convert μgm/m^3 to PPM (particle per million), the conversion factor depends on the molecular weight of the chemical, atmospheric temperature and pressure. Typically, conversions for chemicals in air are made assuming a pressure of 1 atmosphere and a temperature of 25 degrees Celsius. For these conditions, Equation (8) is utilized to convert the units of concentration from μgm/m^3 to ppm, considering the molecular weight of NO_2 is 46.05.

$$\text{Concentration (ppm)} = 0.02445 \times \text{concentration}\left(\mu\text{gm/m}^3\right)/\text{molecular weight}$$
$$= 0.02445 \times (112312/46.05) = 59.631\text{ppm} \tag{8}$$

Taking into consideration the minimum monetary values of pollution which are listed in Table 7

$$\text{Pollution cost} = (43186) \times 12.0 + (706) \times 8.9 + (59.631) \times 6.0 = 524,873\ \text{L.E/person/year}$$

6 BRIEF

The above case example demonstrates the significant contribution of metro projects in saving the social costs resulted from the monetary values of morbidity and mortality. This example aims at providing the policymakers with the necessary information to set priorities.

7 CONCLUSIONS

One of the more troublesome problems, both practical and ethical, is that of valuing the social costs of pollution. This study aims at calculating the social costs of air pollution. It calculates the costs of health damages associated with air pollution. It estimates the expected value of physical impacts of air pollution (morbidity and mortality). This estimation is performed in two steps. The first step is identification and measurement of health impacts. The second is that once impacts have been determined, it is often necessary to estimate monetary values for the associated morbidity (illness) and mortality (death). The estimation of expected values of morbidity and mortality is desirable for evaluating the benefits of specific pollution control policies. This estimation aims at providing the policymakers with the necessary information to set priorities. To this end, Cairo metro project presented as an example of a green project that can save a considerable amount of social costs.

REFERENCES

Brouwer, R. 2000. Environmental value transfer: state of the art and future prospects. Ecol Econ., 32 (1), 137–152.

CIA. 2007. The World Fact book. https://www.cia.gov/cia/publications/factbook/.

Cifuentes, L.A., Escobari, J., Prieto J.J. 2000. Valuing mortality risks reductions at present and at an advanced age: Results from a contingent valuation study in Chile. Proceedings of EAERE Conference, Rethymnon, Crete, Greece.

Cropper, M.L., Krupnick, A.J. 1989. Social costs of chronic heart and lung disease. University of Maryland and Resources for the Future, Washington, D.C., US. http://yosemite.epa.gov/ee/epa/eerm.nsf/vwAN/EE-0120-02.pdf/$file/EE0120-02.pdf.

Detels, R., Tashkin D. P, Sayre J.W., Rokaw, S.N., Massy, F.J., Coulson A.H., Wegman, D.H. 1991. The UCLA population studies of COPD: X. A cohort study changes in respiratory function associated with chronic exposure to SOx, NOx, and hydrocarbons. Am J Public Health, 81(3), 350–359.

Dockery, D.W., Speizer, F.E., Stram, D.O., Ware, J.H., Spengler, J.D., Ferris, B.G. 1989. Effects of inhalable particles on respiratory health of children. Am Rev Respir Dis., 139(3), 587–594.

European Commission DG. 2005. Environment. Methodology for the Cost-Benefit Analysis of the CAFE Program. Health Impact Assessment, AEAT/ED51014/Methodology Volume 2: Issue 2, AEA Technology Environment, UK.AEAT, UK.

Fairley, D. 1990. The relationship of daily mortality to suspended particulates in Santa Clara County, 1980-1986, Environ Health Perspect, 89, 159–168.

Hatzakis, A., Katsouyanni, K., Kalandidi, A., Day, N., Trichopoulos, D. 1986. Short-term effects of air pollution on mortality in Athens. Int J Epidemiol. 23(5), 957–967.

Japan International Cooperation Agency. 2009. JICA Preparatory Survey on Greater Cairo Metro Line no.4 Feasibility Study Report.

Japan International Cooperation Agency, 2002. JICA Transportation Master Plan and Feasibility Study of Urban Transport Projects in Greater Cairo Region in the Arab Republic of Egypt.

Jones, L.M., Loomes, G., Rowlatt, P., Spackman, M., Jones, S. 1998. Valuation of deaths from air pollution. Report to the Department of Environment, Transport and the Regions and the Department of Trade and Industry, NERA and CASPAR, London, UK.

Krupnick, A., Alberini, A., Cropper, M., Simon, N., Brien, B., Goeree, R., Heintzelman, M. Age, 2000. health and the willingness to pay for mortality risk reductions: A contingent valuation survey of Ontario residents. J Risk Uncertainty. 24(2): 161–186

Krupnick, A.J., Harrington, W., Ostro, B. Ambient ozone and acute health effects: evidence from daily data. J Environ Econ Manage. 1990, 18(1), 1–18.

Le Net, M. 1994. Le prix de la vie humaine: calcul par la méthode des préférences individuelle. Commissariat Général du plan.

Mazumdar, S., Schimnmel, H., Higgins, I.T. 1982. Relationship of daily mortality to air pollution: An analysis of 14 London winters. 1958/59-1971/72. Arch Environ Health. 37, 213–220.

Ostro, B. 1992. Environmental Pollution and Health. Lancet. 340(8829), 1220–1221.

Ostro, B. 1994. Estimating Health Effects of Air Pollution: a Method with an Application to Jakarta, Policy Research Department. Working Paper 1301, World Bank, Washington DC, US.

Ostro, B.D. 1983. The effects of air pollution on work loss and morbidity. J Environ Econ Manage. 10(4), 371–382.

Ozkaynak, H., Thurston, G.D. 1987. Associations between 1980 U.S. mortality rates and alternative measures of airborne particle concentration. Risk Analysis. 7(4), 449–461

Plagiannakos, T., Parker, J. 1988. An assessment of air pollution effects on human health in Ontario. Ontario Hydro, Canada.

Pope, C.A. 1991. Respiratory hospital admissions associated with PM10 pollution in Utah, Salt Lake and Cache Valleys. Arch Environ Health. 46(2), 90–97.

Rowe, R., Chestnut, L.G., Peterson, D.C., Miller, C., 1986. The benefits of air pollution control in California. Energy and resource Consultants, Boulder for California Air resources Board.

Rowe, R., Chestnut, V. 1985. Oxidants and Asthmatics in Los Angeles: A Benefits Analysis. Report to Office of Policy Analysis, US EPA, EPA-230-09-86-018. Washington, DC, US.

Samet, J.M., Speizer, F.E., Bishop, Y., Spengler, J.D., Ferris, B.G. 1981. The relationship between air pollution and emergency room visits in an industrial cornniury, J. Air Pollut. Contr. Assoc. 31(3), 236–240.

Schwartz, J. 1991. Particulate air pollution and daily mortality in Detroit. Environ Re. 56(2), 204–213.

Schwartz, J., Dockery, D.W. 1992. Increase mortality in Philadelphia associated with daily air pollution concentrations. Am Rev Respir Dis., 145(3), 600–604.

Schwartz, J., Dockery, D.W. 1992. Particulate air pollution and daily mortality in Steubenville, Ohio. Am J Epidemiol. 135(1), 12–20.

Schwartz, J., Hasselblad, V., Pitcher, H. 1988. Air pollution and morbidity: a further analysis of the Los Angeles student nurses data. J. Air Pollut Control Assoc. 38(2), 158–162.

Schwartz, J., Zeger, S. 1991. Passive smoking, air pollution and acute respiratory symptoms in a diary study of student nurses. Am Rev Respir Dis. 141(1), 62–67.

Tong Siak Henn. 2004. Demographics and Country Profile - Per Capita Income. http://siakhenn.tripod.com/capita.html.

UNICEF 2005. Egypt Statistics. http://www.unicef.org/infobycountry/egypt_statistics.htm/#31

Viscusi, W.K., Wesley, A.M., Huber, J. 1991. Pricing environmental health risks: Survey assessments of risk-risk and risk-dollar trade-offs for chronic bronchitis. J Environ Econ Manage. 21(1), 32–51.

Weather Forecast. 2009. http://www.climatetemp.info/egypt/cairo.html.

Whittemore, A.S., Korn, E.L. 1980. Asthma and air pollution in the Los Angeles area. Am J Public Health. 70(7), 687–696.

World Health Organization (WHO). 2000. Table 1: Numbers and rates of registered deaths, Egypt. http://apps.who.int/whosis/database/mort/table1_process.cfm.

Zhou, Y. 2005. Economic analysis of selected environmental issues in China International Max Planck Research School on Earth System Modeling, Hamburg University.

Tunnels and Underground Cities: Engineering and Innovation meet Archaeology, Architecture and Art, Volume 2: Environment sustainability in underground construction – Peila, Viggiani & Celestino (Eds)
© 2020 Taylor & Francis Group, London, ISBN 978-0-367-46579-7

Compliant reuse of Terzo Valico excavation material: Design and operations

N. Meistro, G. Parisi, A. Scuderi, S. Pistorio & S. Genito
COCIV, Consorzio Collegamenti Integrati Veloci, Genoa, Italy

ABSTRACT: The Terzo Valico railway line project produces excavation material for around 14.000.000 cubic meters. The Italian Ministry Decree n. 161/2012 regulates the management of surplus excavated material. This latter requires a specific design plan called "excavated soils use plan" (also reported as "plan of use") to define methods, means, routing and final destinations of the excavated material in excess. This must be submitted to the Italian Ministry of the Environment for its approval. Excavation materials management strategy for Terzo Valico is reported in the present paper. Quarry sites reclamation is a technique mostly employed, which is beneficial for the environment and landscape of territories involved by the new line works.

1 PROJECT DESCRIPTION

The railway Milan–Genoa "Terzo Valico" belongs to the High Speed/High Capacity Italian system and to the European railway corridor Genoa – Rotterdam.

The new line will improve the connection from the port of Genoa with the hinterland of the Po Valley and northern Europe.

The "Terzo Valico" project is 53km long, 36 km of which in underground between Piemonte and Liguria, crossing the provinces of Genoa and Alessandria, through the territory of 12 Municipalities. To the South, the new railway will be connected with the Genoa railway junction and the harbour by the Voltri and Fegino interconnections. To the North, the railway layout crosses Liguria Apennines with Valico tunnel (27 km long), then running outside in the municipality of Arquata Scrivia continuing towards the plain of Novi Ligure; here, the railway will connect existing Genoa- Turin rail line (for the traffic flows in the direction of Turin and Novara – Sempione) and Tortona – Piacenza –Milan rail line (for the traffic flows in the direction of Milan- Gotthard). under passing, with the 7km long Serravalle Tunnel, the territory of Serravalle Scrivia (Figure 1).

Here, after connecting with the historic line Turin directed (Novi Ligure interconnection), the route continues outdoors in the artificial tunnel, up to Tortona where it links the existing line Milan directed. Client is the company RFI "Rete Ferroviaria Italiana", while COCIV Consortium is General Contractor (GC) for the design and construction of the project.

2 EXCAVATED SOILS USE PLAN IN TERZO VALICO

The construction of the Terzo Valico, mainly consisting of underground excavation and earthworks which amount to nearly 14.000.000 cubic meters, requires a coordinated plan in order to allow an effective management of the excavation surplus. In fact, only 15% of the excavated materials is reused within the project, leaving around 12.000.000 cubic meters to be conferred at suitable sites for requalification purposes.

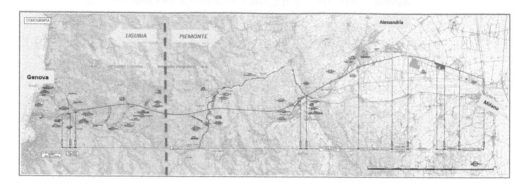

Figure 1.   Terzo Valico project layout.

The excavation material surplus is produced both in Liguria (37%) and in Piemonte (63%) (Figure 2).

Out of the excavation material surplus produced up to date, 37% was transferred to authorized sites in Liguria and 24% in Piemonte (Figure 3).

The Ministerial Decree n. 161 of 10/08/2012 is the applicable regulation, which defines the procedures for the correct reuse of excavated material. These require a specific design plan called "Excavated soils use plan", also reported as "Plan of Use" (in Italian PdU), whose

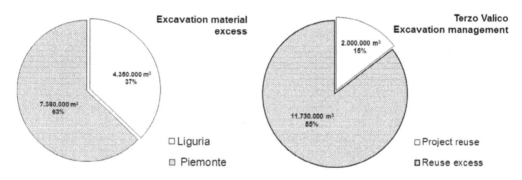

Figure 2.   Excavation and reuse volumes in Terzo Valico (COCIV, 2018, unpubl.).

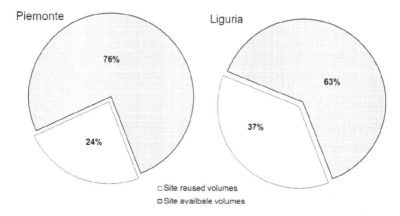

Figure 3.   Liguria and Piemonte region excavation material reuse progress (COCIV, 2018, unpubl.).

purpose is to define the final destination of the excavation surplus on the basis of its environmental compatibility with the recipient site.

In the Plan of Use, each production site is assigned one or more recipient site and the respective amount of material, the methods of characterization of the soil and traceability of the material to ascertain the environmental compatibility the routes and the means of transport.

The Plan of Use for the Terzo Valico was prepared by COCIV in its quality of General Contractor, and was approved by the Minister for the Environment. The field monitoring is performed and granted by the "Regional Agencies for the Environment".

The approved sites indicated in the Plan of Use, can only be operated after the acquisition of the necessary authorizations. Hence, the approval of the Plan of Use is not in itself sufficient to initiate the conferment, which is subject to administrative procedure, one for each of the sites.

The Plan of Use for the Terzo Valico mainly envisages the reuse of the excavation surplus for morphological and environmental upgrading of quarries, previously exploited for the production of aggregates.

The operability of a single destination site is subject to different administrative and procedural aspects, which depend on the type of requalification foreseen, and the current authorization status of the site. Moreover, the formalities required by the sector regulations in force in the Region in which the site falls are also to be taken into account.

3 LIGURIA AND PIEMONTE SITES LOCAL ENVIRONMENTAL REQUIREMENTS

In the very first phases of the work, the material excavated in Liguria was intended to be used for infrastructural works in the port of Genoa as well as in the redevelopment of exploited quarries, subsequently to the signing of an Agreement between the local Administrations, the disposal sites Operators and the General Contractor.

The last update of the approved Plan of Use partially confirms the sites quoted in the aforementioned Agreement, lately signed in 2011, and identifies new locations as a replacement for those that are no longer available or incompatible with the construction schedule.

COCIV activity in Liguria, is limited to the transport of excavated material to the site, while the final placing within the quarry is performed by the site operator, as the holder of the authorization of the redevelopment project.

As in Liguria, also in Piemonte the excavated material is used for reclamation of existing quarry sites. In addition, the so called "open/close quarries" are meant to receive suitable excavated material from Terzo Valico, in quantity sufficient to backfill the quarry after the extraction of aggregates for the Terzo Valico itself.

In Piemonte, administrative procedures to implement in order to grant the operability of a site are different from those in Liguria and actually more complex. In fact, Piemonte Region developed a specific regulation (Regional Law. 30/99) for the use of quarries in public works construction. This law requires that a public work is provided with an additional "Litoid Materials Repository Plan" indicating the quarries supplying aggregates as well as the sites to be reclaimed through the excavated material surplus.

Therefore, a quarry site in Piemonte already included into the Plan of Use, may be authorized for environmental morphological recovery only after being approved according to the Regional Law no. 23/2016. This latter, also regulates the mining and requalification permits for quarry sites. Those permits that must be obtained necessarily by the General Contractor in case of public works.

4 ENVIRONMENTAL RECLAMATION OF QUARRY SITES

Such a use of excavated material surplus from the Terzo Valico, determines an opportunity to improve the landscape through the elimination of discontinuity and degradation caused by previous mining activities.

It should also be considered that quarries requalification are performed under the environmental monitoring from National Governmental authorities, leading to a great advantage both in environmental protection and quality fields.

4.1 *Reclaiming of pit quarries and quarry lakes*

The requalification of a quarry pit consists in backfilling it at ground level, in order to restore the original morphology allowing an agricultural use.

Another reclamation possibility is given by a quarry lake. The lake originates from an industrial stones extraction in an aquifer area and a properly designed refill can be made.

Great attention is paid to environmental protection during the quarries requalification design and their authorization process, carrying out in-depth studies and specialized studies aimed at investigating potentially impacted matrices. In this regard, the hydrogeological and geochemical study carried out in Tortona Municipality site, "Cascina Pecorara", quarry design is summarized.

The study purpose was to ascertain the stratum diffusion of those metals potentially occurring in the reusable excavation material, such as chromium, nickel and cobalt (Cr, Ni, Co), followed

Figure 4. Requalification of the quarry pit C.na Clara e Buona - Alessandria (above) and of the quarry lake C.na Bolla - Alessandria (below) (COCIV pictures, 2018, unpubl.).

by an impact assessment on the existing drainage wells, originally placed for irrigation or domestic use.

Therefore, specific chemical examines have been carried out, both at excavation material so-lid matrices and at site ground water level. Conceptual hydrogeological model thus obtained and solid-liquid matrices chemical data processing, allowed the correlation between metals concentration in the excavation material and their occurrence in the stratum, assumed filtration dissolution and leaching phenomena in the aquifer.

Chemical elaboration and conceptual hydrogeological model results have been used for a further implementation of a numerical predictive model of groundwater flow and mass transport, to forecast pollutants transmission scenarios, referred to site receptors (irrigation or potable wells).

The Figure 5 shows Cr potential diffusion in the aquifer, where it can be seen that the wells around the quarry (PT, PZ in Figure 5), are not affected by this metal dissolution, and that the Cr concentration is lower than the regulatory threshold in water human use. Similar results were found in eluates of the other metals investigated, Ni and Cb.

4.2 Slope quarry reclaiming

The slope quarry reclaiming is a more complex intervention compared to a morphologic reco-very of an extractive pit. The slopes are stepped and the excavated material is then placed according to a slopes succession with a proper inclined profile, to guarantee the stability. The following figures show the slope upgrading rendering designed for Cava Vecchie Fornaci site, in Genoa Municipality (Figure 6).

Another example of slope quarry reclaiming is the no longer active "Ex Cava Cementir" in Voltaggio Municipality (Alessandria province). The design is made particularly complex due to a considerable quarry front highnesses, as well as a placement nearby a torrent named Tor-rente Lemme. The environmental recovery is implemented by COCIV and it is made possible by reinforced soils placement at the foot of the slab, as Lemme Torrent protection (Figure 7).

In addition to the environmental requirements to which the quarry reclaiming plans must undergo, another crucial aspect regards the Italian Mining Police rules compliance, aimed at protecting the health & safety of quarry workers. For instance, in Ex Cava Cementir, COCIV

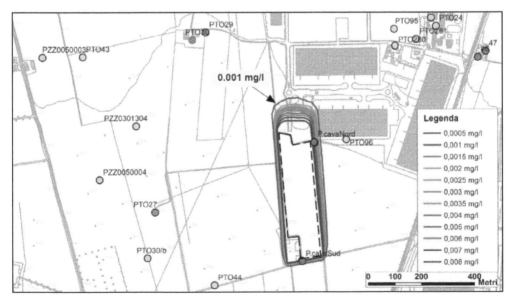

Figure 5.   Chromium potential diffusion in the aquifer below C.na Pecorara - Tortona quarry (COCIV, 2018, unpubl.).

Figure 6. Status of the Cava Vecchie Fornaci site and reclaiming rendering design - Plan and elevation view (COCIVpictures, 2016, unpubl.).

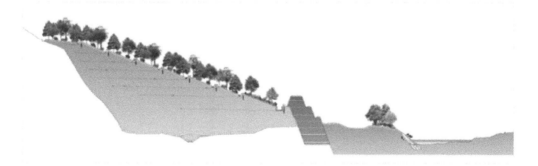

Figure 7. Section view of the quarry morphological recovery at Cava Cementir, Voltaggio (Alessandria province) (COCIV, unpubl.).

has put in place a slope topographic monitoring system which, by surveying the displacement of some appropriately chosen points, allows to assess any movement significance thus to forewarn any collapsing risk.

Aforesaid survey system consists in a topographic station able to detect and record points and cornerstones position in a network made of by optical prisms (Figure 8).

The measurement frequency is set with a cycle of double readings on prisms, usually performed every 2 hours, increasable when needed.

For each reading cycle, angles and distances measurements on prisms are acquired, plus point coordinates are calculated. In a monitoring point, the variances between the gauged coordinates with the former measurement cycle and the initial ones define the displacement vector (Δx, Δy, Δz).

Up until now, no significant movement has been registered (Figure 9).

Figure 8. Topographic monitoring points placement in the Ex Cava Cementir - Voltaggio (AL) (COCIVpicture, 2017, unpubl.).

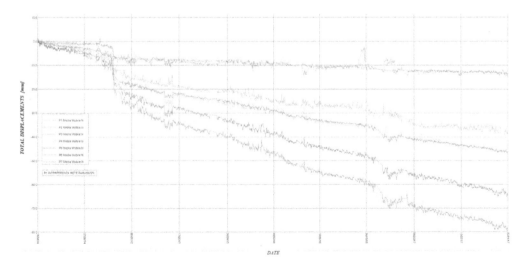

Figure 9. Total displacements of the topographic monitoring points recorded in the Ex Cava Cementir - Voltaggio (AL) (COCIV, 2017, unpubl.).

Knowing the extent over time of displacements, collapsing risk assessment is made possible. With the measurements series recorded to date, potential areas impacted by eventual landslides and related attention thresholds to be adopted during working progress, are in agreement with Piemonte Region's Mining Police Office.

The monitoring system installed is set up to manage alarm situations (SMS, email or acoustic alarm) also, to be activated if the assigned critical levels are exceeded.

5 TERZO VALICO ORGANIZATION FOR MANAGEMENT OF EXCAVATION MATERIAL

Excavated soils reuse management of a huge project such as the Terzo Valico, is not easy and implicates an *ad hoc* organization in term of people and processes. Essential aspects to be checked and documented constantly are the environmental characterization of the excavation materials and their traceability from production sites to destination ones. In particular, COCIV is well organised with a specific and adequate structure, made up of environmental competent field personnel, looking after the project plan management. In particular, an overall coordinator supervises excavated materials characterization and traceability. This figure is supported by two operational coordinators, one for each Region. Moreover, for each "production site" (operatio-nal site) and for each "disposal site", an employee deals with the operational control of excavation material. Other professionals work at the material management, i.e. geologists, for instance, performing the lithological surveys of the tunnel excavation fronts, to determine the sampling materials frequencies.

Considering that in the Project "Excavated soils use plan" encompasses 6 "production sites" and 17 "disposal sites" provided in Liguria Region, and 12 "production sites" and 16 "disposal sites are expected in Piemonte Region", it is easy to understand how complexity the entire ma-nagement is.

5.1 *Environmental characterization of excavation material*

The project executioner (as a "Producer") ascertains the environmental suitability of soils to the planned destination. Regarding the characterization, some specific sampling protocols have been prepared concerning both traditional excavation mode and mechanized one, with TBM. Material samples taken are analyzed by accredited laboratories. These have carried out a ring-test of analysis methods intercalibration with the control authorities (regional Environmental protection agencies, i.e. ARPA Piemonte and ARPA Liguria), in order to use a common me-thodology, compliant with the law.

On the basis of the geological profile of the project, natural asbestos mineralization occurring is possible, in the excavated material. This aspect is cared by adopting a characterization on the asbestos content, according to the specific "Asbestos Management Protocol" of the Terzo Valico project. These activities were entrusted to university laboratories (DISTAV Genoa University "Chemistry and Industrial Chemistry Department" and DIATI laboratory of Polytechnic University of Turin) registered by the Italian Ministry of Health asbestos laboratories list (according to the Ministerial Decree 06/09/1994). In this regard, it should be noted that, to optimize the efficiency and logistics of characterization activities, the Terzo Valico is the first Italian construction site where a university task force laboratory has been set up (Polytechnic University of Turin).

Even these institutes, of well-established expertise in asbestos sector, have performed the intercalibration of analytical methods with the aforesaid control authorities in order to use shared methods.

Another basic aspect of material characterization is surfactant content determination whereas used for tunnels excavation with TBM-EPB machine. To ensure an environmentally sound management of these materials, in fact, despite the existing environmental legislation does prescribe an *ad hoc* study, COCIV has entrusted the "Mario Negri" Institute of Pharmacological

Research (I.R.C.C.S.), internationally body with proven experience in ecotoxicological research field, an experimental study about ecotoxicology and degradation of conditioned excavated muck. Approved by the Ministry of the Environment and tested in-site in a Liguria tunneling work, the study was deepened afterward to upgrade surfactants analytical methods.

Regarding the TBM excavation in Piemonte region project area, University of Genoa conducted further investigations aimed to improve surfactants analytical methods. Also these analysis methods have been subject of a specific ring-test/intercomparison trial with the control bo-dies laboratories.

Ultimately, the excavation material Producer verifies the disposal by the destination sites while any regional ARPA monitors the compliant management of excavation materials with their on-site inspections and characterizations.

5.2 *Excavation material traceability*

During the excavated material transportation phase, it is crucial to ensure their effective traceability to the destination sites, in order to verify the compliance with planned origins and destinations, at any time. Thus, each vehicle journey is accompanied by a delivery note (in Italian: D.D.T.) which contains details on the origin, destination and environmental characterization of supplied material. Given the complex implementation of the excavated soils and subsoil management, COCIV has set up a specific Geographic Information System (GIS) for recording and monitoring the handled material characterization and traceability. This system is called "TerreSc@" and it allows to record the whole of significant management data, including the excavation phase survey, the environmental characterization information and the transport data.

Moreover, for a more effective transport control, all the vehicles have been equipped with a tracking and telematic itineraries recording G.P.S.; this system, called "W.A.Y." (Where Are You), delivers data on a web platform, allowing an in real time remote monitoring of each vehicle position and a routes checking, for any transport.

From TerreSc@ and W.A.Y. data crossing, a comprehensive both documental and spatio-temporal traceability of Terzo Valico excavation material management is given.

W.A.Y. system allows to monitor any construction site transit that in a given time interval can affect a road network. It is possible, in fact, to extract traffic reports on particular sections of traffic, to observe the flow induced by a working site, in order to prevent any risk of public road congestion.

6 CONCLUSION

This paper points out the management of the Terzo Valico' excavation material surplus, according to the project "Excavated soils use plan", compliant to Ministry Decree n. 161/ 2012. In particular, the Plan implementation complexity has been outlined, both in relation to administrative requirements concerning the sites opening (Liguria and Piemonte Regions disparities) and the operative organization necessary for such a complex work.

Project disposal sites are mostly abandoned quarries (pit quarries, lakes quarry and sloping quarries), suitable for receiving excavation materials, according to an authorized reclamation design. As well as design is implemented according to the regional technical regulations, each authorized quarry reclamation work is subject to Terzo Valico environmental control as a public project; this is certainly in favor of a superior reclamation environmental protection and quality.

The compliant management of excavation materials requires a complex organization consisting of specialized personnel. Also COCIV has set up an information system for a complete traceability of excavated materials, namely "TerreSc@", allowing to record the excavation surveys data, environmental characterization information and destination site transportation. In addition, for a more effective control of excavation material transfers, vehicles have been

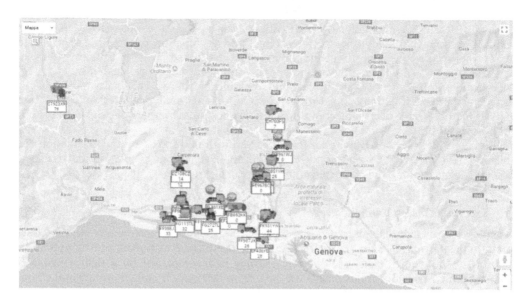

Figure 10.   WAY web: location of active vehicles (COCIV, 2018, unpubl. - Map data ©2018 Google).

all equipped with G.P.S. devices, for itineraries tracking and telematic recording, by means of a system named "W.A.Y.". This system allows the real time monitoring of any vehicle position and the routes checking. Plus, W.A.Y. is a tool for site Terzo Valico induced traffic monitoring as it allows to survey those flows affecting a road net-work portion in a given time interval.

Ultimately, the Terzo Valico construction, mainly based on excavation of long tunnels sections, cannot proceed without effective management of the excavation material surplus. To date, in Italy, there is no other project with such a large quantity of excavation materials to deal with, in a complex and ramified context, even for administrative, normative, geographic, geological and geomorphological dissimilarities, encountered by the project territory.

Furthermore, the Terzo Valico excavation materials management fulfills the railway construction requirements, while renewing those territories previously interested by mining activities, removing discontinuity and degradation.

REFERENCES

COCIV, 2015. "Linea Ferroviaria AV/AC Terzo Valico - Gli aspetti gestionali-operativi del marino del "Terzo Valico"". INERTIA, Ferrara.

Piemonte region, 2017. "Deliberazione della Giunta Regionale 18 luglio 2017, n. 1-5386 - LL.RR. 30/1999 e 23/2016. Piano di reperimento dei materiali litoidi finalizzato alla realizzazione della Li-nea ferroviaria AV/AC Milano-Genova "Terzo Valico dei Giovi". Approvazione dell'aggiornamento 2017". Turin.

Ministry for the Environment, 2017. "DVA protocollo n. 0000309 del 31/10/2017 del Ministero dell'Ambiente e della Tutela de Territorio e del Mare, Direzione generale per le Valutazioni Ambientali – Aggiornamento Piano di Utilizzo, ai sensi dell'art. 8 del D.M. 161/2012; ID Fascicolo: 3324; Allegato: Parere CTVA n. 2530 del 20 ottobre 2017". Rome.

COCIV, 2017. "Linea Ferroviaria AV/AC Terzo Valico – La gestione come sottoprodotto dei materiali da scavo contenenti amianto in matrice minerale". ECOMONDO, Rimini.

COCIV, 2018. "Linea Ferroviaria AV/AC Terzo Valico – La gestione del materiale condizionato con tensioattivi proveniente da scavo meccanizzato nel rispetto dell'ambiente". XVII National Congress of the Environment and Cultural Heritage - University of Genoa, Genoa.

Tunnels and Underground Cities: Engineering and Innovation meet Archaeology,
Architecture and Art, Volume 2: Environment sustainability in
underground construction – Peila, Viggiani & Celestino (Eds)
© 2020 Taylor & Francis Group, London, ISBN 978-0-367-46579-7

Management of tunneling machines excavation material

N. Meistro, G. Parisi & C. Zippo
COCIV, Consorzio Collegamenti Integrati Veloci, Genoa, Italy

L. Captini
Salini Impregilo S.p.A., Milan, Italy

ABSTRACT: The new railway line extends for about 53 km, 36 of which in underground and, partially, excavated by TBM-EPB mechanized method. Such a method requires the usage of ground conditioning agents to improve the muck fluidity. The excavated material is then dumped on the ground and disposed in layers, a conventional industrial practice which allows both drying and the natural bio-degradation of the additives below the reference threshold as defined by specific ecotoxicological trials.

Once the material is compliant with the environmental quality requirements and the aforesaid thresholds it can be managed as a reusable resulting material in accordance to the Ministerial Decree 161/2012.

1   PROJECT DESCRIPTION

The railway Milan–Genoa "Terzo Valico" belongs to the High Speed/High Capacity Italian system and to the European railway corridor Genoa – Rotterdam.

The new line will improve the connection from the port of Genoa with the hinterland of the Po Valley and northern Europe.

The "Terzo Valico" project is 53km long, 36 km of which in underground between Piemonte and Liguria, crossing the provinces of Genoa and Alessandria, through the territory of 12 Municipalities.

To the South, the new railway will be connected with the Genoa railway junction and the harbour by the Voltri and Fegino interconnections.

To the North, the railway layout crosses Liguria Apennines with Valico tunnel (27km long), then running outside in the municipality of Arquata Scrivia continuing towards the

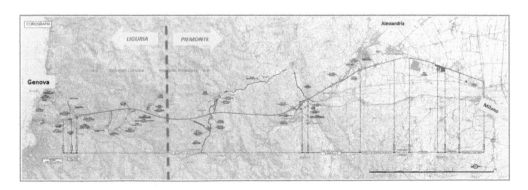

Figure 1.   Terzo Valico project layout.

plain of Novi Ligure; here, the railway willconnect existing Genoa- Turin rail line (for the traffic flows in the direction of Turin and Novara – Sempione) and Tortona – Piacenza –Milan rail line (for the traffic flows in the direction of Milan- Gotthard). under passing, with the 7km long Serravalle Tunnel, the territory of Serravalle Scrivia (Figure 1).

Here, after connecting with the historic line Turin directed (Novi Ligure interconnection), the route continues outdoors in the artificial tunnel, up to Tortona where it links the existing line Milan directed. Client is the company RFI "Rete Ferroviaria Italiana", while COCIV Consortium is General Contractor (GC) for the design and construction of the project.

2 SURFACTANT CONDITIONED EXCAVATION MATERIALS "TERZO VALICO" MANAGEMENT

TBM-EPB cutter allows the full mechanization of both tunnel excavation and covering phases. Digging occurs through the rotation of a shield, whose anterior extremity is equipped with the rock cutting gears. Underground soil material is get away from the excavation room to the back of the shield by a screw conveyor system (Figure 2).

Then, excavation debris ("muck") are directed to "accumulation tanks" in the site work area with a conveyor belt system (machine belt and continuous belt).

In order to reduce the shield friction in favor of an increased removal effectiveness of excavation material, conditioning addition in necessary by a water solution of additives set up surfactant components.

Although all surfactants selected for the operational phase are degradable in aerobic environment, their depletion needs time to happen.

According to the national legislation applicable to the project, the excavation material containing also the aforesaid additives used to allow the mechanized excavation, can be managed in exclusion to the waste regime, in full compliance with National rules.

In particular, excavation material can be reused wether the content of polluting substances in the excavation soil is below the "contamination threshold concentration" reported with reference to a specific reclamation land destination, in Legislative Decree n. 152/2006, Part IV, Annex 5, Table n. 1, Columns A and B or as naturally occurring contamination values.

Nevertheless, the law does not define expressly a concentration threshold for additives conditioning substances in muck, thus leading a few disputes.

Despite a lack of specific regulatory limit, for these parameters also, COCIV wanted to determine a reference threshold intended as the maximum concentration of surfactants allowed in the excavated material without this implying a danger to the environment.

This paper addresses the whole investigation process which led to the management of TBM-EPB muck respecting the environment.

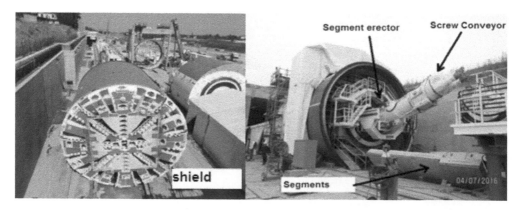

Figure 2.   TBM components (COCIV picture).

This paper concerns surfactants additives reference thresholds studies, surfactants analisys and methods procedures to define a proper operative control in excavation material management.

2.1 The ecotoxicological study

With the aim of determine surfactants additives reference thresholds, an in deep experimental study was commissioned at the "Mario Negri" Institute of Pharmacological Research (IRCCS), an internationally body with proven experience in ecotoxicological research field.

The study was aimed to verify, before the start of excavation operations, in how long additives degradation process would takes place and to thoroughly assess any environmental ecotoxicological effect.

The research was developed in two phases encompassing at first a thorough chemical properties review of products followed by a laboratory scale plus an actual samples calibration toxicity study. For each phase, description and results are hereafter showed.

Phase 1 – Analysis of additives available technical and scientific information.

In a careful review of commercial products technical data sheets, chemical structures of components were analyzed and further reinforced by computational data arising from an *in silico* toxicity predictions in aquatic – terrestrial environment (QSAR modeling and Read Across analysis).

Products chemical characterization was carried out with liquid chromatography high resolution – Fourier transform-based mass spectrometry (HPLC-FTMS). Those tests pointed out the principal components of each investigated conditioning agents.

At laboratory scale, the products main chemical components were examined, targeting degradability time. To assess degradation pathway, a number of samples for every product were prepared and tested likewise. Trials have been carried out with OECD standard soil samples and actual subsoil samples. Sample preparation encompassed spiking of soil with such doses simulating predicted excavation conditioning.

Samples conditioning agents concentration was measured at different time intervals with a semi-quantitative analysis, using pure original products measure as reference standard.

Investigated conditioning agents act similarly in terms of degradation of all components. For OECD soils, degradation is complete after about 8 days yet (Figure 3), while in real subsoil samples, degradation is slower (Figure 4).

Ecotoxicological properties laboratory essays were conducted on conditioning agents surfactants main components by an *in silico* modeling prediction of toxicological properties. At this stage, a structural analogues exam was accomplished with a special regard to U. S. Authorities documented ones. A toxicological/structural comparison approach has been used too, according to procedures established by Italian Environmental Protection Agency.

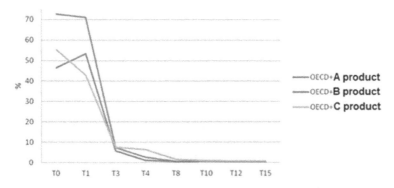

Figure 3. Degradability of main anionic components in spiked OECD soils (COCIV & Mario Negri, 2017, unpubl.).

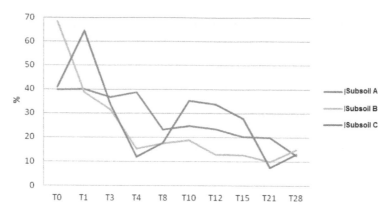

Figure 4. Degradability of main components anionic in spiked real subsoil (COCIV & Mario Negri, 2017, unpubl.).

In silico analysis with two different software, showed that some component would have potential aquatic toxicity (LC_{50} in fish and *Daphnia*).

All components degradation is quiet quick.

Regarding mutagenesis, "Negative" result is always yielded (4 models: 3 in the VEGA platform and 1 T.E.S.T.).

A chemical-structural similarity analysis of identified molecules (test set) was carried out with a database of molecules (training set) that present legislative reference values in a number of national and international regulations.

The similarity analysis, performed with the "TOX Match" software, resulted in a matrix where 10 molecules show high similarity with *sulfonated anionic surfactants* (SLES and LAS) and 1 (molecule 9) reveal it with some *glycols*.

Crossing chemical similarity data with molecules encompassed in analyzed documents, reference thresholds values have been pinpointed.

In particular, according to the main compounds identified, the surfactant concentration reference threshold for two of the three products tested, was set up at 100 mg/kg and the third product tested surfactant concentration reference threshold was set up at 200 mg/kg.

Phase 2. Additives toxicity tests were carried out on standardized animal and plant models, as much representative as possible of living organisms. For these studies, the following biological models and methods are described.

- *Lepidium sativum* (watercress), *Cucumis sativum* (cucumber) and *Sorghum saccharatum* (sorghum) are the higher plants whose seeds are used to measure short time compound effects in soil environment. Germination inhibition and root elongation in higher plants is measured on products conditioned soil plantation (ISO 11269-1).

- *Eisenia andrei* (earthworm): standard species used for acute toxicity tests. The test consists in keeping the organisms in contact with the conditioned medium and evaluating their mortality over time. This test was carried out at first adding the compound maximum concentrations foreseen (limit test), then conducting a more in deep assessment of dose/toxicity relationship.

- *Daphnia magna*: the freshwater crustacean was used to assess toxicity in water environment. The test consists in assessing the crustacean immobilization at 48 hours of exposure to the compound, according to the guidelines of ISO 6341. Based on the results of the studies carried out and the international references acquired, the surfactant concentration reference threshold for two of the three products tested, was set up at 100 mg/kg and the third product tested surfactant concentration reference threshold was set up at 200 mg/kg. These thresholds represent the maximum quantity of product that can be present in the excavation material so that it can be reused as a by-product.

2.2 Analytical method for surfactants measurement in subsoil

To assess environmental reference threshold compliance, a laboratories intercomparison procedure has involved Genoa University "Chemistry and Industrial Chemistry Department", Piemonte Environmental Protection Agency in Alessandria Department (territorial control unit) and "Mario Negri".

Furthermore, this ring test was launched to reach an analytical method agreement to measure surfactants, in order to allow environmental operational controls on muck, compatible with excavation progress needs.

The analytical method was ratified by Italian Ministry of Environmental and by the "Environmental Monitoring Centre". It is used nowadays to assess reference thresholds compliance as determined by "Mario Negri" institute.

2.3 Radimero sitework TBM-EPB muck sampling

In production site, samplings from TBM-EPB sitework belt conveyor for "process verification" analytical trials are carried out to (1) verify the content of surfactants in the excavation material with respect to the reference thresholds; (2) assess real laboratory results delivery time; to (3) to asses degradation time of in-site excavation material surfactants.

Starting and final additive contents were detected in subsoil samples. Additives concentrations were measured in every sample at n. 0, 1, 3, 6, 8, 10 day.

In 8–9 days, degradation process bring surfactants concentration levels below the reference thresholds.

To comply with successive reuse requirements, the excavation material must undertake an ordinary industrial practice. The muck is layered and stockpiled, to reach an as much as possible efficient material oxygenation and natural degradation level.

Considering the limited space available at the excavation worksites, this practice is performed at the Cascina Romanellotta intermediate deposit site identified in the project "Excavated soils use plan".

Required degradation times to comply with the reference thresholds, as verified in-site and the production foreseen for each single TBM-EPB machine, conducted to size the spaces required for the management of the conditioned excavation material. In particular, for each TBM-EPB a surface of about 15.000m^2 were provided, including logistic and service vehicles maneuvering space.

2.4 Surfactants conditioned excavation material site management in Terzo Valico sitework

At the intermediate deposit site, the muck is stockpiled, lying in sectors forming daily piles, for a maximum high of 1 meter and a width depending of the daily admission volume, to reach an as much as possible efficient material oxygenation and natural degradation level. Furthermore, to speed up the drying and oxygenation, every pile is turned once per day.

Each pile is properly labeled, with posters and tracked with registration form. Care is taken to avoid any contact with other piles.

Sampling and analisys are replicated until surfactants concentration is under the reference thresholds. Concentration is certified by Genoa University.

Each Decree 161 analyte conformity is assessed too. If material is compliant both with contamination threshold concentration and reference thresholds, it can be transferred to final disposal site. Otherwise, if threshold concentrations are exceeded, muck is managed as waste. These are "conformity verification" steps.

If material yields contamination threshold concentration only, the intermediate deposit degradation process is carried on until surfactants compliance is fully reached.

The overall demucking process, operational control, management and disposal to final sites is represented in figure 5.

3 ENVIRONMENTAL OPERATIONAL PROTECTION IN INTERMEDIATE DISPOSAL SITE "CASCINA ROMANELLOTTA"

The intermediate disposal site "Cascina Romanellotta", set up, in the municipality area of Pozzolo Formigaro (Alessandria Province), nearby the construction siteworks is properly equipped with environmental protection of the surrounding environmental components.

Intermediate Deposit is divided in n. 4 sectors, i. e. n. 1 for every excavating TBM-EPB machine.

Underground water and aquifer leaching risks were prevented by a suitable waterproofing of the entire site, including service areas. It has been applied a double waterproofing layer with a geomembrane one coupled with a bentonite geocomposite tier, forming a waterproofing package coat of high efficiency and reliability.

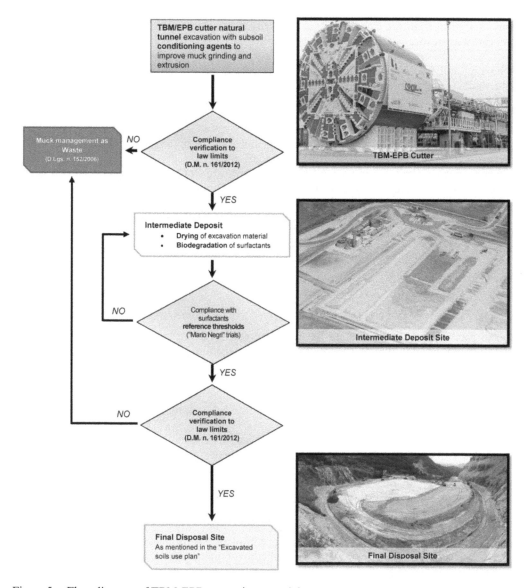

Figure 5.   Flow diagram of TBM-EPB excavation material management.

A local network of 7 piezometers is installed to have a feedback about ground water quality evolution during site management. Monitoring findings reported a water quality comparable with the one before the work site setting up.

Leaching waters are collected and then conveyed through an intermediate deposit perimeter channel toward a chemical-physical treatment plant ensuring any contamination removal, in full compliance with the regulatory standard discharge limits (Figures 6, 7 and 8).

Figure 6.   waterproofing leaching water channel in intermediate deposit in Cascina Romanellotta.

Figure 7.   Chemical-physical leached water treatment plant in intermediate excavation material deposition area – Subsoil and water treatment areas.

Figure 8.   Chemical-physical leached water treatment plant in intermediate excavation material deposition area and waterproof channels and ponds.

Regarding the impact on air, the muck does not release dust because of its high humidity; nevertheless, if the disposal process is prolonged significantly after dehumidification of muck, any airborne dust release is prevented by means of cover sheets or others.

Lastly, in order to limit the visual impacts on surrounding areas, with respect to inhabited zones, site shields as dunes of vegetal soil from the superficial site preparation, are grassed over time.

4   CONCLUSION

The paper deals with management of surfactants conditioned excavated materials produced by the mechanized tunneling activities of new high-speed railway line "Terzo Valico".

During the TBM-EPB excavation, to ensure the machine operation, sufficient consistency of muck must be assured to uniformly transfer the pressure to tunnel excavation face and to facilitate its extraction with a screw conveyor. The conditioning of material takes place through an ad hoc dosed fluidifying agent containing biodegradable foaming surfactants.

The extracted material is not immediately suitable with final disposal site and it requires a maturation/aeration process to allow a natural degradation of compounds contained in such foaming agents. This process, framed among the ordinary industrial practice envisioned in Decree 161, takes place laying the muck on the ground in an intermediate deposit area properly designed for this purpose.

Furthermore, before reuse, the environmental quality requirements are accomplished if "[...] the average composition of the whole mass does not enclose concentrations of pollutants above the envisaged maximum standard threshold" (Decree 161, article 1, paragraph 1, letter b)). Italian current regulations do not explicitly set a concentration standard for additives in TBM-EPB muck.

In this regard, COCIV commissioned to the "Mario Negri" Institute the additives reference thresholds definition as a surfactants standard concentration level of unpolluted and not toxic for the environment conditioned muck.

In site and laboratories studies brought to a degradation time definition, that allowing the dimensional design of disposal areas for excavation subsoil material management.

Chemical and toxicological results were enclosed in the project "Excavated soils use plan" procedure and then approved by Italian Ministry. Analytical method to measure total surfactants concentration to compare with reference thresholds is the result of an intercomparison phase among Contractor laboratories and Piemonte Environmental Agency and such method allows to quickly assess environmental compatibility reflecting consistently the project progress. The analytical method identified has been ratified by the Authorities.

In site test protocols allowed to refine operational control procedures for environmental management of reusable subsoil.

In the light of all the achievements conditioned muck is subjected to an ordinary industrial practice on intermediate disposal site and, after that, it is tested in order to verify its environmental compatibility: if it encounters both Italian legal requirements (i.e. contamination threshold concentration compliance) and concentration criteria pointed out in "Mario Negri" Institute study (i.e. reference threshold conformity), it can be allocated to final disposal site provided in the project tunnel material extraction plan.

REFERENCES

COCIV, 2017. "Elementi riepilogativi relativi allo scavo con TBM e alla gestione dei materiali in Piemonte. (Aggiornamento a seguito della riunione del 01/ 02/17presso ARPA Piemonte – Dipartimento di Alessandria relativa alla metodica per la ricerca dei tensioattivi)". Genoa.

COCIV, 2017. "Linea ferroviaria ad alta velocità Milano-Genoa Terzo Valico dei Giovi. Procedura per la caratterizzazione del materiale da scavo condizionato nell'ambito del Piano di Utilizzo del Terzo Valico". Genoa.

COCIV, 2017. "Procedura per la caratterizzazione del materiale di scavo condizionato nell'ambito del Piano di Utilizzo del Terzo Valico – "Mario Negri" Ecotoxicological study annex". Genoa.

COCIV, 2017. "Procedura per la caratterizzazione del materiale di scavo condizionato nell'ambito del Piano di Utilizzo del Terzo Valico – Surfactant agents laboratory assessment annex". Genoa.

Ministry Decree n. 161, 10/08/2012. "Regolamento recante la disciplina dell'utilizzazione delle terre e rocce da scavo".

Tunnels and Underground Cities: Engineering and Innovation meet Archaeology,
Architecture and Art, Volume 2: Environment sustainability in
underground construction – Peila, Viggiani & Celestino (Eds)
© 2020 Taylor & Francis Group, London, ISBN 978-0-367-46579-7

Stockholm bypass: Environmental sustainability in the excavation of the Lovön rock tunnels

G.L. Menchini & S. Piscitelli
CMC Cooperativa Muratori Cementisti Ravenna, Italy

ABSTRACT: Stockholm bypass is the biggest infrastructure ever designed in the Nordics. The paper describes the most critical section, length 7 km, running below Lovön Island and Malaren lake, respectively a relevant natural reserve and Stockholm's drinking water reservoir. The main challenge is minimizing the impact of mucking out the 9 million tons of rocks coming from the tunnels excavation. Strong focus is given to water protection. The systematic pre-grouting prevents any possible impact on the lake and the water table; wastewaters coming from the tunnels, from the rock materials stockpiling and from the equipment parking areas are kept separate, collected and diverted to avoid even the minimal contamination.

1 INTRODUCTION

Stockholm is the fastest growing capital in Europe. Stockholm bypass is the biggest infrastructure ever designed in the Nordics, it will adsorb the upcoming shock in traffic, that the city will face due to the continuous demographic growth. The only sustainable way to build such a major infrastructure, located in the heart of one of the greenest city in Europe, is to go underground; 18 km out of 21 are underground.

The paper describes the largest and most critical contracts in the Stockholm bypass, Lovön Rock Tunnels North and South.

The Project features 14 km of Main Tunnels – two parallel tubes 7 km each - having a typical section of 128 m^2 and 6 km of underground Ramps having a typical section of 70 m^2.

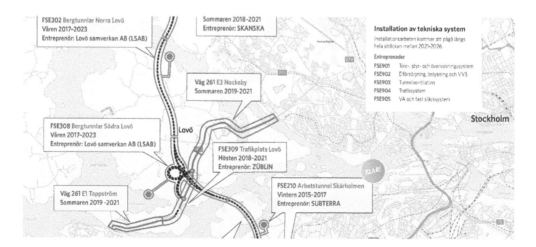

Figure 1.   The Stockholm Bypass – Lovön Rock Tunnels.

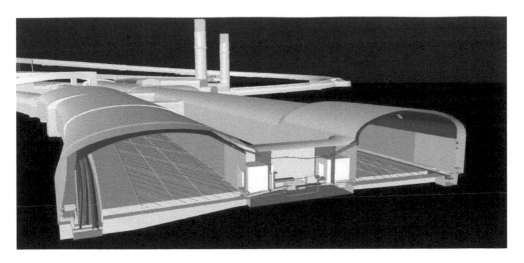

Figure 2.   The Stockholm Bypass – Lovön Rock Tunnels – Main Tunnel Typical Section.

The Main Tunnels rage from 65m to 100m depth and pass mainly below the island of Lovön and, in two different sections, below lake Malaren, where the tunnels connect with the two adjacent Contracts of the Bypass.

To access the Main Tunnel the Contractor has excavated two temporary access tunnels, with an overall length of about 1,8 km (850 m in Site 302 and 950 m in Site FSE 308), with the typical section being 77m^2. Eight ventilation shafts, 10 meter diameter, 40m to 100m deep, complete the Project.

The aboveground site installations are massive featuring, in each site, a stationary crusher, a stockpiling with tripper and feeding tunnel, a conveyor belt, a temporary pier, a batching plant, a water treatment plant and multiple facilities such as offices, warehouses, workshops, tents, etc.

Figure 3.   Site Installations at Lovön Site FSE 302.

The excavation, performed with the Drill and Blast technique in gneiss/granite rocks, will then be followed by tunnel lining, preparation for the tunnel road pavement and installation system.

One of the biggest challenges for the project is the context, Lovön Island is a relevant natural reserve and a major cultural heritage site, home to the King's Residence Drottingholm, facing Lake Malaren, the primary source of drinking water for Stockholm.

With these premises, a full commitment to environmental sustainability has been the top condition for the award of the Project, therefore attention toward adopting the best precautions and mitigation measures has been extremely high during both the design and construction phases.

In order to contain the CO_2 emission and minimize the impact to the traffic in the island, the 9 million tons of excavated material are crushed on site to the commercial fraction 0–150, transported by ship, from temporary piers built by the Contractor, to receiving piers and then reused in the construction industry.

Systematic pre-grouting is performed before the excavation to prevent changes in the level of the underground water. Waters coming from tunnel activities, rock stockpile, site platforms and paved areas are collected, diverted to a water treatment plant, purified and then recirculated in the system, while the overflows or wastage are, instead, convoyed to the existing sewage system.

2 REDUCED CO2 EMISSIONS AND TRAFFIC IN THE SURROUNDINGS

The result of the excavation is a total of around 9 million tons of rock material which the Contractor is encouraged to reuse on site (temporary roads, permanent roads in the tunnel, aggregates, etc.), to comply to this requirement, the construction of an additional crusher, dedicated to the production of aggregates of various dimensions, has been planned on site.

However, most of the excavated material, in commercial 0 – 150 mm fraction, has been already sold by the client, therefore it will be crushed and transported outside the Island of Lovön using two different logistics depending on the site.

FSE 302 Tunnel North, is reachable only through an old unpaved road, not available to heavy traffic, making necessary to find an alternative way of transportation for materials and equipment, identified in waterways through ferries provided by the Contractor. This al

FSE 308 Tunnel South is, instead, reachable by land through Ekerövägen, that will be completely upgraded and refurbished during the execution of the project.

Figure 4.   Crusher at Site FSE 308.

2.1 Site Installations

The Contractor has built in each of the two sites: one temporary pier, a conveyor belt, a temporary stockpiling featuring a tripper and a feeding tunnel and a stationary crusher.

After every round the excavated material is transported with dumpers to the hopper of the stationary crusher which, after filtering the fraction below 150mm, transfers it first to a primary crusher and then to a secondary crusher. Together they can process the full quantity of rock coming from the excavation in order not to hinder the progresses.

From the secondary crusher the material. in fraction 0–150 mm, is transferred to a 18m high tripper, who distribute the excavated material in a temporary stockpiling. The stockpiling is the buffer who disconnect the excavation progress from the maritime transportation, thus increasing the flexibility of the mucking out system.

The stockpiled material is automatically loaded on the conveyor belt from a feeding tunnel located at the bottom of the stockpiling, its live storing capacity 6.000 m^3, however its overall capacity reaches about 18.000 m^3.

The conveyor belts are 650 m long in Site 302 and about 2 km long in site FSE 308. A system of covers and trays, as well as closed hoppers, prevent the spreading of the material dust and water. At the end of the belt, installed directly on the floating temporary pier, there is a swinger, adjustable in both the horizontal and vertical plan, which performs the loading of ships and self -propelled barges. The average loading capacity of the system is 500 to 600 ton per hour.

Figure 5.   Loading operation at the temporary pier at Site FSE 302.

2.2 Maritime Transportation

A fleet of self-discharging ships/barges transports the material from the two sites to the three receiving ports indicated by the Client. The travel time is about 3 to 3,5 hours. Once at destination the material is discharged with an excavator, installed on board, into the hoppers of the conveyor belt at the receiving piers. The average discharging capacity of the on-board excavator is 300 ton per hour.

Considering the draft, the limitation along the route, and the loading/discharging capacity, it has been concluded that the optimum capacity of the ship/barge ranges from 2.000 tons to 3.000 tons. With operations being normally conducted 24/6, the 2.000 tons ships perform 7 to 8 trips per week and the 3.000 tons barges around 6 trips per week.

To boost the production, the Contractor has deployed five excavation team in total: two excavation teams in Tunnel North FSE 302, working on four fronts, and three excavation teams in Tunnel South FSE 308 working on 10 fronts.

In order to not slow down the progress of the excavation, while at the production pick point, the Contractor shall transport, on a weekly basis, up to 35.000 tons of material from Site FSE 302 and up to 45.000 tons of material from Site FSE 308.

The average production, together with the excavation period has been, however, calculated based on the production plan and time schedule. Therefore, the Contractor, has engaged two different shipping companies, that together deploy up to three ships with capacity between 2.000 tons and 2.500 tons and two self-propelled barges with 3.000 tons capacity.

One ship (2.500 tons), supported, after a few months, by second ship (2.000 tons), has been used during the excavation of the access tunnels. A third ship (2.500 tons) has been added when the Ramps and the Main Tunnel have been reached, and finally two additional barges (3.000 tons) have been deployed when all excavation fronts have become available.

3 PROTECTION OF THE GROUND WATER LEVEL: SYSTEMATIC PRE-GROUTING

The purpose of tunnel grouting is to prevent changes in groundwater levels. Environment may get harmed if those levels change considerably, reducing the amounts of water pumping out of a tunnel also affects long-term operating costs. It is further considered that leakages may wash out the material filling the rock fractures and erode the rock itself, possibly having a long-term effect on the strength and stability of rock structures. The flow of water also induces forces and change in pressure, which modifies the natural effective stress and may have immediate effect on the strength and stability of rock structures. All grouting is done in compliance with the Client's design, instructions and requirements within the Contract documents.

Pre-grouting is systematic before any excavation. The length of pre-grouting fans is typically 25meters, with at least 6 meters of overlap.

Figure 6. As Built 3 D model – Grouting Overlap.

The quantity of water leaking into the tunnel is constantly measured at specific points where measuring dams are built; according to the Contract the leaking water shall not exceed 4 l per minute per 100 meter of tunnel. Post-grouting is performed only if required and instructed by the Client, based on the leaking water data.

Grouting units and grouting mix are approved by Client before work start and all the workers responsible for the grouting shall be certified by the client.

According to the typical sections, included in the contract documents, in the Main Tunnel there are three different classes of pre-grouting:

- Class A: 8 holes + 8 holes
- Class B: 16 holes + 16 holes
- Class C: 32 holes + 32 holes

Figure 7. As Built 3D Model - Grouting Section.

However, according to the longitudinal profiles, included in the contract documents, only class B and C are applicable to FSE 302 and FSE 308 tunnels; there is not Class A pre-grouting.

3.1 Drilling

Drilling rigs are equipped with data collecting devices for recording drilling, MWD (Measurements While Drilling). Grouting boreholes is drilled with automatic drilling parameters registration (MWD): feed pressure, rotation speed, drilling speed, percussion pressure, rotational pressure, flush water pressure, flushing water flow, hole depth, time.

In all pre-grouting classes, drilling is performed in two rounds. After the first round MWD data are immediately saved to Rockma program, the Client's geology is informed by SMS and e-mail. From these data the Client's Geologist decides if extra grouting holes are needed and, in this case, their locations. Observations of leaking grouting holes is notified to the geologist in connection with the transmission of MWD data. If no notice of additional grouting drilling has been received within one hour after MWD data has been submitted and remaining holes according to typing drawings are drilled, grouting may begin. During the calibration the corresponding time is three hours.

Figure 8.   Three Boomer Drill Rig at site FSE 308.

3.2   *Calibration of MWD equipment*

To increase the reliability of the system, calibration of MWD equipment is performed for at least eight grouting screens, for at least 6 working weeks from the date when drilling rig was put into operation. When calibrating MWD equipment, water loss measurements are performed.

Water loss measurements is carried out with double packers in sections along the entire borehole. The distance between packers is 3 m. Water loss measurement is performed on the first five grouting screens in 10 evenly spaced grouting holes per screen.

The water loss measurements are performed in the initial grouting holes according to typical drawings, at an overpressure of 1.0 MPa for at least 2 minutes with a stable flow for each measurement section.

3.3   *Grouting*

The Contract consider three different mixtures

Table 1.   Grouting mix features.

	Mixture 1 Normal	Mixture 2 Alternative	Mixture 3 Sealing coarser cracks
bicritical	<75 μm	<90 μm	<140 μm
Bmin	<45 μm	<45 μm	<90 μm
Float Limit	1–2 Pa	2–6 Pa	> 8 Pa
Viscosity	10–30 mPas	10–50 mPas	> 30 Pas

Grouting begins with Mix 1. Downward holes are filled with grout from the bottom. When grouting, the connection between the grout holes is observed. Packers are placed 1–2 m in to the hole. Holes with big leaks and holes with connections are grouted first. During MWD equipment calibration, only one hole per pump is grouted.

The grouting pressures are shown below where, rock mass coverings also refer to possible soil coverage.

- Rock Mass < 5 m = Grouting Pressure 0.5 MPa
- Rock Mass 5 - 15 m = Grouting Pressure 1 MPa

- Rock Mass 50 - 75 m = Grouting Pressure 2.5 MPa
- Rock Mass > 75 m = Grouting Pressure 5 MPa

After completion of the hole filling, the grouting must start with high flow to achieve the specified grouting pressure as quickly as possible and to be performed for the entire hole length. The grouting pressure should then be kept constant throughout the grouting time. Grouting must last for 15 minutes for each hole, after the specified grouting pressure has been reached and is stable.

Grouting of holes where the flow does not reach 2 l/min, within 5 minutes, can be interrupted after reaching the specified grouting pressure. When grouting several holes per pump, flow per pump applies. Grouting must be interrupted when the specified maximum volume is grouted, even if the prescribed design time is not reached. Grouted volume is excluding hole and hose filling. The specified maximum volume refers to 25 m long grouting holes. For single hole holes, the maximum volume must be 500 l. For multi-hole design of two holes, the maximum volume is 750 l, and for multi-hole design of three holes, the maximum volume must be 900 l. For multi-hole design of more than three holes, volume criteria must be determined and reported in the working preparation.

4 SUSTAINABLE WATER MANAGEMENT

Overexploitation of clean water is a great concern, due to the increased water demand connected a growth in population, technological development and climate change. Besides that, pollution from Nitrogen in the groundwater reservoir and water basin is a great concern.

Tunnel operations have a significant demand for water within the tunnel due to its use in the drilling machinery. On the other hand, one of the main environmental problem of the drill and blast tunnel excavation method, is the presence of high content of Nitrogen (NH_3) contained in the emulsion and left in the excavated rocks. For this reason, the Contractor developed and implemented an advanced system where greatest part of the water is collected and, after the required treatment, reused on site. The implementation of an innovative design for the collection of storm water and wastewater, together with an advanced solution for the treatment of water on site, leads to a considerable reduction of water pollution and depletion, together with a reduction of cost for handling of water for the society.

At site FSE 308 and FSE 302 the following type of water are collected and treated before being sent to the existing sewage system.

- Water from the tunnel activities, where high concentration of nitrogen and cement is present due to the use of emulsion from blasting and shotcrete
- Water drained from the stockpile, where high concentration of nitrogen is present in the mucked rocks
- Water leakage from platforms and paved roads (areas near offices and industrial buildings) that may be exposed to oil spill from passing/parked cars
- Water coming from all paved areas exposed to high pollution risk (storage of crushed shaft masses, equipment drainage, crushing plant, dosing plant, workshops)

4.1 Calculation for dimensioning the Water treatment plant

Regarding the calculation of flows of industrial water, the water consumption required for the following uses has been considered:

- Dewatering Tunnel: 38m³/h (Each Drill rig uses 300l/m)
- Crusher: 2 l/s
- Batching Plant: 5 l/s
- Vehicle Washing: 5 l/s
- Workshop: 1 l/s

Since the greatest concentration of pollutants are found in the first part of the runoff, the system has been designed to treat this first storm water runoff, related to areas of interest of the project. In detailed, the design data required for the size calculation of flows corresponds to the following:

- 228 litres/s for ten minutes, duration with a time to return in ten years.
- All flows have been determined based on this precipitation intensity.
- For the calculation of flows, the "rational" method was used.

The formula for calculating the maximum flows is directly derived from the kinematic method:

$$Q = \frac{C \cdot i \cdot A}{3600000}$$

where: Q = maximum flow (in m³/s); A = surface of the water catchment (in m²); C = output coefficient; i = measured precipitation (mm/h)

The calculation of the rational formula is correct under the following assumptions:

- uniform precipitation intensity in space and constant over time
- constant outflow coefficient during the event and irrespective of the intensity of precipitation
- permanent linear model of conversion of inputs and outputs
- initial water flow equal to zero.
- the coefficient C is a parameter which is less than 1, considering the total water losses in the basin (inland soil, retention in superficial soil boundaries) which causes the flow to be lower than the total precipitation flow. This coefficient has been determined by calculating the weighted average of the coefficients for each type of surface to which the pool belongs.

Below are reported the achieved values.

- Industrial Area (Drainage Area 6.500 m³ – runoff coef. 0.80) = 0.119 mc/s
- Muck Storage (Drainage Area 2.360 m³ – runoff coef. 0.80) = 0.078 mc/s
- Industrial Area (Drainage Area 2.910 m³ – runoff coef. 0.42) = 0.028 mc/s

As it can be seen, the greatest part of the runoff comes from the paved areas (industrial area) and the clay membrane area, 0,119 m³/s (water collected from the muck storage) while the roads have a lower contribution, equal to 0,028. Indeed, the greatest part of the roads are built with permeable area and water is discharged to the nature, since no pollution is expected from these areas.

4.2 *Layout and characteristics of the catchment areas*

In the general layout showed below it is possible to identify, through several colors, the different water flows coming from the different areas of the construction site. The catchment areas and the pipe network system have been developed in order to collect all the polluted water and divert it to the water treatment plant, minimizing the risk of spread of pollutants in the surrounding environment and maximize the reutilization of water, after dedicated treatment, on site. As it can be seen, water is diverted from the stockpile (green color), tunnel (orange color) and catchment areas/paved surfaces (red color) to the water treatment plant. Then, the treated water is pumped again to the main water storage container next to the tunnel. From here, the water is used again in the tunnel and pumped back to the water treatment plant, reaching a closed cycle of water and a considerably saving a large amount of water from the lake.

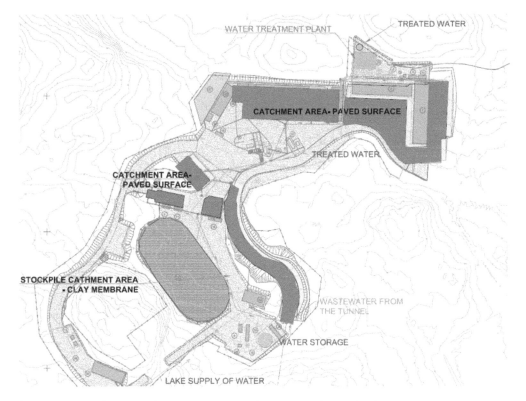

Figure 9.   General Layout of catchment areas and wastewater flow at the site.

In the following subchapters, each catchment system is described in detail.

4.1.1 *Tunnel wastewater collection system*

In the tunnel, a high quantity of water is discharged due to the work activities and, partially, due to the leakage from the groundwater. This water is then mixed with sediments from exploded rocks, emulsions and concrete used in the tunnel during the shotcrete and explosive charging. Several sedimentation tanks are located along the tunnel. At the entrance of the tunnel a larger sedimentation tank is located. From here, the water is then diverted to the Water treatment plant.

4.1.2 *Stockpile Wastewater collection system*

The new material excavated from the tunnel is transported continuously to the stockpile. The mucked rocks retain a high concentration of Nitrogen from the emulsion used as explosive. During the rainfall event, the nitrogen is diluted into the water and its runoff could end into natural environment, with considerable damage to the ecosystem, such as eutrophication issues. For this reason, the stockpile is provided with a bentonitic clay membrane layer underneath it.

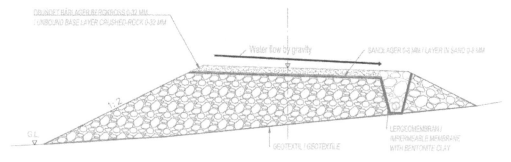

Figure 10.   General Layout of catchment areas and wastewater flow at the site.

The bentonitic membrane consists in a layer of material characterized by suitably low diffusion coefficient, equal to 1.58×10^{-9} m/s. Here, water drains into a pipe and, by gravity, it is collected to a manhole. From the manhole, water is subsequently diverted to the water treatment plant. The pipe network has been designed to handle the maximum runoff level of the catchment area (approximately 5000m^2). Besides that, the bentonitic membrane acts as a retainment system for pollutants present in the water, avoiding their discharge in the groundwater or surrounding environment even in case of filtration.

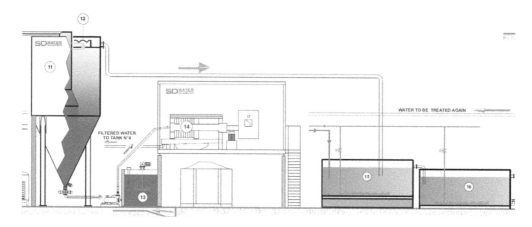

Figure 11.   Layout of the clarifies and last two treatment tanks.

The clean water then flows to a containment tank with a volume located outside the water treatment plant. Here, water is temporary stored before being diverted to the main storage area next to the tunnel where water is used for tunnel activities.

Overflow is instead diverted to the sewage system, therefore no spillage into the nature is allowed. All the information regarding processed water, instantaneous values and problems/alarms are accessible through a dedicated software where the plant can be controlled and monitored on real time and action can be taken directly from remote. The set of Data can be visualized in a three-variable graph to facilitate the readability of the result.

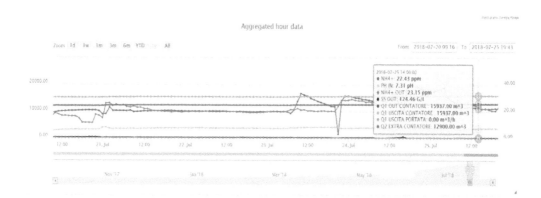

Figure 12.   Graphical layout of the monitoring system of the Water Treatment Plant software.

5   CURRENT STATUS OF THE WORK

The works are performed by Lovön Samverkan AB (LSAB), the Swedish company which CMC and its JV partner have established after the signature of the Contract with Trafikverket.

The access tunnels have been completed during Summer 2018 and the excavation of the main tunnels and ramp tunnels is now proceeding at full speed. Five excavation teams, working 24/6, are deployed to excavate simultaneously up to 14 fronts. The excavation will be completed at the beginning of 2021, while the overall contract completion date is 3rd March 2023.

Figure 13.   Tunnel Works at Lovön site FSE 308.

6   CONCLUSION

The execution of Lovön Rock Tunnels is a successful experience which proves that through accurate planning and professional performance, it is possible to ensure the environmental sustainability of a large infrastructure project, even when performed in the most sensitive and protected areas.

*Tunnels and Underground Cities: Engineering and Innovation meet Archaeology,
Architecture and Art, Volume 2: Environment sustainability in
underground construction – Peila, Viggiani & Celestino (Eds)*
© 2020 Taylor & Francis Group, London, ISBN 978-0-367-46579-7

Utilization of microtunnelling technology for construction of the new gas pipeline between Azerbaijan and Europe

D. Mognol
Bessac, Toulouse, France

ABSTRACT: To reduce Europe's dependency on Russian gas, the European Union has decided to construct the Southern Gas Corridor to transport Caspian Gas from Azerbaijan to Europe. The Southern Gas Corridor is composed by three pipelines: the South Caucasus Pipeline eXtension, the Trans ANAtolian Pipeline, and the Trans Adriatic Pipeline. We have constructed six microtunnel sections over 400 meters' length for a total of 3630 meters of microtunnel between 2015 and 2018 on this project leaded by major Oil&Gas Companies. Indeed, the microtunneling technology presents several advantages over other construction technologies to allow installation of pipeline on the most challenging sections of the projects with minimal disturbance and maximal efficiency. Microtunneling provides a reliable, eco-friendly and competitive technology for challenging geological crossings. However, construction of microtunnel for pipeline installation presents certain specificities to be considered to lead to the success of the operation.

1 INTRODUCTION

1.1 *The Southern Gas Corridor*

Gas supply to Europe is strategic issue for the European Union. To reduce Europe's dependency on Russian gas and diversify the supply sources, European Commission has decided the construction of the Southern Gas Corridor (SGC) in 2008. The Southern Gas Corridor is aimed to provide a mean of transportation for the natural gas of the Caspian Sea in Azerbaijan (Shah Deniz gas field) to the European gas network in Italy.

The Southern Gas Corridor project consists in the construction of three gas pipeline in the south of Europe (See Figure 1) for a total 3500 kilometers of 48″ pipeline:

– The South Caucasus Pipeline extension (SCPx) going through Azerbaijan and Georgia
– The Trans-Anatolian Pipeline (TANAP) crossing Turkey
– The Trans-Adriatic Pipeline (TAP) through Greece Albania and Italy

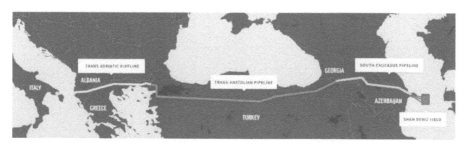

Figure 1.   The Southern Gas Corridor.

1.2 *Pipeline microtunnels contracts*

We completed EPC contracts for the design and construction of 6 challenging micro-tunnels and pipeline installation on SCPx and TAP projects under strict safety and environmental regulations.

Six microtunnels for a total of 3630 meters have been excavated in three different countries on the Southern Gas Corridor:

- One 605 meters' river crossing DN1800 microtunnel in Georgia (Kura East (Ge) river crossing) on SCPx project
- One 1010 meters' river crossing DN1800 microtunnel in Azerbaijan (Kura West (Az) river crossing) on SCPx project
- Two 600 & 560 meters DN1800 microtunnels to cross landslides and faulty areas in Corovode, Albania for TAP project
  - Two 435- & 420-meters' river crossing DN1600 microtunnels in KP114 & KP139 of Albanian part of TAP project.

Each of these crossings presents specifics that have been considered from the design phase of the projects: ground properties, pipeline characteristics, pipeline installation methodology, pipeline riser installation, etc...

Geologies encountered ranged from hard sedimentary rock in mountain areas to loose uncemented alluviums with boulders in the river crossings.

In addition, EPC projects included the construction of the shafts to allow microtunnel construction, pipeline and pipe riser installation, grouting of the annular void between the tunnel lining and the pipeline.

A wide range of construction methodologies have been deployed to successfully complete these projects, among which: shaft sinking, secant piles boring, traditional shoring (timbering), underwater concreting, deep well dewatering, shotcrete, sheet piling, diamond wire concrete cutting, and other civil works.

2 PROJECTS PRESENTATION

2.1 *Crossings on SCPx*

2.1.1 *Kura East River Crossing (Georgia)*
Kura East Microtunnel is a river crossing of the Kura river situated near Rustavi in Georgia. The scope of work included the reparation and deepening of a segmental shaft built and abandoned 10 years before, the construction of the reception pit, the construction of microtunnel, installation of the 605m pipeline in one string, pipeline riser installation, the final pipeline grouting and backfilling of the shaft.

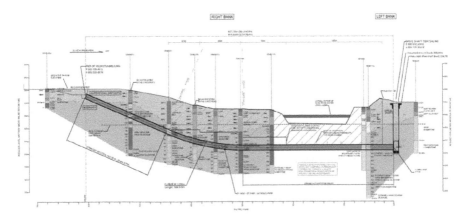

Figure 2. Kura East river crossing route.

477

The microtunnel is 605 meters: 302 meters of horizontal section, followed (in the direction of drilling) by 150 meters of curve at 1700m radius and then a straight slope at 9% for the last 153 meters.

The microtunnel is layed in the bedrock layer consisting of an alternance of extremily weal to strong mudstone, siltstone, and sandstone. The microtunnel route has been chosen such as to limit the probability to cross the superficial deposit of the Kura River, which depth was unknow under the river.

2.1.2 *Kura West River Crossing (Azerbaijan)*

Kura West Microtunnel is a river crossing of the Kura river situated in the region of Agstafa in Azerbaijan. The scope of work included the construction of a secant pile shaft, the construction of microtunnel, the construction of the reception pit, installation of pipeline in two strings, pipeline riser installation, the final pipeline grouting and backfilling of the shaft.

The microtunnel is 1010 meters: 490 meters of straight section with a 0,6% slope, followed (in the direction of drilling) by 10 meters of curve at 1700m radius and a horizontal section of 219m, then by 148 meters of curve at 1700m radius and a straight slope at 9% for the last 143 meters.

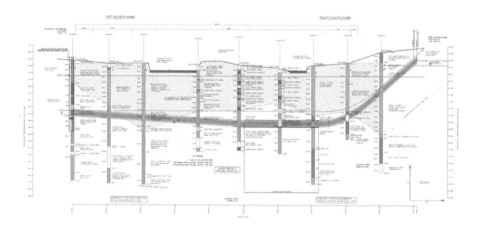

Figure 3. Kura West river crossing route.

Along its path, the microtunnel crosses different geological layers: Clay (25%), Gravels and Sand (45%), and Gravels and Cobbles (30%).

The main challenge of this microtunnel was the quantity of cobbles encountered on the total length of the drive combined with the difficulty to access the cutting chamber to inspect and change the cutting tools if necessary (high permeabilities).

2.2 *Crossings on the TAP*

2.2.1 *East Corovode Microtunnel (Albania)*

East Corovode Microtunnel is a 560meters straight microtunnel with a 2% slope to allow pipeline installation under a landslide, for which the traditional open-trench methodology to lay down the pipeline presented high risks. The scope on this project includes the drive shaft and reception shaft conception and construction, tunnel design and construction, pipeline insertion and grouting.

The geology encountered is Flysch bedrock (alternance of sandstone and siltstone layers) from very weathered to sound bedrock with Uncompressive Strength up to 100MPa.

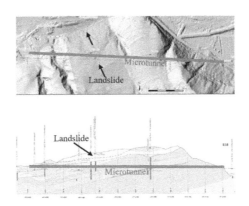

Figure 4. East Corovode Microtunnel project.

2.2.2 *West Corovode Tunnel (Albania)*

West Corovode Microtunnel is a 600meters straight microtunnel to allow pipeline installation under steep mountain slope and avoid the destruction of a newly built road to Corovode. The scope on this project includes the drive shaft and reception shaft conception and construction, tunnel design and construction, pipeline insertion and grouting.

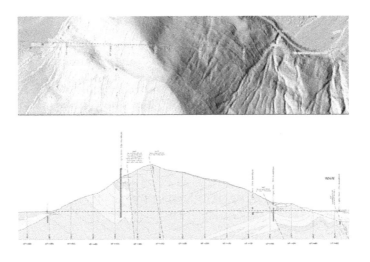

Figure 5. West Corovode Microtunnel project.

The geology encountered is Flysch bedrock (alternance of sandstone and siltstone layers) from very weathered to sound bedrock with Unconfined Compressive Strength up to 100MPa.

2.2.3 *KP139 River Crossing (Albania)*

Initially planned as open cut crossing, the construction methodology of this crossing has been changed to microtunnelling to limit the environmental impact and to get rid of the limited time periods during which open-cut crossings were feasible (4 months throughout the year).

KP139 crossing is located near the locality of Berat, Albania. It is a 420 meters' river crossing going through alluvial deposit of Osumi river. The drive consists in a straight and horizontal tunnel. The drive begins from a 14meters deep segmental shaft, the first 20 meters of tunnel are drilled in the Flysch bedrock, then the drive goes through in the alluviums of Osumi river for 400meters. Break-out is performed in a sheet pile shaft on the other bank of the river.

479

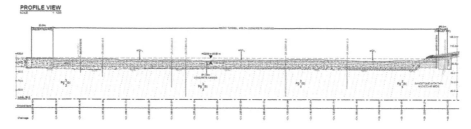

Figure 6a. KP139 river crossing route.

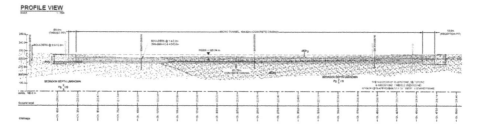

Figure 6b. KP114 river crossing route.

2.2.4 *KP114 River Crossing (Albania)*

As for KP139 crossing, this crossing was initially planned as open cut crossing, the construction methodology of this crossing has been changed to microtunnelling to limit the environmental impact and to get rid of the limited time periods during which open-cut crossings were feasible (4 months throughout the year).

KP114 crossing is located near the locality of Corovode, Albania. It is a 435 meters river crossing going through alluvial deposit of Osumi river. The drive consists in a straight and horizontal tunnel. The drive begins from a 12 meters deep secant piles shaft, the first 100 meters of tunnel are drilled in a alluvial fan containing large boulders, then the drive goes through in the alluviums of Osumi river for 325meters. Break-out is performed in a segmental shaft on the other bank of the river.

3 CHALLENGING PROJECTS FROM ENGINEERING PHASE. . .

3.1 *Shafts design*

The shafts do not play a structural role in the permanent work, nevertheless, their design should allow to perform the successive phases of the work. The main challenges to be considered from the design phase to perform a rational design are:

- Geometrical requirements and load cases during shaft construction and microtunnel construction
- Geometrical requirements for pipeline string and riser installation
- Structural stability during shaft opening (window cutting for riser installation)

For instance, on Kura East river crossing (Georgia), the segmental shaft already in place did not present sufficient strength to be cut opened without any additional measures. Thus, a special reinforcement concrete structure has been designed at the top of the segmental shaft to ensure the stability of the shaft during the opening of the shaft window to allow the installation of the pipeline riser section.

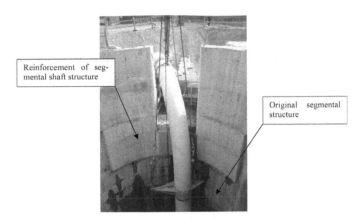

Figure 7. Reinforcement of segmental shaft structure.

3.2 *Microtunneling engineering*

3.2.1 *Microtunnel alignment choice*

Although tunnel route design is generally chosen at FEED phase by the client, it remains a sensitive point to be looked at during detailed design phase, it should be reviewed such as:

– To limit the depth of the shafts (to simplify further pipeline installation)
– Allow MTBM to pass through most desirable geological layers for tunneling
– Allow pipeline installation according its mechanical characteristics (diameter, maximal allowable pipeline radius,...)

On Kura West river crossing the FEED route proposed was found not to be optimal. Indeed, the initial route had long lengths where it was possible to encounter cobbles. To limit this risk, we proposed a variation, accepted by the client, allowing to reduce the lengths where the excavation should have been performed through superficial deposits (containing cobbles and being highly permeable).

3.2.2 *Choice of equipment*

Considering the long length of the microtunnel in SCPx and the difficult geological conditions, we have anticipated potential problems that could occur and jeopardize the success of the project:

– High wear of cutting tool, cutting wheel & crusher cone with difficulties to perform maintenance in highly permeable grounds
– Possibility to encounter boulders
– Risks of blockage of the cutting head

To reduce these anticipated risks, some modifications have been performed on the MTBM equipment:

– The cone crusher have been modified to increase the crushing gap, thus allowing the crusher to work less and decreasing risks of blockage of the cutter head
– Slurry lines on the return have been changed from 6″ to 8″ as a result of the crushing cone modification
– A secondary hydraulic crusher has been installed in the tunnel in case necessity
– A cutter head has been specially design for this project (hardox coating, openings more important than in the original rock cutter head, crushing gap modification)

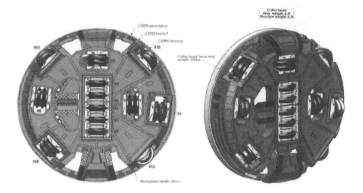

Figure 8.   New cutter head design for Kura West project by our Design Office.

3.3   *Pipeline installation design*

3.3.1   *Pipeline string installation*

The design of the pipeline string installation is a challenging engineering work that includes calculations at different stage of the project:

– Design and calculation of the external and internal tunnel roller supports that should be designed to be harmless to the pipeline coating
– Calculation of the pipeline stresses and deformations during all the phases of the project, including insertion, hydrotest, grouting and operation phases, these calculations should be performed according to the standards of the project
– Calculation of temporary structures used for this operation, among which the winch holding structure, pulling head, holdback head, pipeline braking collar, anchoring...

3.3.2   *Pipeline riser*

The pipeline risers are the part of the pipeline that is used to raise the pipeline level from the bottom of the shaft up to the normal installation depth in surface. The geometry of the pipeline risers composed by two elbows, their dimensions (more than 15*15meters) and weights (20tons) require to carefully prepare the riser installation operation.

Figure 9.   Pipeline riser installation sketch.

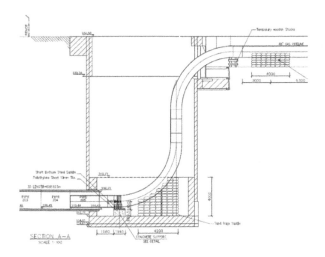

Figure 10. Riser lifting collar design and load test.

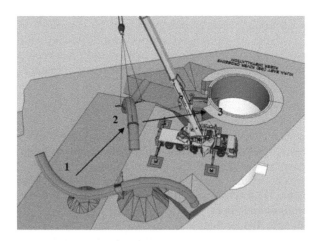

Figure 11. 3D modelling of riser installation.

We developed a specific lifting collar to perform this complicated lifting operation within the strict safety requirements of SCPx project.

In addition, as the last step of preparation, the operation has been modelized in 3D.

4 BUT ALSO DURING CONSTRUCTION

Construction phase often presents unforeseen conditions or deviations from what have been planned during the engineering phase, these news conditions can be initiated by the client, the site conditions, third parties, laws or other factors. In all the cases, the success of the job lies in the ability for the Contractor to adapt quickly to these new hypotheses and all optimize processes.

4.1 *Guidance*

Deviations of the MTBM from its centerline can be of various reasons, these projects with long drive lengths and changes of geologies are at risk. Indeed, for drives longer than 300/350meters the laser system should be replaced by other systems to guarantee correct precision (for instance HWL and gyrocompass), these systems are a bit less accurate and less robust in terms of process. Moreover, changes of geologies, mixed face excavation, or orientation of geological layers can have an impact.

The final objective of the tunnel is the installation of the pipeline section through it. The pipeline is made of thick metallic pipes welded one to another. The resistive stiffness of the pipeline string to bending is extremely important on small intervals, but this stiffness and the pipeline own weight allows bending on longer intervals (for a 48″ pipe bending becomes noticeable on lengths over 30/40 meters depending on thickness).

This behavior has an impact on how the guidance of the MTBM should be managed, or more accurately how the deviations to the project centerline should be managed. Indeed, if the difference of diameter between the pipeline and the tunnel allows some deviations, for the client the absolute position of the pipeline is not as important as the insertion itself. Thus, deviations should be corrected slowly avoiding as much as possible "bumps" or quick changes of direction in the tunnel.

Guidance issues being identified as a critical for these projects, we have put in place a process and control measures to master as much as possible the guidance:

– Identification of at-risk areas for guidance (discontinuities)
– An experienced in-house surveyor has been deployed on the projects
– Tests and calibration of each device used for guidance
– Trigger values of deviations with corresponding actions for MTBM operators
– Position survey each 30/50 meters
– In case of serious variation, a new alignment should be computed in agreement with the pipeline design office to ensure pipeline installation can be performed successfully
– Before pipeline installation an iterative pipeline calculation has been performed to guarantee that the pipeline can be installed within the technical requirements

4.2 Geology

Microtunnels for pipelines installations are used in challenging geological areas where standard installation methodologies cannot be used. Microtunnels are generally used to cross rivers as they allow to perform the works with the minimal disturbance to the environment and to get rid of limitations such as water level of the rivers (and thus the season). But, as performed in Corovode in Albania, Microtunnel can also be used to perform crossings under landslides or to cross too steep mountains slopes.

As for all microtunnels, the geology influences a wide range of choices made from the design phase, but when the time of construction arrives, unforeseen conditions are legion and adaptions should be made.

On KP139 and KP114 river crossings the foreseen geology was mainly composed of alluviums with approx. 40% of gravels 40% of sand 10% of silts and 10% of clay. The first unforeseen conditions appeared at the excavation of the reception shaft where a boulder layer have been found at the level of the tunnel alignment. Second has been the quantity of boulders found in the alluvial fan of KP114. Following this discovery, it has been decided to:

– Mobilized a rock cutter head
– Use triple disc cutters with carbide reinforced bits
– Mobilize equipment for hyperbaric operations
– Plan a mitigation plan in case of difficulties with boulders (removal from surface)

Moreover, on all river crossings, rapid changes in alluviums composition forced our team to adapt the desanding unit parameters and slurry characteristics to optimize the process. Indeed, the ground conditions could change from 100% clay/silts to 100% gravels within few centimeters. This has been managed as follow:

– For intervals corresponding to the foreseen geology the excavation fluid stayed with normal parameters (around 70sec of Marsh viscosity)

Figure 12. Boulders found in reception shaft of KP139 crossing and in KP114 drive shaft at level of MTBM.

- For intervals of clay, the excavation fluid was made lighter (even simply water) to allow better advance speeds
- For intervals composed mainly of gravels and cobbles, a thicker fluid have been used (marsh viscosity over a few minutes) to be able to confine the front and not to block the cutter head
- As these intervals could change several times a day, additional buffer tanks of slurry have been used to be able to change quickly the excavation fluid from one to another
- At last the desanding have been adapted to the new conditions of viscosity, hydrocyclones output diameters modified for high density slurry

4.3 Pipeline installations

The installation of large and long pipelines into tunnels is as well a challenging activity that is performed in common with the Pipeline Contractor teams. The main idea of the pipeline section installation consists in pulling the pipeline string from the opposite side of the tunnel on rollers supports by using a powerful winch system to pull a cable and the pipeline attached to it.

Figure 13. Pipeline installation works, SCPx, Georgia.

Two main issues can appear during the pipeline installation:

- The pipeline can begin a rotation along its own axis
- If the installation is made in a slope, the pipeline should be holdback

The rotation of the pipeline along its own axis has not an important influence on the surface as the pipeline slides on static rubber rollers. At the entrance of the tunnel, specific collars equipped with rollers are installed on the pipeline. It is thus important to prevent any important rotation to ensure that the roller collars. To this purpose, a sideboom was using a sling to apply a torque opposite to the actual sense of rotation.

485

Figure 14. Anti-rotation of pipeline during pulling.

In Azerbaijan and in Georgia the pipeline has been installed in a slope of 9°. Due to this important slope the risk of the pipeline sliding by itself in the tunnel has been considered. In order to mitigate this risk, a holdback solution has been developed by using 2 heavy machinery. In addition, during nights and weekend, a brake was installed at the entrance of the tunnel to prevent any movement of the pipeline.

Figure 15. Holdback at the pipeline section rear (left) and pipeline brake installed at tunnel entrance (right).

5 CONCLUSION

Microtunnels for pipeline sections installation are challenging projects which includes a wide variety of works and require extensive attention and experience during both engineering and construction phase.

Most of these projects are performed in remote areas where other construction methodologies cannot be foreseen. Microtunneling remains an effective solution for all types of crossings that can be met in pipeline project.

BESSAC by its international positioning, its long experience, its engineering offices and its scientific approach to all technical issues is able to complete successfully all the challenges met in this type of projects.

*Tunnels and Underground Cities: Engineering and Innovation meet Archaeology,
Architecture and Art, Volume 2: Environment sustainability in
underground construction – Peila, Viggiani & Celestino (Eds)
© 2020 Taylor & Francis Group, London, ISBN 978-0-367-46579-7*

Changes in tunnelling in the Big Data Era – The new Milan-Genoa high speed/high capacity

G. Morandini, A. Marcenaro, H.M. Dahir, E.G. Caci, A. Ferrari & G. Petito
Italferr SpA, Roma, Italy

ABSTRACT: Today, more than ever before, the use but, above all, the capacity to manage high volumes of data measured and collected during the implementation of large infrastructure projects, has become decisive. Data analysis is becoming bigger and bigger: to the more traditional production and tunnelling progress data, we must now add environmental monitoring data, such as the concentration of airborne asbestos fibres, and the data characterising spoil materials, geo-mechanical monitoring data, aboveground civil structure monitoring data. The paper investigates the issues, with reference to the High Speed and High Capacity Terzo Valico dei Giovi Line, connecting Genoa and Milan, 53 km long, of which 37 km running underground, in two tunnels excavated according to both the conventional and mechanised models. There is a special focus on the management of the soil and rocks excavated by means of TBM-EPB (closed shield), produced during the construction of the Polcevera lateral access tunnel, preliminary to the excavation of the railway tunnels. The results of the analyses provide useful information for both design purposes and for guiding the choice of the most suitable solutions to be adopted.

1 INTRODUCTION

1.1 *General description of the infrastructure*

The Terzo Valico project concerns the construction of a new rail infrastructure connecting the port of Genoa and the northern Italian rail network, and represents the Mediterranean terminus of the Rhine-Alpine Corridor, one of the main corridors of the strategic trans-European transport network.

Its construction is part of the EU strategy for integrating the European transport networks (TEN-T) and for strengthening the role of rail transport in the development of freight traffic between the large ports of northern Europe and the Mediterranean, and between western and eastern Europe.

The new "Terzo Valico dei Giovi" line is 53 km long, overall, with 37 km of the line running in tunnels, and affects 12 local authorities in the provinces of Genoa and Alessandria and the regions of Liguria and Piemonte. The interconnection points of Voltri and Fegino Junction will be connected, southwards, with the rail systems of the Genoa Node – regarding which important functional upgrading and strengthening works are being undertaken – as well as with the port facilities of Voltri and the city's historic centre.

To the north, from Novi Ligure, the line joins the existing Genoa – Turin line (in the case of traffic flows to Turin and Novara – Sempione) and the Tortona – Piacenza line (for traffic directed to Milan San Gottardo).

2 TUNNEL DATA MANAGEMENT

2.1 *Tunnel monitoring – IT data management platform*

The purpose of monitoring is to verify the validity of design forecasts and to adapt the work in progress, based on a systematic comparison, during construction, of the forecasts and the performance/behaviour of the soil and rock formations concerned by the tunnelling and of the tunnel lining structures. In particular, in the case of underground works, monitoring plays a fundamental role for guiding and overseeing the project decisions in the construction stage, in relation to the variability contained in the Guidelines for the application of the typical sections.

The monitoring process features three main types of activities:

– geotechnical/structural instrumental monitoring; (inclinometers; load cells; strain gauge bars)
– topographical monitoring; (convergence; surface levelling)
– geo-mechanical monitoring. (tunnelling face surveys)

In order to manage this immense amount of data, Cociv and Italferr have jointly prepared a new monitoring platform, comprising different access policies, with dedicated areas for each Contractor, with regard to the relevant WBS, areas dedicated to Cociv for the WBS carried out directly, and which enables Italferr, in relation to its Supervision of Works and High Surveillance functions, to display all the documents approved by the Cociv SPT. Furthermore, the system can highlight the achievement of the threshold values, which are timeously signalled in red. Exploring the item, the comment is displayed showing the value and the related threshold achieved. The software also sends an email to a predefined list of users, containing all the details and comments on the exceedance of the thresholds that need to be kept under control.

Staff members carry out the monitoring by means of daily site inspections (geo-mechanical), while others analyse the data in the back-office; the latter compare the monitoring data with the Execution Design data, looking into them based on the results of the inspections on the ground and validating the data from an engineering point of view, making it available on the monitoring data IT platform within 24 hours from their measurement in the field.

2.1.1 *Conventional tunnelling method*

Conventional tunnelling (drill&blast and/or mechanical systems) is the most used tunnelling method in the case of the Terzo Valico tunnels (about 60%).

The measurement of the convergence of the tunnel walls, the surveys carried out at the tunnel face and the extrusiometric measurements of the tunnel face are the principal elements that are analysed to assess compliance with the design forecasts during the tunnelling stage. An automated navigation system is used for the Valico tunnelling, for accurately surveying the progress of the tunnel and ensuring compliance with the design tunnelling profile, thus accelerating the tunnelling process despite the limited visibility due to the presence of dust. The system is therefore capable of operating independently, which improves the safety of the workers, whose presence is not required in the proximity of the tunnel face.

The aim is to maximise and speed up the tunnelling progress, consistently with the design profile, in all conditions, calculating the tunnelling data for analysis purposes to ensure that

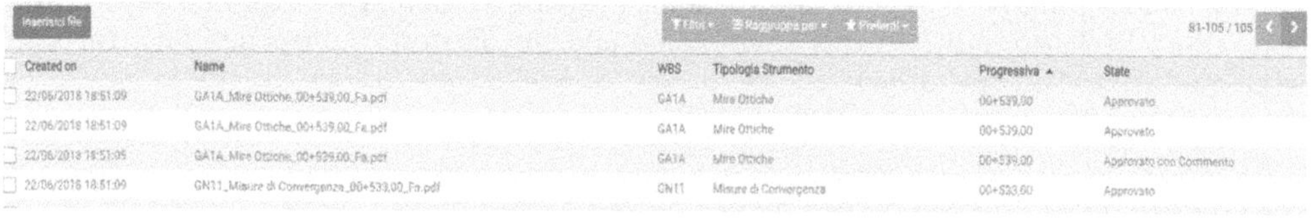

Figure 1.   Platform for managing the geomechanical monitoring data.

the tunnelling effectively takes place according to the design specifications, and to avoid wasting time with re-processing activities.

Thanks to this assistance, the workers can excavate the tunnel with an accurate profile, which means time and cost savings, in terms of the concrete used to line the walls, for example.

The data is transmitted by radio signal, the sensors measure the position and movement of the cutting arm, relative to the machine, using the WiFi interface; the data is then transferred to the site office for assessment and filing.

2.1.2 *Mechanised tunnelling method: machine data management*

The Serravalle Tunnel, the terminal section of the Valico Tunnel and the Polcevera lateral access tunnel to the Valico Tunnel have all been excavated using a tunnel-boring machine (TBM).

During the progress of a tunnel excavated with a TBM, it is essential to continuously monitor the process in every stage, high-sensibility multi-sensor networks are capable of providing detailed data in real time. This data is then used to automate the processes and minimise risks, and the combination with appropriate communication and tracking systems ensures that all the workers are constantly networked, can be located at all times and alerted during both their routine daily activities and in emergency conditions. It is possible to provide a radio cover without solution of continuity, as well as telephone connections, video-surveillance, access control and person/vehicle location, emergency alarm and fire protection. The data collected during the tunnelling process are stored in a database to be used for studying the work progress.

Furthermore, the machine components are also monitored, in order to programme maintenance activities and thus minimise machine dead times, thus also cutting down costs.

2.2 *Monitoring aboveground civil structures*

In areas with low coverage and the presence of aboveground civil structures, it is essential to ensure attentive and customised monitoring, capable of ensuring high levels of safety. The objective is to identify the risks to buildings in built-up and residential areas, in the initial stage, to prevent and accidents and damage. A monitoring system will be used along the entire Terzo Valico line for monitoring the retail area of the Factory Outlet in Serravalle Scrivia (AL), which is characterised by the presence of above- and underground structures and infrastructures.

The interferences between the tunnelling and all pre-existing structures and utilities are analysed by means of continuous measuring systems, through the installation of monitoring instruments for assessing any geometric displacements and the developments in the existing fractures. When a design threshold is exceeded, an alert is automatically sent out by email or SMS text message, to ensure that the appropriate measures are immediately taken.

In modern-day geosensor networks the sensors are capable of automatically collecting and transmitting the values measured, filing them in a central database. This monitoring process is completely automatic, with frequent measurements, providing accurate and punctual data relating to the entire process and, therefore, forecasting developments, and any discrepancies compared to the design values, in useful time.

The data analyses and quality control of the measured values are carried out automatically. The automatic reporting process includes the production of tables and graphs, as well as design drawings and survey diagrams. The data may be constantly viewed, worldwide, via the web platform, by means of password-controlled access, with the possibility of differentiating the access policies by type of user.

3 ENVIRONMENTAL DATA MANAGEMENT

3.1 *Environmental monitoring*

In railway projects, the attention towards the environment translates into constant environmental monitoring, aimed at verifying the actual occurrence of the forecasted impacts and the effectiveness of the environmental mitigation measures put into place, in order to survey and promptly manage any environmental emergencies.

In the case of the construction site of the Terzo Valico dei Giovi railway line, these monitoring activities are the responsibility of the General Contractor, which measures various environmental components, in accordance with the Environmental Monitoring Plan approved by the Entities. Specifically, the significant characteristic components of the project environment are the atmosphere, surface waters, underground water, soil, vegetation and flora, fauna and ecosystems, noise, vibrations, landscape, physical state of the places and social environment. The General Contractor appropriately validates and enters the results of the monitoring campaigns into a GIS database (called SIGMAP and developed by Italferr), for recording all the monitoring figures for the relevant environmental components, and promptly notifying the data to the competent Entities. In such a complex context as the Terzo Valico construction site, it is obvious how the environmental communication policies represent an instrument for providing a complete and transparent picture of the site activities, and for constantly informing the local communities about the state of environmental quality of the areas concerned by the construction activities, the project-related environmental mitigation measures, and the environmental monitoring activities carried out before, during and after the construction.

The particularly sensitive environmental issues related to the Terzo Valico site undoubtedly include the management of airborne asbestos, which has required the definition of specific procedures for making sure that the tunnelling work is carried out in the highest possible safety conditions, and preventing any risks for the workers and the people living in the proximity of the sites. Such procedures define the monitoring and sampling activities, and their frequency for the different tunnelling methods, the alert levels in the case of hazard, the monitoring points of the airborne asbestos identified in the areas subject to exposure and based on the nature of the places.

The adoption of the "Asbestos Risk Management Protocol", in relation to the tunnelling activeities for the Terzo Valico project, ensures the proper management of the relevant "Asbestos Risk", providing the persons concerned with the tools for carrying out their work in the best and safest way, while at the same time enabling the supervisory bodies, the ARPA environmental protection agencies and the local communities to understand the findings and the manner of progress of the construction work.

3.2 *Monitoring with SIGMAP – Geographical Information System for Monitoring the Environment and Projects*

In order to provide a clear, complete and transparent picture of the environmental activities carried out, for the purpose of more effectively managing the monitoring data and speedily consulting and retrieving all the available information, in relation to the specific project works, Italferr has designed a web portal, which has been made available to the GC and its suppliers, so that they may provide their effective contribution. Understanding how the project is perceived by the people living in the affected areas is essential to direct the most expedient stakeholder communication and involvement actions, according to a fruitful participation approach and for the purpose of effectively sharing the decisions taken. Italferr's environmental database is called SIGMAP and, based on a GIS web portal, it enables the centralisation, filing, analysis and downloading of both the local geographical and the mapping data, allowing the consultation of thematic maps, in particular for Design, Monitoring and Remediation.

Suitably reorganised, the data is made available to the public through specific websites, which illustrate the state of environmental quality of the project area, the monitoring results prior to the implementation of the project/during the construction work/after completion of the works, and the related environmental mitigation and compensatory works.

Thanks to web platforms such as SIGMAP, the collected data may be mined for general purposes, instead of being restricted to a single project, providing the basis for a reliable analysis.

Figure 2. Italferr's SIGMAP portal.

3.3 *The environmental management of conditioned tunnelling spoil materials*

Environmentally speaking, the applicable regulations provide for the use of specific "additives for mechanised tunnelling", for the management of spoil materials as a "by-product", although the regulations do not specify a concentration limit for foam agents.

In consideration of the above, ad hoc studies and research work has been carried out, in relation to the Terzo Valico project work, in relation to the environmental compatibility and the possible hazards for the environment associated with the use of conditioning additives for tunnelling with TBM/EPB. After attentive laboratory tests, the "maximum pollutant concentration limits" have been defined, in relation to the project, for three different brands of conditioning agents used in connection with the tunnelling work. The ecotoxicological research has defined, for the tested foam agents, the relevant Reference Thresholds (RT), below which it has been scientifically proven that the conditioned spoil materials are not ecotoxic.

As provided in the approved Utilisation Plan, before being transferred to the final disposal sites, the conditioned spoil is treated – according to normal industrial practice – by being spread out on the ground to dry, undergoing a maturation process to ensure that it attains the best handling characteristics and the best humidity levels, to foster the natural biodegradation of the additives used during the tunnelling. Only at the end of the production cycle, downstream from the treatment according to normal industrial practice, can the spoil materials be correctly classified as a by-product, subject to conformity checks, with regard to the final disposal site, by means of environmental characterisation by the producer, aimed at verifying the concentration of the parameters provided by the environmental regulations and the Reference Thresholds (RT), as defined in relation to the above mentioned studies.

Therefore, the conformity of the spoil is checked at the intermediate disposal site. The checks for tax purposes are then carried out on the spoil materials verified as conforming, by the control authorities, from the moment they leave the intermediate disposal site until they reach the final disposal site.

The entire spoil management and disposal process, to the intermediate and final disposal sites, is subject to checks by the competent control authorities and to sample checks by Italferr's Supervision of Works team.

3.4 *Monitoring Spoil Soils and Rocks – Terresca WEB*

The entire spoil soils/rocks management process – from the production site to the final disposal site – in relation to the spoil materials produced at the Terzo Valico de Giovi sites, is managed through a web platform called Terresc@. This tool, used for extracting customised reports for analysing the data relating to spoil materials managed as by-products of the tunnelling, has been made available by the Ministry of the Environment and the competent authorities, to ensure the transparency of the processes and of the management of all the spoil materials.

The platform's database contains the principal information relating to the balance of spoil soils and rocks (surveys at the tunnel face, transportation logbooks, transportation documents, transport communications adjusted pursuant to D.M 161/2012 and analytical characterisations of the materials).

Besides the database, the system also provides geo-referenced maps showing the location of the production and storage sites relating to the Terzo Valico dei Giovi project, and by interacting with the maps users can retrieve information about the excavated and transported amount of spoil materials, in the case of production sites, or the amount of spoil stored at the storage sites.

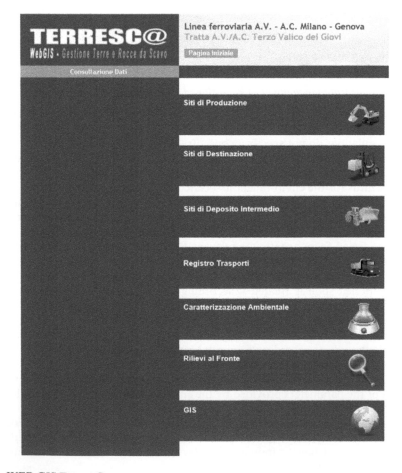

Figure 3.   WEB GIS Terresc@.

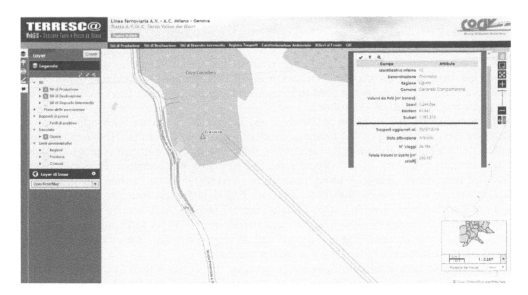

Figure 4.   WEB GIS sito di produzione.

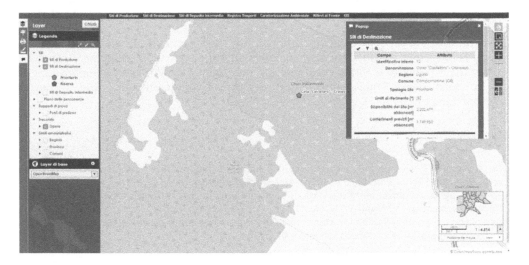

Figure 5.   WEB GIS storage site.

4   CONCLUSIONS

4.1   *The role of Italferr*

For the management and control of the complex site for the Terzo Valico project, Italferr has deployed, on the ground, besides the Supervision of Works team, other specialised engineering units for the tunnelling operations and environmental monitoring and control.

The staff of the Field Operations, Safety and Commissioning unit carries out control activities on the ground, on a daily basis, in support of the Supervision of jointly Works, verifying the monitoring activities carried out by the GC and, at the same time, analysing the data recorded in the web platforms, checking the accuracy of the data and compliance with the threshold values.

493

4.2 *The key elements of successful management*

Big data is like large quantities of bricks or cement stored in a warehouse; taken individually, they are insufficient for building anything. What makes the difference is what can be done with them once one has mastered the software for analysing the big data. The constant growth of this data increases the need to streamline them, so that they can be efficiently used and speedily and effectively consulted.

The construction sector, given its inherent characteristics, is slower than other sectors in analysing data, compared to the financial or retail sectors, for example. However, managers could rapidly obtain a summary of the quality, safety, workforce and equipment data, which could streamline the task of identifying risks and assessing performance. Problems may be identified before they emerge, leading to significant savings, in terms of time and costs, in the implementation of corrective measures. By recording data in real time, and then using the data for modelling the possible implications on current projects, it would be possible to make more accurate assessments. The analysis can provide more useful information on the risk levels, before a certain threshold is exceeded and an alert generated, it may also offer further information, which conventional systems cannot guarantee.

In the case of such large and complex projects as the Terzo Valico line – which require many years for construction – the local and nationwide economic and political activities may significantly affect the construction time and costs; therefore, the next challenge is building the capacity to take these aspects into account as well.

Big Data have the potential to provide solutions to all these problems, if the trend of a continuous collaboration between the construction sector and technological development continues.

Over the next five years, Big Data is set to radically transform both the construction process and the construction contract business. A bright future awaits those companies capable of building the capacity to embrace data analyses and the new technologies for innovating the way we build infrastrutures, and Italferr has already unquestionably and fortunately embarked upon this path.

REFERENCES

ARPAL Agenzia Regionale per la Protezione dell'Ambiente Ligure. 2016, *Technical-scientific support activities at the "Terzo Vali dei Giovi" Environmental Observatory*.

Ministry of the Environment and Protection of Land and Sea, Technical Commission for Environmental Impact Assessment – VIA and VAS, *Plan of use of excavation materials*.

Geographical Information System for Monitoring the Environment and Projects; https://sigmap.italferr.it.

WebGis – Management of excavated soil and rocks, *Terresc@*

Tunnels and Underground Cities: Engineering and Innovation meet Archaeology,
Architecture and Art, Volume 2: Environment sustainability in
underground construction – Peila, Viggiani & Celestino (Eds)
© 2020 Taylor & Francis Group, London, ISBN 978-0-367-46579-7

Autarkic aggregate supply with recycled tunnel spoil at the Brenner Base Tunnel

R. Murr, T. Cordes, M. Hofmann & K. Bergmeister
Brenner Base Tunnel BBT-SE, Innsbruck, Austria

ABSTRACT: In the execution of the lot E52 of the *Brenner Base Tunnel*, environmental challenges were mastered for reduceing the environmental impact and sparing resources. According to the geological preliminary investigations, the *Bündner Schist* which formerly should be deposited was released for recycling on the basis of intensive geological and engineering investigations. After beginning the construction phase, the project partners planned cooperatively to implement the recycling of tunnel spoil not previously used in Austria. Within a few months, the necessary steps of sampling, trial production, laboratory tests for decisions and technical development were carried out. The complexity of this topic is presented and the measures for quality control and adaption of process cycles are displayed. With the implementation of tunnel spoil recycling at *"Erkundungslos Padastertal + Wolf II"*, BBT-SE mastered important steps for planning and concrete execution of further construction lots as well as an innovative path for the construction industry.

1 INTRODUCTION

The *Brenner Base Tunnel* is the key structure of the trans-European North-South link between *Helsinki* and *Valletta (Malta)*. The construction of the North-South corridor is part of the *TEN Strategic Plan* to cope with the increasing movement of goods and people. The strategic plan gives priority to the development of the rail infrastructure, the *Brenner Base Tunnel* is subject to the highest funding from the *European Union* due to its importance.

Lot *Wolf II* is part of the *Brenner Base Tunnel* and essentially comprises the following construction measures:

- 4 000 m access tunnel, excavation area 120 m², via which the central construction site logistics for the subsequent main construction lot must be ensured
- Connecting cavern, branch cavern, fan cavern, excavation areas 230–350 m²
- 950 m spoil removal tunnel, excavation area approx. 60 m²
- 180 m cross-connection tunnel from access tunnel to exploration tunnel
- 1 140 m diversion tunnel for *Padasterbach*, excavation area 26 m²
- 360 m diversion tunnel for *Padasterbach* in open construction method
- 145 m transverse drainage tunnel, excavation area 10 m²
- 700 m inner lining *Padastertunnel*
- 1 000 m inner lining *Saxenertunnel*, connection tunnel to motorway A13
- Hydraulic structures (bed load barriers, intake structures, fish ponds etc.) in the *Padastertal*
- Landfill construction of the landfill *Padastertal*

All tunnels of this lot were excavated by blasting. The contract value of the construction lot was approx. 104 million euros, construction began on December 4th 2013.

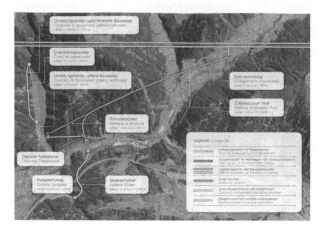

Figure 1.   Lot Wolf II, E52 (picture: BBT SE).

2   RECYCLING OF TUNNEL SPOIL FOR AGGREGATE PRODUCTION

On the basis of the Waste Management Law (*Abfallwirtschaftsgesetz - AWG*), with the target of "recycling of waste" (see AWG §1 and §8 (3)), the Brenner Base Tunnel BBT SE commissioned studies involving the *University BOKU Vienna* and external experts to examine the main lithologies of the Austrian project area (*Innsbruck Quartzphyllite* and *Bündner Schist*) for the possibility of processing aggregates for concrete and filter material. Since no mining operator in Austria used *Bündner Schist* for concrete at that time, its suitability as aggregate was unclear. Investigations for *Bündner Schist* (see e.g. (Voit, K 2013)) have shown that Schist prepared by crushing, screening and washing is suitable for the aggregate production for shotcrete as well as for structural concretes. In contrast, the suitability of the *Innsbruck Quartzphyllite*, which is to be found in the northern project area, as aggregate for structural concretes could not be proven.

In the course of the construction of the lot *Wolf II*, innovative steps were taken to optimise material logistics and valuable experience was gained in the use of prepared tunnel spoil as aggregate for concrete production. Mainly shotcrete (almost 100 000 m³) was produced out of tunnel spoil.

Processing and recycling of tunnel spoil brought the following positive aspects for the project:

- Use of processed tunnel spoil and thus cost reduction for inert materials to be purchased externally
- Reduction of landfill volume
- Avoid any transport of inert materials outside of the site
- Compliance with *AWG* requirement that tunnel spoil is recycled
- Improvement of the life cycle assessment (e.g. CO2, SO2, non-renewable energy and mineral resources according to (*BBT*-intern, 2015)

3   RECYCLING CONCEPT

With regard to processing and production of the fractions, an optimal concept suitable for construction site had to be found.

The *Bündner Schist* is characterised above all by its good fissile strength along the shearing surfaces and its flat appearance. In order to produce suitable aggregates for concrete production, the individual crushing stages were therefore selected primarily to achieve the required grain shape.

For this purpose and to obtain fractions for working out mix designs two processing trials were carried out in December 2013 and March 2014 with the following plant constellation (see (Voit, K 2013)):

- Test series 1: Jaw crusher - impact mill
- Test series 2: Jaw crusher - Impact mill - Vertical mill

Test series 1 showed that the production of aggregates from *Bündner Schist* is basically possible, but that it is difficult to achieve the required grain shape. It was therefore decided to additionally use a third crushing stage in the form of a vertical mill. Vertical mills are mainly used to achieve as cubic a grain shape as possible and subsequently as low an LA value as possible.

Based on findings from the tests, the planning of the plant in *Wolf* was started in March 2014 and should meet the following requirements (see also (Bergmeister, K., Kogler, H., Murr, R., Cordes, T. & Arnold R., 2016)):

- low space requirement
- rapid availability
- flexible system constellation
- low treatment losses
- acceptable investment costs

3.1 Low Space Requirement

The complete planning and implementation took place parallel to the ongoing excavation. The available space in the *Padastertal* was too small to install all three crushing stages on the *BE Padastertal* site (see Figure 2). The pre-screening system was therefore installed in the fan cavern. A further aspect was the all-season availability and the largely independent operation of the pre-separation situated under the surface.

3.2 Rapid Availability

The period from the decision to implement to full commissioning was only four and a half months. By comparison, mining operations usually have a period of two years.

3.3 Flexible System Constellation

The plant constellation was chosen in such a way that it would also have been possible to relocate the plant locally, if necessary during construction operations. The choice of a mobile intermediate screen after the second crushing stage was made in order to enable the production of building materials for landfill construction (e.g. filter gravel) during downtimes of wet processing (e.g. in high winter) and thus loss of production of aggregates for concrete.

A decisive factor for the planning of the processing of the tunnel excavation was the conversion of the processing plant. Depending on the construction phases, large quantities of sand had to be produced for shotcrete, coarse grains for inner shell concrete and larger quantities for filter layers of the landfill. The processing concept therefore took into account a rapid

Figure 2.   Overview of material recycling plant in *Padastertal*.

changeover of the crushing stages in order to be able to react to changes in demand with a short-term back-crushing of surplus granulations.

3.4 Low Treatment Losses

The choice of the separation cut of the pre-screening is a compromise between the highest possible percentage of material recycling on the one hand and the elimination of the less suitable or unsuitable proportions such as layers rich in mica and graphite on the other. The separation cut at 32 mm proved to be advantageous. A similar behavior was found when the suitable grain shape was achieved, since the grain shape was improved by repeated crushing, but higher processing losses were generated. The achievement of a sufficient grain shape with as low a proportion of blowdown materials as possible was achieved during operation by regular adjustment of the speed of the impact mill and vertical mill.

3.5 Acceptable Investment Costs

The planned short service life required low investment costs. Both used equipment was purchased to ensure faster availability and, in the case of wet processing, new equipment to guarantee high quality. By producing in advance, a shutdown of the wet processing in high winter could be planned. Otherwise, considerably higher investments in the form of a winter-proof housing would have been necessary.

The following plant scheme has been provided for:

- Jaw crusher
- Pre-screening at 32 mm
- Impact mill
- Intermediate screening (for a possible production of filter gravel 16/32)
- Vertical mill
- Wet sieving 0/4, 4/8, 8/22 mm
- Rinsing sand 0/4 mm in bucket wheel

3.6 Construction Sequence

From the working face, the tunnel excavated material was transported to the fan cavern for coarse processing, crushed by jaw crushers and pre-screened at 32 mm. The further crushing process took place in the *Padastertal* processing area using impact mills and vertical mills to produce the coarse and fine fractions combined with an improvement in grain shape and LA value.

For shotcrete production with a sand content of approx. 70%, the surplus granulations were broken back both in the impact mill and in the vertical mill.

4 PREPARATION OF THE MIX DESIGN

For manufacturing the shotcrete lining, a shotcrete of the grade SpC25/30(120)/II/J2/XC4/GK8 was used in the *Wolf*, which was mixed at the *Wolf* construction site mixing plant. The concrete strength class was determined on the basis of optimisations based on the geology encountered and the deformations measured. With regard to durability, great importance was attached to a high structural density, which was intensified by testing the water penetration depth of drill cores from grouting boxes and the structure.

The formulation was developed in two parallel steps:

- Binder paste and mortar tests with different setting accelerators (EB) without reference to *Bündner Schist*
- Mix design of concrete, fresh concrete and gunning tests with prepared Bündner Schist and the binder combinations of the binder paste and mortar tests

4.1 Binderpaste- and Mortar Tests

In order to optimise the choice of setting accelerator and to improve the coordination between setting accelerator and binder combination, tests were carried out in the laboratory to determine the setting times, compressive strength development at a young age (3h to 24h) and the loss of strength at the age of 7 days compared to the unaccelerated mortar. The tests were carried out in accordance with the guideline on shotcrete (Austrian Society for Construction Technology öbv, 2009), the results of which are shown in Figures 3 to 5.

When determining the setting acceleration, it was found that the required setting acceleration was achieved with significantly different dosing heights for the various products. During the test, the mixing in of some combinations of accelerator and binder very quickly led to clumping and inhomogeneities in the accelerated binder paste due to the high reactivity. It was also shown that accelerator products which allowed high early strengths within the young age (tested up to 24h) showed a higher loss of strength compared to the non-accelerated zero mortar.

Based on this, a binder combination of a CEM I 52.5 N and a CEM II/A-M 42.5 N in a ratio of 50% to 50% and the addition of a hydraulically reactive additive type II (AHWZ) was selected in the application, which significantly reduced lump formation and the occurrence of

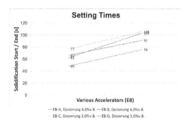

Figure 3. Determination of the setting acceleration of binder paste with different setting accelerators Products.

Figure 4. Determination of the compressive strength development of accelerated binder paste in cylindrical test specimens with different setting accelerators Products.

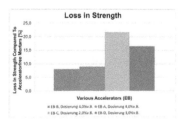

Figure 5. Determination of the loss of strength of accelerated mortar prisms with different accelerated setting products at the age of 7 days.

inhomogeneities. The drop in strength determined on the mortar prism at the age of 7 days was between 8% and 22%, depending on the product, compared with the unaccelerated zero mortar, see Figure 5.

4.2 *Concrete Mix Design, Fresh Concrete and Shotcrete Tests*

Based on a "standard" shotcrete formulation with 420 kg/m³ binder, the binder content was increased due to the higher water requirement and the influence of the phyllosilicates. At the same time, new admixtures had to be used to introduce fine air into the fresh concrete and to improve consistency. The graphite content in the *Bündner Schist* was very low in absolute terms - examinations of the crushed raw material revealed graphite contents of between 0.4% and 0.5% - but reduced the effectiveness of the air entraining agent to generate fine air in the fresh concrete. The fine air was used specifically to improve workability, reduce tackiness and increase concrete stability.

In shotcrete tests, the phenomenon known from earlier times clearly showed that at EB doses which ensured that the J2 curve was reached at a very young age (approx. 3 to 6 hours), a strength deficit compared to products with a low early strength performance was caused in the further course. This phenomenon could be controlled to some extent with different products and different dosages, but could not be completely eliminated. It was also shown that different EB products reacted differently sensitively to EB dosage changes with the binder combination used; i.e. for some products a small change in dosage showed a clear effect - mainly for EB products with strong strength development at an early age - some products behaved rather well.

These observations and the results of the binder paste and mortar tests led to an optimised concrete formulation and the use of the EB-C solidification accelerator, which largely met the requirements of driving and the construction site:

Table 1. The mix designs.

Binder kg/m³	Batch kg/m³	Aggregat kg/m³	Batch	Water batch	Additive
CEM I 52.5 R	210	Sand 0/4	1119	220	Superplasticizer
CEM II/A-M	210	Coarse 4/8	488		Consistency holder
(S-L) 42.5 N					
AHWZ[1]	50				Air-entraining agent

1 Admixture Type II, Chargeable Hydraulically Active Admixture (AHWZ)

5 EXCAVATION ACCOMPAINING TESTS

The shotcrete SpC25/30(120)/II/J2/XC4/GK8 was used in the period from December 2014 to April 2016 with a compacted monitoring to ensure the uniformity of the paving quality. The formulation of the sprayed concrete is listed in Table 1 (see also (*BBT*, 2016)). The evaluation of the installed shotcrete quality was carried out using the available test documents:

- Recipe check from 18th December 2014
- Conformity tests in the period between December 2014 and April 2016
- Identity tests between December 2014 and April 2016
- Component testing (drill core removal)
- Additional tests (Youngs-modulus test, frost resistance test XF3)

5.1 *Conformity Tests*

The formulation showed an increased water demand due to the processed sand fraction of 0/4 mm. The evaluation of almost 200 conformity testing showed a satisfactory, very homogeneous result for the shotcrete quality with slight fluctuations in air content and a clear drop in compressive strength to 76% on average from shotcrete to zero concrete. The early strength curve J2 showed a slight underrun in individual samples at an age of approx. 3–9 hours. A higher EB dosage leads to a reduction of the strength development at an age of >28 d.

The evaluation of the shotcrete test according to *ÖVBB guideline shotcrete* (öbv, 2009) and ÖNORM EN 13791 according to Table 2 led to a confirmation of the required compressive strength:

Table 2.   Evaluation of the tests according to (öbv, 2009) and ÖNORM EN 13791 (BBT, 2016)).

SpC25/30 Minimum Characteristic Compressive strength fck,is	Demanded (MPa)	Attained Conformity testing	Identity testing including
Drill Core (L/D=1/1)	(MPa)	(MPa)	Structural testing
	26		
From the mean value criterion Estimated characteristic Compressive strength fck,is		26.8	26.2
From the single value criterion Estimated characteristic compressive strength fck,is		29.5	30.0

5.2   *Identity Testing Including Structural Testing*

The evaluation of the ID tests also confirmed the requirements with a slightly higher compressive strength. Based on these results, a total of 35 drilled cores out of the construction were taken and additionally tested. Of these, 4 drill cores that were not intact had to be excluded from the tests. The measurement of the microstructure density on a total of 22 drill cores from the component confirmed compliance with the highest requirement XC4, the evaluation of the compressive strength as an average and the required shotcrete quality according to ÖNORM EN 13791, see Tables 2 and 3.

In addition, the development of compressive strength and Young's modulus was determined during the ID tests. The result in Figure 6 shows the development of a relatively low Young's modulus after 28 days. For comparison purposes, the temporal curves according to Model Code 2010 are shown. However, this property can basically be classified as advantageous for use in tunnel construction due to the resulting more flexible shell. The reduction of the modulus of elasticity results from the use of the prepared aggregates.

Table 3.   Test results of structural density and compressive strength.

Parameter (mm)	Structural density (MPa)	Compressive strength on drill cores
Number of testing results	22	31
Mean	29.0	29.7
Standard deviation	6.9	2.4

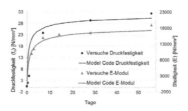

Figure 6.   Development of Young's modulus and compressive strength.

6 STUDIES TO OPTIMICE MIX DESIGN

Based on the experience gained, investigations were carried out for further recipe optimisation. Since it had to be established during the execution that the water demand of the processed aggregates was very high despite adjustment of the grading curve and optimised concrete admixture, the possibility of reducing the water demand by partial to complete replacement of the processed aggregate fraction 0/4 was investigated. In the investigations, the percentages of the tunnel excavation sand 0/4 were gradually replaced by a calcitic sand 0/4. If the sand is replaced, it is increasingly necessary to adjust the additive dosages or products for the individual mixtures (e.g. also the air-entraining agents). The following mixtures were compared, whereby the binder remained unchanged:

Table 4. The mix design.

Mix Design	Binder 1[1]	Batch kg/m³
Cement 1	CEM II/A-S 42.5R	210
Cement 2	CEM I 52.5 R	210
AHWZ[1]	AHWZ	50
Additive	Superplasticizer	
Additive	Consistency holder	
Additive	Air-entraining agent	

1) Admixture Type II, Chargeable Hydraulically Active Admixture (AHWZ)

The proportion of aggregates remained uniform in the investigations, but with different compositions of the individual fractions:

Table 5. Division of the fractions into the overall sieve line.

Fraction (mm)	NM (%)	M20 (%)	M35 (%)	M50 (%)
Calcitic sand 0/4	0	20	35	50
Tunnel excavationsand 0/4	70	50	35	20
Gravel 4/8	30	30	30	30

6.1 Fresh Concrete Testing Results

The following table shows the results of the fresh concrete tests of the different mixtures.

6.2 Compressive Strength

It is known that the compressive strength increases with decreasing air content or decreasing porosity (see e.g. (*Wesche, K.*, 1993)). Since the investigated compounds had a very different air content, the test results were converted to a uniform fictitious air content of 5.0% in order to compare the individual compressive strengths. As a simple approximation, the empirical approach was chosen in this paper that with an increase in the air content of 1%, the

Table 6. Fresh Concrete Results.

Parameter	NM	M20	M35	M50
Watercontent (kg/m³)	224	214	210	189
W/B-ratio (-)	0.49	0.45	0.44	0.40
Density (kg/m³)	2110	2234	2288	2132
Slump after 10 min (mm)	613	565	580	555
Slump after 60 mini (mm)	495	440	420	475
Air content (%)	9.5	6.4	4.6	10.9

compressive strength decreases by 5%. Under this assumption, the following table shows the compressive strength for the existing air content of the test specimens and the compressive strength related to the calculated air content of 5.0%.

It can be seen that with the increasing replacement of tunnel spoil 0/4 by calcitic sand 0/4, the compressive strength of the concrete also increases and that this is mainly expressed in the possibility of water saving. The following figure illustrates this relationship:

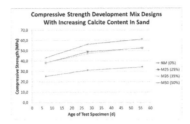

Figure 7.   Concrete Strength Curve for NM, M20, M35 and M50 Mixes.

7   EFFECT OF MATERIAL PREPARATION ON LANDFILL VOLUME

7.1   *Share of shotcrete in landfill volume*

A reliable forecast of the required landfill volume is of decisive importance for a project such as the *Brenner Base Tunnel*, because the planning and approval of landfills takes a long time due to requirements and procedures. In total, the required landfill volume in the Austrian project area amounts to approx. 12 million m³. Changes in the percentage range lead to large absolute increases or decreases in quantities.

At first glance, the calculation of the landfill volume as the product of the excavation cross-section with the tunnel length appears to be trivial. However, a closer look at the mass balances reveals a different picture. The deposited material does not only consist of the excavated rock but also of shotcrete (from rebound and temporary securing e.g. at the working face). In addition, the water content generally changes. These effects are relevant and must therefore be taken into account in the mass balance.

The mass ratio of shotcrete to rock in the landfill can be estimated on the basis of a fictive circular tunnel assuming a density of rock of 2.7 to/m³ and shotcrete of 2.2 to/m³ as follows:

Table 7.   Estimation of the proportion of shotcrete in the landfill.

Radius Soffit (m)	Excavated area Orbastrally (m²)	Thickness of shotcrete (m)	Rebound (m)	Pitch length (%)	(m)
4.0	50.3	0.15	0.05	10	1.4

Volume per pitch			Mass per pitch		
Rock (m³)	Reveal (m³)	Orbastrally (m³)	Rock (to)	Sprayed concrete (to)	Ratio (%)
70.4	5.4	2.5	190.0	7.0	4

This simple estimation results in a mass ratio of shotcrete to rock in the landfill of approx. 4%. A similar value resulted from a back-calculation of the actual masses for the lot *H33 Tulfes-Pfons*. The proportion of rebound is relatively low (approx. 1%), the more important part is the shotcrete used to secure the face of the village.

Table 8. Ratio of excavated to recycled rock volume for lot *Wolf II*.

Year (-)	Quantity excavated (to)	Quantity recycled (to)	Mass percentage (%)
2013	20.500	0	0
2014	764.000	0	0
2015	600.000	140.000	23
2016	341.250	100.000	29
2017	20.500	0	0
Total	**1.746.250**	**240.000**	**14**

7.2 *Reduction of landfill volume through material preparation*

It is assumed that the reduction of the landfill volume is equal to the mass ratio of the processed rock to the excavated rock (a closer look reveals a slightly non-linear correlation).

The quantities excavated and processed in lot Wolf II are summarized in the following table.

It follows from this that without material processing, the landfill volume would have been approx. 30% higher in 2016 and approximately 15% higher in the total lot.

8 CONCLUSION

This article gives an overview of the developed processing concept and a description of the current shotcrete technology with the experience of the material processing of *Bündner Schist* as aggregates for concrete production in the construction of the Brenner Base Tunnel. In the successful pilot project lot *Wolf II*, it was shown that the processed *Bündner Schist* is well suited as aggregate for shotcrete and structural concrete. The processed tunnel spoil was used to successfully supply aggregates for 144 000 m³ of concrete. Furthermore, in concrete tests it was determined that the concrete strength can be increased by replacing tunnel spoil 0/4 with calcitic sand 0/4. The recylcling of tunnel spoil improve the sustainability of the project by reducing the environmental impact and sparing resources.

REFERENCES

BBT-internal report, 2015, E52 Life Cycle Assessment (LCA), Ökobilanzstudie Variantenuntersuchung Spritzbeton.
BBT-internal report, 2016, quality evaluation sprayed concrete, Wolf E52 intern
Bergmeister, K. & Kogler, H. & Murr, R. & Cordes, T. & Arnold R., 2016: Brenner Basistunnel Innovationen zur Aufbereitung des Tunnelausbruchmaterials und Optimierung der Betonzusammensetzung. Zement+Beton 1, Page 48–57
Bundesministerium Nachhaltigkeit und Tourismus, 2002, Abfallwirtschaftsgesetz AWG
Österreichische Vereinigung für Beton und Bautechnik, 2009, Guideline Shotcrete.
Swiflty Green, 2015, Report "Infrastructure spatial planning and environmental effects" https://www.swiftlygreen.eu
Voit K. 2013: Einsatz und Optimierung von Tunnelausbruchsmaterial des Brenner Basistunnels, Phd-Thesis, University of Natural Resources and Life Sciences, Vienna
Wesche, K. 1993: Baustoffe für tragende Bauteile, Band 2, Beton und Mauerwerk, Bauverlag GmbH. Wiesbaden und Berlin.
Cordes, T. & Hofmann, M. & Murr, R. & Bergmeister, K. 2018: Aktuelle Entwicklungen der Spritzbetontechnologie und Spritzbetonbauweise am Brenner Basistunnel, Spritzbeton – Tagung 2018, Alpbach.

Tunnels and Underground Cities: Engineering and Innovation meet Archaeology,
Architecture and Art, Volume 2: Environment sustainability in
underground construction – Peila, Viggiani & Celestino (Eds)
© 2020 Taylor & Francis Group, London, ISBN 978-0-367-46579-7

Environmental risk assessment of conditioned soil: Some Italian case studies

S. Padulosi, F. Martelli, A. Sciotti, D.F. Putzu & M. Filippone
ITALFERR S.p.A., Rome, Italy

G. Mininni
CNR Water Research Institute, Monterotondo, Rome, Italy

ABSTRACT: Excavated soil (ES) produced by mechanised tunnelling (EPB-TBM) can be qualified as waste or alternatively as a by-product, depending on its characteristics and the conditions for utilisation. Tunnel excavation produces huge volumes of ES. Managing it as waste might be so expensive to dramatically impair the feasibility of major infrastructure projects. Different site-specific experimental studies were carried out on conditioning products and soil/product mixtures to assess the eco-compatibility of ES and therefore the possibility to manage them as a by-product. Firstly, laboratory tests were carried out to evaluate the less impacting conditioning products and the related dosage for an optimal soil conditioning. An environmental site-specific study was then carried out to simulate operating conditions testing different conditioned soil samples for modelling the real cases. Eco-compatibility of conditioned soil samples strictly depends on the time needed for biodegradation of conditioning agents.

The paper presents a number of significant Italian case studies.

1 INTRODUCTION

The use of EPB (Earth Pressure Balance) technology for mechanised tunnelling in soft ground is widespread throughout most of the world, particularly in urban areas where it enables the surface settlement caused by underground excavation to be minimised, thereby preventing damage to buildings. The use of this technology is spreading quickly, even in non-urban areas, due to its advantages in terms of worker safety, excavation rate, and continuous working.

The operating principle is based on using the excavated soil to counterbalance the earth pressure and stabilise the tunnel face. Natural soil is conditioned with foams and/or polymer, and during excavation is homogenised to form a soft, plastic paste with low permeability that is pressurized by the TBM thrust cylinders as support for the tunnel face.

Tunnel construction generates large quantities of excavated soil (ES) that can be qualified either as waste or as a by-product, depending on conditioned excavated soil characteristics and the conditions for its utilisation. In Italy, managing ES as waste might be so expensive that, in some cases, economic feasibility of civil works is impaired. ES is instead a resource to be used for multiple purposes like recovery of degraded areas, such as derelict quarries and brownfield sites.

The use of additives is a key factor for correct operation of EPB-TBM technology but may affect the options for managing the ES. Conditioning the spoil must render tunnelling easy assuring that excavated soil can be used in compliance with environmental protection standards.

The management of soils produced during the construction of infrastructure works can be described as a project within the project.

Currently, the management of ES is regulated in Italy by Decree 120/2017, which impose limit concentrations for some contaminants (metals, asbestos, heavy hydrocarbons, BTEX and PAH) for managing these materials as by-products. This condition is also applicable to soils that have been conditioned with additives, as in the case of EPB-TBM excavation. In the case of additives containing pollutants that are not listed in the Decree mentioned above, the designer is required to submit technical documentation to enable the responsible authorities to verify that the excavated soils do not represent a risk for the environment and human health. However, Decree 120/2017 does not yet clarify how to assess the eco-compatibility of conditioned soils.

In order to overcome this gap in the Italian regulations, which to date do not specify threshold values for anionic surfactants and other components of additives which are generally used (e.g. polymer and lubricants), Italferr began a successful collaboration with the Italian National Research Council (CNR), as well as a number of universities and other institutions, and developed a design approach which might be a considerable starting point for further discussions.

Currently, the eco-compatibility of conditioned soil is assessed by making site-specific studies based on eco-toxicological and biodegradation tests on the soil/additive mixture.

The natural degradation of foaming agents is an important factor for making a realistic environmental risk assessment. In fact, temporarily storing treated soils for a suitable period can bring the material into compliance with legal specifications for by-products. Since the degradation rate depends on a number of factors like soil characteristics (e.g. granular size, mineralogy, etc.), chemical characteristics of products, treatment ratio and microbial activity of soil, the eco-compatibility study is typically site-specific.

2 SOIL CONDITIONING AT THE DESIGN STAGE

2.1 *The importance of soil conditioning and foam properties in EPB-TBM tunnelling*

Conditioning the soil with additives for EPB-TBM tunnelling has several benefits: it improves the stability of the tunnel face, reduces the internal friction angle, enhances the spoil fluidity (which facilitates extracting and transporting it to destination), reduces wear of the cutters and cutter head, inhibits clogging events that prevent suspending the excavation, etc.

The most common products used for soil conditioning are water, a foam produced mixing air and an aqueous solution of a foaming agent, and a polymer solution not always needed.

The foaming products provided by manufacturers are used in solution at concentration (C_f) of 0.5–6%, being 2% that typically used.

Foaming products are commercialized in water solution of surfactants and other chemical substances. Their chemical formula, not full-known due to industrial secrecy, might have direct impact on water and soil and therefore influences the eco-toxicological properties of ES and its acceptability as by-product. As reported in safety data sheets, most foaming agents contain anionic surfactants at concentrations typically between 5% and 30%. Surfactants are used in various industrial applications (Langmaack & Feng 2015).

There are many chemically different products on the market. Some contain polymers for adsorbing free water and increasing plasticity of the spoil, or high dispersing capacity polymers (anticlogging polymer), as appropriate.

Under some conditions (foam generator, air flow, liquid solution flow and product concentration), different products generate foams with different characteristics (Merritt 2004). They differ not only in terms of foam expansion ratio (FER), i.e. the ratio between the foam volume and the volume of liquid solution, but also in terms of the stability of the foam over time.

2.2 *Methods for defining foam quantity*

Based on case studies, EFNARC (the European Federation for Specialist Construction Chemicals and Concrete Systems, representing the producers and appliers of specialist building products for concrete) has published guidelines on the use of foam, indicating a possible range of conditioning parameters: FIR (the Foam Injection Ratio, i.e. the amount of foam as a percentage of the soil volume in situ), FER (the Foam Expansion Ratio) and C_f (the concentration of surfactant agent in water). However, EFNARC stresses that the operating conditions of the EPB-TBM may or may not confirm the real effectiveness of the product with the selected soil conditioning parameters (EFNARC 2005). Various authors have shown that the quantity of foam used (FIR) is often greater than that predicted (Shinouda et al. 2013, Thewes et al. 2012).

At the design stage, no standard tests exist for defining the soil conditioning parameters and the additive doses. However, laboratory testing is generally used to estimate the parameters to be used during tunnelling: e.g. a foam half-life test to verify foam stability over time, a slump test to measure the 'plasticity' or 'workability' of conditioned granular soils, or a mortar flow table-test to verify the plasticity of conditioned soil. Other tests can be used depending on the characteristics of the soil, the hydrogeological conditions and the support pressure required (Merritt 2015, Peila 2011).

2.3 *Soil conditioning study at the design stage*

The currently applicable regulations require an assessment of the amount of conditioning agents to be made at an early stage, so that a project can be developed for managing the excavated material.

The soil conditioning study begins by analysing the geotechnical tunnel profile and has the purpose of identifying the most representative lithotypes, based on several factors, such as:

- lithotype predominance: the quantity of the various lithotypes to be excavated;
- product chemical: the chemical characteristics of the additives potentially needed for conditioning the soil;
- soil granulometry: the granulometry of the soil (e.g. coarse-grained soils could favour the drainage and removal of liquids including dissolved surfactants or polymers).

The lithotypes used for soil conditioning tests and environmental laboratory tests must be carefully evaluated. Once selected, soil samples of most representative lithotypes should be collected.

A study of soil conditioning based on laboratory tests carried out at the design stage presents some critical issues (Martelli et al. 2017): e.g. it is not always possible to obtain a large amount of soil before starting excavation. In addition, experimental work can only be carried out after the tunnelling technique has been selected, i.e. at the final design stage, thereby increasing cost and time. When a full conditioning study cannot be carried out, the recommendations and suggestions of the product manufacturers may be used.

Based on the test results or the information provided by the manufacturers, the highest additive dosage can be estimated, i.e. the reference value for the environmental risk assessment of conditioned soils. It is important to highlight that this assumed product dosage is a design hypothesis based on the knowledge available at a given time. Before beginning excavation, this hypothesis should be confirmed or updated based on the proper expertise of the Contractor, on additional information from a new geognostic campaign, on the availability of innovative new products, etc.

3 ASSESSING THE TOXIC EFFECTS OF CONDITIONED SOILS

3.1 *Soil conditioning and the selection of samples for studying environmental compatibility*

The management of soil produced by EPB-TBM tunnelling as by-product depends on previous testing to assess eco-toxicity of foaming agents and anti-clogging additives and polymer (if required).

The eco-compatibility study should consider:

- the intrinsic eco-toxicological features of the commercial products, which will depend on the main components and the quantities thereof;
- the quantity of product used to treat soil, i.e. TR: Treatment Ratio, l/m^3 (litres of product per cubic metre of in situ soil). These parameters depend on the lithotype and the product biodegradability;
- the destination site for the by-products and their possible interaction with the ground and with underground/surface water.

The environmental risk assessment focuses on lithotypes that require a higher TR, i.e. the 'worst case scenarios' to be taken as reference in the eco-toxicological study.

As a precautionary measure, the eco-toxicological study can consider a further increase of TR compared to the value calculated by geotechnical testing, to take into account any additional dosing of the foaming agents due to unexpected geological or operational conditions encountered during the works.

3.2 The ecotoxicological study at the microcosmic scale

The ecotoxicological study is carried out on treated and untreated soil samples and elutriate.

Biodegradation tests are carried out on microcosms of soils coming from the excavation site and treated with the foaming agents and anti-clogging additives (if needed) that were specified at the design stage. The microcosms tested in the laboratory reproduce the environmental conditions (temperature, light, humidity) specified at the design stage, in order to simulate the soil maturation over time (i.e. biodegradation process). If biodegradation of the anionic surfactants is enough high that residual concentrations are not of concern both from environmental and health point of view, ES can be classified as by-products.

Laboratory tests are carried out on target organisms, as representatives of both terrestrial and aquatic environments. The selection of the organisms depends on the destination site for the by-product and on the possibility that the treated soils may interact with the ground or with underground/surface water.

With the aim of assessing the degradation of the foaming agent, the eco-toxicological tests at different time steps are carried out (biodegradation time t=0, 7, 14 and 28 days).

The tests used include the following: acute toxicity with bioluminescent bacterium *Vibrio fischeri*, seed germination and primary root growth of *Lepidium sativum*, acute toxicity with earthworm *Eisenia foetida*, Fish Embryo Toxicity (FET) on *Danio rerio*, immobilisation on *Daphnia magna*, mortality rate and growth inhibition of crustaceans *Heterocypris incongruens*.

Chemical analyses are also needed to assess the residual concentrations of foaming agents within soils and water that may affect environmental eco-compatibility of soil and acceptability from sanitary point of view. Such concentrations mainly depend on lithological and mineralogical characteristics of the soils and on the biodegradation time.

4   THE DESIGN AND CONSTRUCTION STAGES

Based on the results of the ecotoxicological tests, the destination of the excavated soils can be designed and the re-use of conditioned soils planned. However, if for the laboratory testing, the mixture soil/additive has no significant ecotoxicological effects on the target organisms beginning from only a few days of maturation, suitable storage basins should be constructed.

In the case of work sites in urban area, this requirement may be not easy to meet due to the unavailability of areas in which to allocate large facilities.

4.1 Eco-toxicity tests at the construction stage

Eco-toxicological tests provide significant information for defining the tests to be carried out during the construction stage in order to confirm the eco-compatibility of the excavated material and agreeing them with the responsible controlling bodies.

5   CASE STUDIES

The results of the soil conditioning and environmental eco-compatibility studies that were carried out as part of the following railway projects are described below:

– The Polcevera Tunnel along the Milan-Genoa high speed railway;
– The Palermo railway junction;

- The Florence railway underpass;
- The new Turin – Lyon railway: tunnels between Avigliana – Orbassano;
- The track-doubling between Apice and Hirpinia (Naples–Bari railway)- the Rocchetta tunnel;
- The track-doubling between Hirpinia and Orsara (Naples–Bari railway) – the Hirpinia tunnel.

For each of these projects, one to three representative lithotypes were selected and one to three products were tested for each lithotype. Figure 1 shows the granulometric curves of each lithotype tested (10 cases). Essentially, they cover the whole applicability range of the EPB-TBM (EFNARC 2005), and for this reason the results can be considered sufficiently significant.

For each case and for each lithotype, the dosages of products suitable for soil conditioning are shown in Figure 2. Moreover, the results of the eco-toxicological studies on conditioned soil, in terms of days of maturation before its reuse as a by-product, are shown.

It should be highlighted that Figure 2 only shows products for which the environmental study has determined that the stockpiling period of the conditioned material, before its re-use as a by-product, is less than or equal to 28 days.

For different lithotypes, a total of 8 foaming agents were tested and only in two cases the use of a polymer was necessary: a polymer for cohesiveness and consistency control in Case 6 (a/b), and an anti-clogging polymer in Case 9.

Dosages of these products, suitable for conditioning the lithotypes, generally range from 0.22 to 2.25 $l/m^3_{in\ situ}$; the biodegradation time, based on the environmental assessment, ranges from 0 days to 28 days. In Figure 2, the cases (5, 8, 9, 10) related to lithotypes mainly at fine grain size are highlighted (light grey background). As can be seen, the biodegradation time heavily depends on the chemical formula of the products rather than on the granulometry or the dosage of the products.

Product no.1 was tested in several cases. The graph in Figure 3 shows that the higher the dosage, the longer the biodegradation time. The data indicates that when TR is lower than 1.3 $l/m^3_{in\ situ}$, the biodegradation time is 7–10 days. For a higher TR, the maturation time considerably increases up to 28-days for a TR of 2.25 $l/m^3_{in\ situ}$.

It is noted that the dosage increment of about 0.5 $l/m^3_{in\ situ}$, from case 4 to case 5, causes an increase of biodegradation time of 7 days, while the same dosage increment from case 5 to case 6 produces an increase of 14 days of biodegradation time.

Cases 5 and 6 refer to the same project (Florence railway underpass): one to the coarsest grain size sample, the other to the finest grain size sample. It is inferred that the lithotype particle size surely affects the biodegradation time: the finer the sample granulometry and the longer the biodegradation time.

It could also be hypothesized that above a certain dosage the increase of additive causes a non-linear increase of the time of biodegradation.

However, all these considerations should be confirmed with further studies.

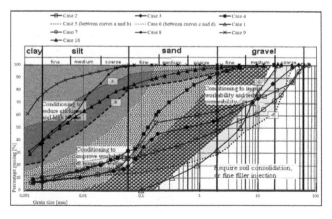

Figure 1. Particle size distribution referred to soil sample tested. In the background: area of EPB application by EFNARC 2005.

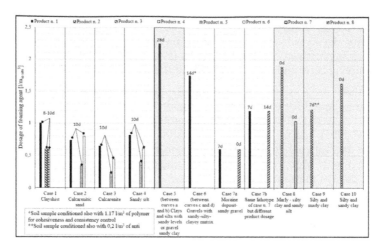

Figure 2.   Foaming agent dosage tested for each lithotype and days of maturation inside storing basins

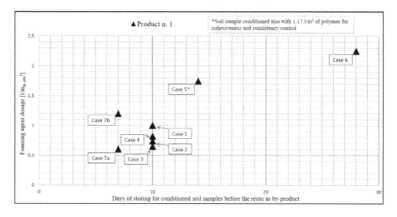

Figure 3.   Dosage of Product 1 for each soil sample tested and days of maturation.

5.1   *Case 1: The Polcevera Tunnel on the Milan – Genoa high-speed railway*

The representative lithotype of the Polcevera Tunnel is clayshist (see Figure 1 – Case 1). In order to identify suitable conditioning parameters, the Polytechnic of Turin carried out a soil conditioning study at laboratory scale, testing three commercial products and finding a Treatment Ratio (TR) of 0.6–1.0 l/m^3 in situ (see Figure 2).

Subsequently the Mario Negri Pharmacological Research Institute carried out the eco-toxicological study by experimentally assessing the Reference Threshold values (RT) referred to the surfactants contained in the treated soil. It was found that the RT ranges from 100 to 200 mg/kg of soil, depending on the type of surfactants and the conditioning products used. Tests show that for surfactant concentration less than RT, the soils are not toxic for any vegetal and animal organism.

Table 1.   Milan – Genoa high-speed railway – Design data and tests to be carried out during the works.

Lithotype	Clayshist (Case 1)
Biodegradation time	8 ÷ 10 days depending on the season
Tests on material excavated by EPB-TBM	Surfactant concentration, *Vibro fischeri*, environmental characterisation
Treatment on excavated soil	Storage in maturation basin
Further requirements by Authorities	Agreement of the sampling methodology with the Regional Agency for the Environment

In table 1 some of the design data is summarised along with the action to be taken on the excavated material during the works, based on the conclusions of the Mario Negri Institute.

5.2 Cases 2, 3, and 4: Palermo railway junction

The representative lithotypes of this project are: calcarenitic sand (Case 2), weakly cemented bioclastic calcarenitic (Case 3) and sandy silt (Case 4) (see Figure 1).

The Polytechnic of Turin carried out the soil conditioning study at laboratory scale, testing three commercial products. A treatment ratio (TR) of $0.2 \div 0.8$ $l/m^3_{in\ situ}$ was determined (see Figure 2).

The Mario Negri Institute assessed the RT, the surfactant degrading time, and the optimal mixture between the three commercial products and the three representative soil samples. The RT was found to be 238 mg/kg of conditioned soil.

In the table below, some of the design data is summarised along with the action to be taken on the excavated material during the works, based on the conclusions of the Mario Negri Institute.

Table 2. Palermo railway junction – Design data and tests to be carried out during the works.

Lithotypes	Calcarenitic sand (Case 2)
	Weakly cemented bioclastic calcarenite (Case 3)
	Sandy silt (Case 4)
Biodegradation time	10 days
Tests on material excavated by EPB-TBM	Surfactant concentration, environmental characterisation
Treatment on excavated soil	Storage in basin for maturation and liming
Further requirements by Authorities	Monitoring of surfactant in groundwater and soil, monitoring of dusts for liming

5.3 Cases 5 and 6: the Florence railway underpass

There are two representative lithotypes of the Florence railway underpass: clays and silts with sandy levels (Case 5) and gravels with sandy-silty-clayey matrix (Case 6). The first is included between curves a and b in Figure 1; the second between curves c and d.

At an early design stage, the CNR verified the soil conditioning tests carried out by the product manufacturers on products and dosages suitable for tunnelling by EPB-TBM. Subsequently the General Contractor (GC) added polymers and fixed a dosage ranging from 1.6 $l/m^3_{in\ situ}$ to 2.25 $l/m^3_{in\ situ}$ (i.e. more than twice the dosages given by the product manufacturers) for each foaming additive and lithotype.

Table 3. Florence railway underpass– Design data and tests to be carried out during the works.

Lithotypes	Clays and silts with sandy levels (Case 5)
	Gravels with sandy-silty-clayey matrix (Case 6)
Biodegradation time	14 (Case 6) and 28 days (Case 5)
Tests on material excavated by EPB-TBM	Grain size, surfactant concentration, *Danio Rerio* (FET), *Vibrio fischeri*, COD, environmental characterisation
Treatment on excavated soil	Storage in basin for maturation
Further requirements by Authorities	Lithological characterisation, environmental monitoring of the water

The chemical and eco-toxicological tests carried out by the CNR on soil samples conditioned with dosages provided by the GC show that, for one of the two foaming products, the toxicity is substantially null after a maturation time of the treated soil of 14 or 28 days, depending on the lithotype.

In the table 3, some of the design data is summarised along with the action to be taken on the excavated material during the works, based on the conclusions of the CNR.

5.4 Case 7: New Turin – Lyon railway: tunnels between Avigliana – Orbassano

The CNR and the 'Bicocca' University of Milan tested two foaming products and one soil containing sandy gravel of moraine deposit (see Figure 1).

The Polytechnic of Turin carried out the soil conditioning study at laboratory scale testing two commercial products and determining a treatment ratio (TR) of 0.6 $l/m^3_{in\ situ}$ (Case 7a) and 1.2 $l/m^3_{in\ situ}$ (case 7b), respectively (see Figure 2).

The tests were carried out on a soil sample mixed only with water ('blank') and two soil samples conditioned with different dosages of products. In the first sample, the FIR (Foam Injection Ratio) was increased by 50% (Case 7a) with respect to 'optimal' conditioning parameters. In the second sample, both the FIR (Foam Injection Ratio) and the FER (Foam Expansion Ratio) were increased by 50% (Case 7b).

The CNR carried out eco-toxicological tests at the microcosm scale. The results show that:

– First product: tests on elutriates and soils (blank and treated) show no significant toxic effects on all tested organisms for both dosages since t=7 days.
– Second product: tests on elutriates and soils (blank and treated) show no significant toxic effects on all tested organisms at the minimum dosage since t=0 days whilst at the maximum dosage the effects are null since t=14 days.

Table 4. New Turin – Lyon railway: tunnels between Avigliana – Orbassano – Design data.

Lithotypes	Moraine: sandy gravel (Case 7)
Biodegradation time	0 – 7 – 14 (depending on product and dosage)

5.5 Case 8: Track-doubling between Apice – Hirpinia: Rocchetta tunnel

The CNR and the 'Bicocca' University of Milan tested two foaming products and one soil containing marly – silty clay and sandy silt (see Figure 1).

The preparation of the representative soil samples from the excavation site and the conditioning tests were carried out by the 'Sapienza' University of Rome, which arranged a soil sample mixed with water only ('blank') and two soil samples conditioned with the maximum doses as per the designer's recommendation: 1.04–1.89 $l/m^3_{in\ situ}$.

The CNR carried out ecotoxicological tests at the microcosm scale. The results show that since t=0, there are no significant toxic effects on all the tested organisms for both products in elutriates and soils ('blank' and treated).

Table 5. Track-doubling between Apice – Hirpinia: Rocchetta tunnel – Design data and tests to be carried out during the works.

Lithotypes	Marly – silty clay and sandy silt (Case 8)
Biodegradation time	0 days
Tests on material excavated by EPB-TBM (only if TR is less than 1.89 $l/m^3_{in\ situ}$ or 1.04 $l/m^3_{in\ situ}$ depending on product)	Monitoring of the Treatment Ratio (TR, $l/m^3_{in\ situ}$), *Vibrio fischeri* on elutriate
Treatment on excavated soil	Storage in basin for maturation
Further requirements by Authorities	Set up a field laboratory, monitor the conditioning parameters, environmental monitoring of the waters, environmental characterisation, lithological characterisation

Based on the experimental study, the CNR suggested carrying out the acute toxicity test with the bacterium *Vibrio fischeri* to assess the eco-compatibility of the excavated material at real scale during the works. This test must be repeated at least three times to calculate the mean (m) and standard deviation (σ). If significant toxic effects are observed, the test must be repeated at further degradation time step of the soil stored in basins.

5.6 *Cases 9 and 10: Track-doubling between Hirpinia – Orsara: Hirpinia tunnel*

In order to select the representative lithotypes and assess the optimal treatment parameters, the Polytechnic University of Turin carried out geotechnical conditioning tests on eleven lithotypes, determining for each lithotype the additives and the soil conditioning parameter suitable for EPB-TBM tunnelling.

Lithotypes requiring the maximum dosage of foaming agent or foaming agent plus polymer were chosen as the most representative soil samples. The dosages were: 1.22 $l/m^3_{in situ}$ of foaming agent and 0.2 $l/m^3_{in situ}$ of anti-clogging polymer for silty and marly clay (Case 9) and 1.63 $l/m^3_{in situ}$ of foaming agent for silty and sandy clay (Case 10).

The environmental risk assessment study by the CNR and the 'Bicocca' University of Milan refers to one foaming product, one anti-clogging additive, and two lithotypes.

The CNR carried out ecotoxicological tests and simultaneously determined the concentration of the main anionic surfactant in both the products in elutriates from soil samples at different maturation times.

As expected, the elutriate produced by soils conditioned with the foaming agent and the anti-clogging additive shows the highest concentration of surfactant at the beginning of the test (t=0). In any case, at t= 7 days, there is a significant reduction of the surfactant concentration and, consequently, no toxicity is longer detected in the elutriates.

Based on the experimental tests, the soil conditioned with the foaming product only shows no toxicity since t=0, whilst the soil that was also treated with the anti-clogging additive shows no toxic effects from t= 7 days.

Table 6. Track-doubling between Hirpinia – Orsara: Hirpinia tunnel– Design data and tests to be carried out during the works.

Lithotypes	Silty and marly clay (Case 9)
	Silty and sandy clay (Case 10)
Biodegradation time	0 – 7 days (depending on lithotype)
Tests on material excavated by EPB-TBM	Not yet available
Treatment on excavated soil	Storage in basin for maturation
Further requirements by Authorities	Not yet available

6 CONCLUSIONS

The Italian regulations have not yet defined procedures and laboratory tests for conditioning studies of soils, or a methodology for eco-toxicological testing. However, thanks to experimental studies carried out by CNR, universities and other institutions, Italferr has identified a design approach that made possible to overcome these gaps.

These studies carried out for several large-scale infrastructure projects show that results are thoroughly affected by different approaches and input data.

As for granulometry, the analysed cases cover the whole applicability range of the EPB-TBM (EFNARC 2005). The experimental conditioning tests result in a foaming agent dosage of less than 2 l/m^3 of in situ volume of soil.

As various authors have shown, the quantity of foam injection ratio (FIR) is often higher than the predicted dosage from experimental tests; the actual product dosages should feed the back-analysis process.

In order to assess the eco-compatibility of ES by EPB-TBM, some eco-toxicological laboratory tests are carried out to verify whether the conditioned soils represent a risk for the environment and human health. Natural biodegradation of the foaming agents for few days is generally sufficient to render ES suitable for subsequent utilization in accordance with law specifications for by-products. In general, the tests show that for each case there is at least one product used for soil conditioning for which within 0–14 days (only in one case within 28 days) the excavated material has the same characteristics of toxicity of the "blank" (i.e. no-toxic).

Because the experimental studies are based on eco-toxicological tests at the microcosm scale, confirmation is required that the mixture of soils/foaming products/anti-clogging additives (if any) is also neither toxic nor eco-toxic at the real scale, i.e. during the works. For this purpose, the excavated material is stored in appropriate basins for collecting the soil samples to be tested. In fact, some boundary conditions (i.e. temperature, rainfall, etc.) differ from those simulated in laboratory. Furthermore, during the excavation, the quantity of the products used to treat the soils, i.e. the Treatment Ratio (TR l/m^3) measured as volume of foaming product per cubic meter of in situ soil, may change due to a lithotype with characteristics slightly different from those tested in the laboratory.

Results at real scale are a useful tool for calibrating experimental studies (e.g. by suggesting new and/or additional laboratory tests), supporting future infrastructure designs and driving improvements in the eco-compatibility of products.

The laboratory tests show that foaming agents persist longer in elutriates than in conditioned soils. As consequence, back-analysis tests should be focussed to identify threshold values for the product components in elutriates rather than in the mixture soil/products.

However, the Italian regulations have not yet defined procedures and laboratory tests for conditioning studies of soils, or a methodology for eco-toxicological testing.

Laboratory and full-scale testing could be more efficaciously used whether the following methodological criteria are precisely fixed by the authorities:

- standard procedures for developing soil conditioning studies at different design stages;
- risk assessment should be site-specific as it is currently for remediation of contaminated soils;
- procedures to test commercial products and the soil/additive mixture;
- standardization of the procedures for assessment of conditioned soils during the works.

Finally, a national database of overall ecotoxicological test results would be very useful.

REFERENCES

EFNARC 2005. *Specification and guidelines for the use of specialist products for mechanised tunnelling (TBM) in soft ground and hard rock.* Farnham, Surrey, UK: EFNARC
Langmaack, L. & Feng, Q. 2005. Soil conditioning for EPB machines: balance of functional and ecological properties. In: Erdem, Y. & Solak, T. (eds.), *Underground space use. Analysis of the past and lessons for the future; Proc.: World Tunnel Congress and 31st ITA Assembly, Istanbul, Turkey, 7–12 May 2005.* Rotterdam: Balkema.
Martelli, F., Pigorini, A., Sciotti, A., Martino, A. & Padulosi, S., 2017. Main issues related to EPB soil conditioning and excavated soil. *Congrès International de l'Aftes, Paris, 3–15 November 2015.*
Merritt, A.S. 2004. Conditioning of clay soils for tunnelling machine screw conveyors. *PhD. Thesis, University of Cambridge.*
Merritt, A. 2015. Soil conditioning for EPB tunnelling: some examples of laboratory testing and field monitoring. *Terre e rocce da scavo nelle opere in sotterraneo: un problema o una opportunità?; Atti del convegno SIG, Samoter, Verona, Italy, 8–11 May 2014.* Milano: SIG (Società Italiana Gallerie)
Peila, D., Borio, L. & Pelizza, S. 2011. Lab test for EPB ground conditioning. *Tunnels & Tunnelling International*, September 2011: 48–50.
Shinouda, M.M., Garahbagh, E.A. & Shinouda, M.M.R. 2013. Untangling the mystery of soil conditioning in EPB tunnelling. In Micheal A. Di Ponio and Chris Dixon (eds.), *Rapid Excavation and Tunneling Conference; Proceedings, Washington, DC, 23–26 June 2013.* Englewood, CO: Society for Mining, Metallurgy & Exploration, Inc.
Thewes, M., Budach, C. & Bezuijen, A. 2012. Foam conditioning in EPB tunnelling. In Giulia Viggiani (ed.); *Proc. of the 7th Int. Symp. on Geotechnical Aspects of Underground Construction in Soft Ground, Rome, Italy, 17–19 May 2011.* London: Taylor & Francis.

*Tunnels and Underground Cities: Engineering and Innovation meet Archaeology,
Architecture and Art, Volume 2: Environment sustainability in
underground construction – Peila, Viggiani & Celestino (Eds)
© 2020 Taylor & Francis Group, London, ISBN 978-0-367-46579-7*

Low energy nobilitation of clay waste from tunnelling

V. Perugini
Experimentations Srl, San Mariano di Corciano (PG), Italy
School of Science and Technology, Geology Division, University of Camerino, Italy

ABSTRACT: Infrastructures and tunnelling works should increase circular economy, maximizing waste reusing, in order to save quarrying, landfill dumping, transport and environmental impact. An integral reusing of tunneling waste can minimize the necessity of waste relocation. Hard rocks can be easily used as aggregate, instead, argillaceous waste commonly requires expensive relocation works. This research is pointed to demonstrate the possibility of producing refined materials from argillaceous tunneling waste for use in concrete, as aggregate, binders, supplementary cementitious materials and geopolymers, by applying a low energy treatment associated with chemical activation. The main advantage of the approach of this study is the potential to reuse larger quantities of clay materials with low treatment costs. It would be very useful also in case of Earth Pressure Balance TBM bored tunnel, where clay soil is often conditioned with surfactant and polymer agents to assure the stability of the excavation and to improve the TBM performance.

1 INTRODUCTION

New technologies in tunnel production, first of all due to the use of Tunnelling Boring Machine (TBM), allow many advantages, as e.g. higher velocity of excavation, lower labor activity, lower environmental and landscape impact. This leads to increase the TBM use in the tunnelling and to the possibility to adopt longer tunnel in the infrastructures planning. Nevertheless, especially in case of Earth Pressure Balance (EPB) and surfactant use, the TBM process determines new environmental issues in the muck managing, whose results rapidly produce a higher amount.

The nature of the rock and geological formations involved in tunnelling works determines different challenges and possibilities of waste reusing. In fact, good mechanical properties of the rock make tunnelling waste suitable for use as concrete aggregate. In this case they require only physical treatments, as e.g. washing, grinding, sieving, separation, etc. Instead, different issues are linked to the argillaceous formations, due the plastic behaviour, water retention, high hydraulic limits and deformability, which make it complicated to manage and relocate them. Moreover, in case of EPB-TBM, the pollution due to the use of surfactant addition to the soil during the excavation, requires storing them in dedicated areas, waiting for the reduction of surfactant activity, before definitive relocation can take place.

Considering the difficulties for long steps of storage, the limitations in the relocation possibilities and landfill dumping costs, treatment of the muck could become convenient also in cases of final vale of the produced material lower than the treatment cost, thanks to the high saving of relocation. Other mainstone to take in account from this solution regard a better landscape safeguarding and environmental sustainability, due to the saving of quarrying of new raw materials for building material industry, lower greenhouse gas emission, lower impact of transports on infrastructure and natural environments during the works. Considering all possible advantages, also soil waste which looks ineligible as building materials, as e.g. for lime, binder, cement, aggregate, addition or supplementary cementitious materials (SCM), should be evaluated in order to detect possible forms of reusing inside or close to the tunnelling work.

This research is focused on the possibility of producing an SCM, to use in concrete, cement, or low grade binder, through available clay waste from one of the largest tunnelling works today in progress in north of Italy. It should also be considered that, the use of SCM as cement replacement, is today one of the most recommended practices for several reasons. It allows the reduction of content of cement in the concrete, or clinker Portland in cement production, which is one of the main causes of CO_2 emissions, being responsible for about 7% of global greenhouse gas emissions (Perugini et al., 2014; Hendriks et al., 1998). Moreover, in most cases, SCM use in concrete determines higher durability, increasing resistance to chemical damage and leading to higher alkali-silica reaction mitigation (ASR). In the same time, it can also lead to better physical performance, such as lower permeability, lower shrinkage, or lower hydration heat.

SCM from argillaceous soils, especially if they contain kaolin, can be produced by means of calcination, in order to transform alumina-silicates in artificial pozzolan, as recommended in the common industrial practice (Chakchouk et al., 2009; Cook, 1986; Johanson and Anderson, 1990). In the case of not strong, or insufficient activation of the pozzolanicity, alternative low strength binders, chemically activated geopolymers, fillers or artificial aggregates, could be produced, especially if they result as adapted to use in the same infrastructure construction ring. (Perugini 2018a; Perugini 2018b). Nevertheless, in some cases, a lower activation could be due to the convenience of a lower energy treatment, which could lead to accepting a lower quality of the final product, in consideration of the saving in the treatment or due to the unavailability of higher energy for treatment. The following research work demonstrates as, in some cases, also for clay without kaolin, low energy of activation could be possible. This solution could result as convenient, offering the possibility to reuse tunnelling waste in infrastructure works through a low cost of treatment, by means calcination at temperatures lower than 600°C, instead of the more common and energy consuming treatment at 750°C or higher, e.g.1000°C.

2 MATERIALS AND METHODS

A representative sample from different points of the same geological formation involved in the tunnelling work, was sampled, premixed, homogenized and assessed regarding the chemical and mineralogical composition, using X ray techniques, such as XRF and XRD.

After chemical and mineralogical characterization, the soil sample was treated by means of calcination for 2 hours, at 400°C and 600°C, obtaining two different calcined, or partially calcined, materials. Following, they were ground to a grinding fineness of 600 m^2/Kg, which was assessed, step by step, until reaching the right value, by means of the Blaine method, in accordance with EN 196-6. For both temperatures of calcination, mineral dissolution and transformations were assessed through X ray diffraction (XRD), as shown in Figure 1. In both cases, chemical activation produced by means of calcination was verified through the Strength Activity Index (SAI), in comparison to fly ash and natural pozzolan, using 25% of cement replacement, following the indication of EN 450-1 (very similar to the ASTM C 618 and C311). As base cement, was used a commercial Portland cement type I class 52,5 R, produced according to EN 197-1, containing more than 95% of clinker Portland, having alkali content of 1%, as shown in Table 1. The same Portland cement was also adopted for the production of two experimental Pozzolanic blended cements, in accordance to EN 197-1, using each calcined soils.

Soil samples show very low loss on ignition (L.O.I.). That treated at 400°C indicates not relevant mineralogical transformations, while that treated at 600°C show evident changing of some phases. Both products, treated at 400 and 600°C, were used in the production of two experimental Pozzolanic cements containing respectively 20% and 50% each of calcined clay, in replacement of the same percentages of Portland cement. Following, in order to evaluate the possibility to increase the strength activity, or hardening contribution, in a chemical way, a sort of hybrid binder OPC-AAC-Geopolymer (ordinary Portland cement - alkali active cement - Geopolymer), containing 50% of calcined clay treated at lower temperature, and

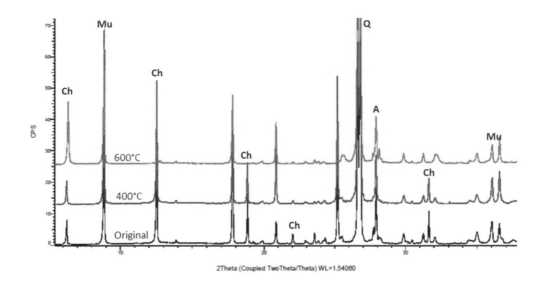

Figure 1. XRD spectra of tunnelling waste sample and mineralogical transformations at temperature of 400, 600; minerals and their dissolving or transformation are indicated by means the symbols: Q = quartz; Ill = illite; Mu = muscovite; Cc = calcite; A = albite; Ch = chlorite/clinochlore. Quartz is the main element, calcite and chlorite are present in lower percentages, similar to mica component muscovite, and plagioclase, as albite.

Table 1. XRF data of tunnelling waste samples, natural pozzolan, fly ash and Portland cement.

	Chemical characterization											
	L.O.I.	SiO_2	Fe_2O_3	Mn_3O_4	TiO_2	P_2O_5	Al_2O_3	CaO	MgO	SO_3	Na_2O	K_2O
Soil waste - Noth Italy	0,70	56,63	8,15	0,24	0,42	0,01	23,37	0,34	2,67	0,50	2,28	4,68
Natural Pozzolan - Borghetto, Italy	8,97	53,04	5,44	0,12	0,59	0,15	18,19	4,67	1,46	0,23	1,63	5,30
Burning coal fly ash - Cerano, Italy	6,93	50,05	4,37	0,04	1,49	1,39	27,50	5,02	1,08	0,39	0,30	1,18
Reference CEM, type I 52,5 R	2,71	19,25	3,08	0,08	0,17	0,07	4,65	64,54	1,09	3,53	0,63	0,57

addition of NaOH as activator, was tested in concrete, in comparison to the other cement containing similar components without NaOH activator.

All experimental cements were compared to the reference Portland cement type I 52,5 R, used as base cement, and four Pozzolanic cements containing analogue percentages of natural pozzolan from Borghetto, (Viterbo - Italy) and fly ash from burning coal from Cerano power plant (Brindisi - Italy), regarding pozzolanicity (in accordance to EN 196-5), as well as for mechanical strength in standard mortar (in accordance to EN 196-1) and in concrete at equivalent cement dosage and conditions (in accordance to EN 12390 and 12350).

Finally, the mentioned experimental hybrid binder (OPC-AAC-Geopolymer), made by adding 0.2% of alkali (as NaOH) to the experimental cements containing 50% of calcined clays for compensation of low Strength Activity Index (SAI) of the calcined clay, especially detected for the sample treated at lower temperature (400°C), was assessed regarding mechanical strength in concrete, in order to verify the additional chemical activation due to the higher alkali content and relative increasing of hardening of the soil calcined at low energy. This conditions could result as very useful considering that the

517

treatment at 400°C allows to resolve the issue of pollution of the soil due to the surfactant use in EPB-TBM process.

Chemical assessment, regarding clay sample, Portland cement, natural pozzolan, fly ash, by means XRF (as alternative method in accordance to EN 196-2), are shown in Table 1.

3 RESULTS AND DISCUSSION

Chemical characterization of the sampled soil waste involved in this experimentation shows a content of silica, alumina and iron oxide higher than pozzolanic materials usually used for Pozzolanic cement production, as evident from the comparison with natural pozzolan from Borghetto and fly ash from burning coal from Cerano power plant (see Table 1).

Mineralogical composition of sampled soil shows quartz as prevalent element, with lower content of chlorite and clinochlore, mica component as muscovite, and plagioclase as albite.

Loss on ignition (LOI) of clay sample, as shown in Table 1, is lower than 1% at 950°C, in accordance to EN 196-2, indicating not presence of water and calcite. Evident mineralogical transformation occurs at temperature higher than 400°C, involving at first chlorite and clinochlore, which are totally decomposed at 600 °C. (Johanson and Anderson, 1990; Kakali et al., 2001; Cook, 1986).

As shown in Table 2, after thermic treatment, soil treated at 600 °C match the prescribed limits of EN 450-1 for strength activity index (SAI) of fly ash from burnig coal (as well as ASTM C 618). In fact, SAI test indicate for this soil, if treatment at temperature at 600°C, a value higher than the limit of 0.75 at 28 days and 0.85 at 90 days. Instead, sample treated at 400°C shows a value of 0.75 at 28 days and a value lower than the limit of 0.85 at 90 days, being 0,84.

Regarding chemical assessment, as shown in Table 2, data match the prescribed parameter of content of silica, alumina and iron oxide, being their sum higher than 70 %. Regarding the values of reactive silica content, the sample treated at 600°C match the EN 197-1 and 450-1 limits of 25%, being 25,4%, while the sample treated at 400°C do not match the minimum value of 25%, showing a content of 22,7%.

The hardening contribution of these calcined clays is confirmed by mechanical strength in standard mortar (in accordance to EN 196-1) of the experimental cements containing 20% of them, as shown in Table 3 and Figure 2.

Tests indicate that compressive strength in standard mortar at 2 and 28 days, for both treatment temperatures, are very similar to the cement containing natural pozzolan and fly ash in the same percentages. Mechanical strength at 2 and 28 days are respectively of 24 MPa and 45.6 MPa for clay calcined at 400°C and 25.5 and 48.9 MPa for the sample calcined at 600°C. Instead, cements containing 50% of them show mechanical strength in standard mortar of 10.9 MPa at 2 days and 35 MPa at 28 days, for clay calcined at 400°C, and 13.9 and 39 MPa for the sample calcined at 600°C.

Table 2.   Strength Activity index (EN 450-1 at 28 and 90 days); reactive SiO_2 (EN 197-1 and EN 450-1); sum of silica+alumina+iron oxide (ASTM C618 and EN 450-1) of natural pozzolan, fly ash and calcined clay.

	SAI test 28 days	SAI test 90 days	Reactive SiO_2 (%)	SiO_2+ Al_2O_3+ Fe_2O_3 (%)
Limits (EN 450-1 and EN 197-1)	0,75	0,85	25	70
Natural Pozzolan	0,78	0,9	37,2	76,6
Fly Ash	0,78	0,89	38,1	81,9
Calcined clay - 400°C	0,75	0,84	22,7	88,2
Calcined clay - 600°C	0,77	0,91	25,4	88,4

Table 3. Mechanical strength in standard mortar (in accordance to EN 196-1) and concrete (EN 12390 and 12350) of cements containing Natural Pozzolan, Fly Ash and Calcined Clay, as clinker replacement, in percentage of 20 and 50%.

CEM	Maturation age (days)	Standard Mortar			Concrete				
		2	7	28	2	7	28	60	180
I 52,5 R		38,0	48,0	58,0	51,8	66,7	80,1	84,5	90,2
IV/A (P) 42,5 R (Natural Pozzolan = 20%)		26,0	37,3	52,7	43,2	53,5	65,0	73,5	83,7
IV/A (V) 42,5 R (Fly Ash = 20%)		25,0	40,1	52,8	41,0	52,2	64,4	74,4	84,9
IV/A (Q) 42,5 R (Calcined clay 400°C = 20%)		24,0	37,0	45,6	35,2	48,9	59,8	67,2	77,3
IV/A (Q) 42,5 R (Calcined clay 600°C = 20%)		25,5	38,0	48,9	42,8	56,7	68,1	74,9	85,7
IV/A (P) 42,5 R (Natural Pozzolan = 50%)		13,1	25,1	38,7	19,1	39,1	51,3	57,5	63,8
IV/A (V) 42,5 R (Fly Ash = 50%)		15,2	27,0	39,3	18,8	39,7	52,7	58,4	64,9
IV/A (Q) 42,5 R (Calcined clay 400°C = 50%)		10,9	22,4	35,0	17,1	35,4	46,9	53,1	59,3
IV/A (Q) 42,5 R (Calcined clay 600°C = 50%)		13,9	25,5	39,0	18,6	38,6	51,0	58,3	66,1
IV/A (Q) 42,5 R (Calcined clay 400°C = 50%) + NaOH		16,9	27,3	40,5	22,0	42,5	53,9	59,7	66,3

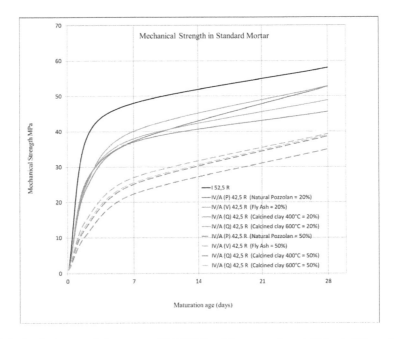

Figure 2. Mechanical strength in mortar (EN 196-1), of four experimental blended Pozzolanic cements containing 20% and 50% of tunnelling waste, calcined at 400°C and 600°C, in comparison to the reference Portland cement type I 52.5 R and four commercial cements containing natural pozzolan and fly ash in similar percentages.

Behaviour in concrete was assessed using a cement dosage of 400 Kg/m³ (as sum of reference Portland cement type I 52.5 R and pozzolanic material separately ground and added to the concrete before mixing), at same workability class (slump 18 cm) and w/c ratio (0.42), adding 1% of superplasticizer Polycarboxylate ether (PCE).

As shown in Table 3 and Figure 3, compressive strength at 28 days of concrete made with cement containing clays calcined at 600°C are equivalent, or higher, than those containing the same percentage of natural pozzolan or fly ash, being 68,1 MPa, for cement containing 20% of calcined clay and 51,0 MPa for the cement containing 50% of the same calcined soil.

Mechanical strength in concrete at long term of this type of calcined soil, treated at 600°C, after 180 days of maturation, confirms the evident good pozzolanicity. In fact, they match perfectly those of concrete containing natural pozzolan and fly ash commonly used, showing a rate of increasing higher than that of the reference Portland cement, type I class 52.5 R, and reaching strength higher than 85 MPa at 180 days for percentage of 20%. Lower strength shows calcined clay treated at 400°C, confirming the lower strength activity index (SAI). Nevertheless, as shown in Figure 5, mechanical strength in concrete could be affected and improved by other aspect, as e.g. alkali content.

Figure 4 shows the pozzolanicity of cements containing 20% and 50% of the soil involved in the experimentation, calcined at 400 °C and 600°C, assessed in accordance to the EN 196-5 and compared to cements containing 20% and 50% of natural pozzolan and fly ash, named IV/A (P) and IV/B (P) in case of use of natural pozzolan, or IV/A (V) and IV/B (V) in case of fly ash.

Data after 15 days indicate than, using high content of calcined soil (e.g. 50% in the cement type IV/B (Q) 32.5 R), also if treated at 400°C, this type of material allows to produce sufficient pozzolanicity, as capability to combine the lime, fixing it by means not soluble end-product, such as calcium silicates and aluminates, and reducing consequently the calcium ion concentration in the cement paste. Instead, the soil sample treated at 600°C shows pozzolanicity perfectly comparable to the natural pozzolan from Borghetto (Italy) and fly ash from Cerano (Italy), which have been successfully used for many years in the production of Pozzolanic cement.

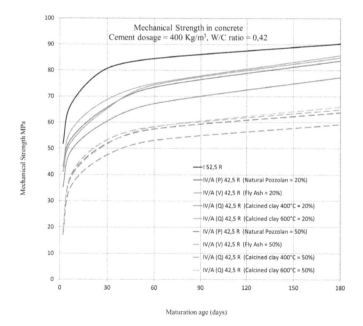

Figure 3. Mechanical strength in concrete, at equivalent workability (in accordance to EN 12390 and EN 12350) of four experimental blended Pozzolanic cements containing 20% and 50% of soil waste, calcined at 400°C and 600°C, in comparison to the reference Portland cement type I 52.5 R and four commercial cements containing natural pozzolan and fly ash in similar percentages.

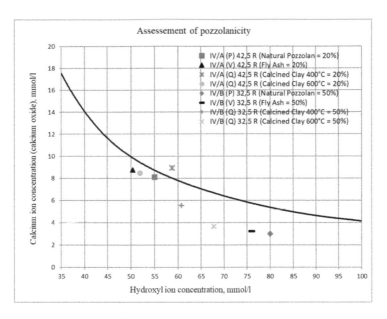

Figure 4. Pozzolanicity tests (EN 196-5) of the Pozzolanic cements containing 20% of calcined clay waste treated at 750 °C and 870 °C, natural pozzolan from Borghetto and fly ash from Cerano power plant, after 15 days.

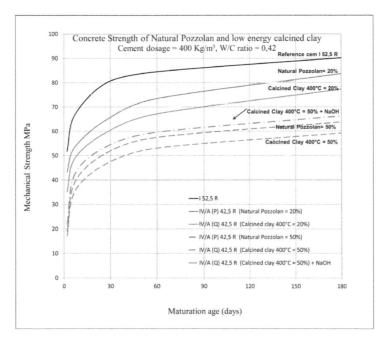

Figure 5. Mechanical strength in concrete, at equivalent workability class (in accordance to EN 12390 and EN 12350) of three experimental blended Pozzolanic cements containing 20% and 50% of soil waste, calcined at 400°C, without and with alkali activation, in comparison to the reference Portland cement type I 52.5 R and two commercial Pozzolanic cements containing the same percentages of natural pozzolan.

Similar to these, it assures sufficient pozzolanicity also using only 20% of addition as clinker Portland replacement. The percentage of 20% is the minimum usually necessary to reach the required reduction of calcium ion concentration under the saturation line, in order to allow the commercialization of blended cement as Pozzolanic cement (e.g. cement type IV/A (Q) 42.5 R).

Regarding the final experimentation, as shown in Figure 5, a sort of hybrid binder (OPC-AAC-Geopolymer), activated by means the addition of 0.2% of alkali (as NaOH) to the experimental cement containing 50% of calcined soil treated at 400°C, was tested for concrete strength comparison, in order to evaluate the possibility to increase the activation in a chemical way and to compensate the low reactivity which the low Strength Activity Index (SAI = 22.7) indicates. The result shows clearly the possibility to improve chemically the reactivity of the cement, also in case of not very reactive calcined soil, allowing the optimization of treatment and reusing cost.

4 CONCLUSION

This study shows the possibility of reusing argillaceous tunnelling waste as pozzolanic material for cement, or supplementary cementitious material (SCM) for concrete, or low grade binder, or e.g. in alkali activated cement (AAC), or similar hybrid binders, as tested in this experimentation.

Finally, reusing of clay waste in tunnelling work leads to increased sustainability in several ways. It allows the reduction of CO_2 emission thanks the possibility to replace cement Portland in the concrete or clinker Portland in the cement, as well as reducing quarrying and landfill dumping. At the same time, the addition to the concrete of pozzolanic materials improve durability of structures and infrastructures. Quality of the products, as addition for cement or supplementary cementitious material for concrete, can be managed considering costs and convenience of the treatment. Decisions should consider relocation cost of the tunnelling waste, surfactant presence from tunnelling process, as well as the necessity and availability of pozzolanic material in the construction works. This is a clear example of circular economy, from which both the natural and urban environment can strongly benefit.

REFERENCES

Perugini V., Paris E., Giuli G., Carroll M. R. 2014. The use of urban ceramic wastes in eco-sustainable durable cement. 34th Cement and Concrete Science Conference 2014. University of Sheffield.

Perugini V. 2018a. Nobiltation of clay waste from tunnelling. 38th Cement and Concrete Science Conference 2018. University of Coventry.

Perugini V. 2018b. Nobiltation of clay waste from tunnelling as building material. Word Tunnelling Conference 2018. Dubai.

Hendriks C.A., Worrell E., Price L., Martin N., Ozawa Meida L., De Jager D., Riemer P. 1998. Emission reduction of greenhouse gases from the cement industry. Proceedings of the 4th International Conference on Greenhouse Gas Control Technologies, Interlaken, Switzerland.

Chakchouk A., Trifi L., Samet B. and Bouaziz S. 2009. Formulation of blended cement: Effect of process variables on clay pozzolanic activity. Construction and Building Materials. 23, 1365–1373.

Cook D.J. 1986. Calcined clay, shale and other soils. In: Cement Replacement Materials. Swamy RN (ed), Surrey University Press, London. 40–72

Johanson and Anderson. 1990. Pozzolanic activity of calcined moler clay. Cement & Concrete Research. 20, 447–452.

Tunnels and Underground Cities: Engineering and Innovation meet Archaeology,
Architecture and Art, Volume 2: Environment sustainability in
underground construction – Peila, Viggiani & Celestino (Eds)
© 2020 Taylor & Francis Group, London, ISBN 978-0-367-46579-7

Large civil works jobsites in densely populated urban areas

F. Poma, S.F. Caruso, F. Ruggiero, V. Bianco & L. Lampiano
COCIV, Consorzio Collegamenti Integrati Veloci, Genova, Italy

ABSTRACT: the construction of new high-speed freight railway line between Milan and Genoa, the "Terzo Valico dei Giovi", implied the installation of 12 operational sites supporting the works development along the project alignment. Fegino jobsite, the southernmost one, serves the works of 12 km of the line, mostly underground, creating a complex system of different type of tunnels and several bypasses. The jobsite, located in Genoa district, has been settled in a heavily urbanized area. Because of the morphology and territory accessibility, the installation of the jobsite for a big infrastructure has been an adverse activity. It has been necessary to face and solve several problems in order to start and carry on works through technical and operational complex solutions that improve the organization of the works and reduce the initial estimated completion time.

1 PROJECT DESCRIPTION

The railway Milan–Genoa, part of the High Speed/High Capacity Italian system (Figure 1), is one of the 30 European priority projects approved by the European Union on April 29th 2014 (No. 24 "Railway axis between Lyon/Genoa – Basel – Duisburg – Rotterdam/Antwerp) as a new European project, so-called "Bridge between two Seas" Genoa – Rotterdam. The new line will improve the connection from the port of Genoa with the hinterland of the Po Valley and northern Europe, with a significant increase in transport capacity, particularly cargo, to meet growing traffic demand.

The "Terzo Valico" project is 53 Km long and is challenging due to the presence of about 36 km of underground works in the complex chain of Appennini located between Piedmont and Liguria. In accordance with the most recent safety standards, the under-ground layout is

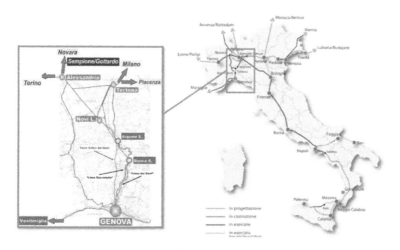

Figure 1. High-speed Italian system.

formed by two single-track tunnels side by side with by-pass every 500 m, safer than one double-track tunnel in the remote event of an accident.

The layout crosses the provinces of Genoa and Alessandria, through the territory of 12 municipalities.

To the South, the new railway will be connected with the Genoa railway junction and the harbor basins of Voltri and the Historic Port by the "Voltri Interconnection" and the "Fegino Interconnection". To the North, in the Novi Ligure plain, the project connects ex-isting Genoa-Turin rail line (for the traffic flows in the direction of Turin and Novara – Sempione) and Tortona – Piacenza –Milan rail line (for the traffic flows in the direction of Milan- Gotthard).

The project crosses Ligure Apennines with Valico tunnel, which is 27 km long, and ex-its outside in the municipality of Arquata Scrivia continuing towards the plain of Novi Ligure under passing, with the 7 km long Serravalle Tunnel, the territory of Serravalle Scrivia (Figure 2). The underground part includes Campasso tunnel, approximately 700 m long and the two "Voltri interconnection" twin tunnels, with a total length of approximate-ly 6 km.

Valico tunnel includes four intermediate adits, both for constructive and safety reasons (Polcevera, Cravasco, Castagnola and Vallemme). After tunnel of Serravalle the main line runs outdoor in cut and cover tunnel, up to the junction to the existing line in Tortona (route to Milan); while a diverging branch line establishes the underground connection to and from Turin on the existing Genoa-Turin line.

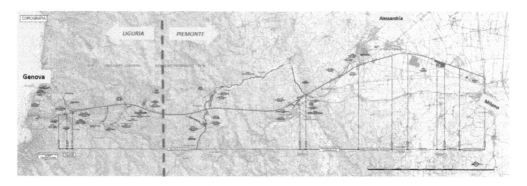

Figure 2.   Terzo Valico project.

From a construction point of view, the most significant works of the Terzo Valico are repre-sented by the following tunnels:

- Campasso tunnel 716 m in length (single-tube double tracks)
- Voltri interconnection even tunnels 2021 m in length (single-tube single track)
- Voltri interconnection odd tunnels 3926 m in length (single-tube single track)
- Valico tunnel 27250 m in length (double tube single track)
- Serravalle tunnel 7094 m in length (double tube single track)
- Adits to the Valico tunnel 7200 m in length
- Cut and cover 2684 m in length
- Novi interconnection even tunnels 1206 m in length (single-tube single track)
- Novi interconnection odd tunnels 958 m in lenght (single-tube single track)

The project standards are: maximum speed on the main line of 250 km/h, a maximum gradient 12,5 ‰, track wheelbase 4,0 – 4,5 m, 3 kV DC power supply and a Type 2 ERTMS signalling system.

2   FEGINO JOBSITE

The Fegino construction yard is the southernmost site of the entire area of the Terzo Valico, located within the urban area of Genoa, in Trasta zone. The site supplies the excavation of five tunnels, for an overall of 12 km, 60% of which has been already carried out:

- The Galleria Campasso (the first one). Is a double-track tunnel that crosses around 700 m of a mountain range;
- The first section of the two single-track high speed/high capacity tunnel of Terzo Valico (about 5.400 km)
- Single-track interconnection tunnels that link the high speed/high capacity tunnels to the port of Voltri. These are about 6.000 m of tunnels which will be linked with the high speed/high capacity tunnels through grafting chambers.

The project includes also several pedestrian bypasses for connection between the even and odd track tunnels. The presence of an high number of tunnels gives to the jobsite a mine-like layout.

Furthermore the project provides other external works like: a new road system, a technological square and the renewal of the railway's plans and other civil works in order to quadruple the capacity of the existing line.

The Fegino yard is divided into two different areas, splitted by the Galleria Campasso: one at the southern entrance and the second located between the north entrance of the city and the southern entrance of the Valico Tunnel.

In the following article, will be also deepened the choice of the realization of a construction shaft in order to reduce the excavation time of the tunnels.

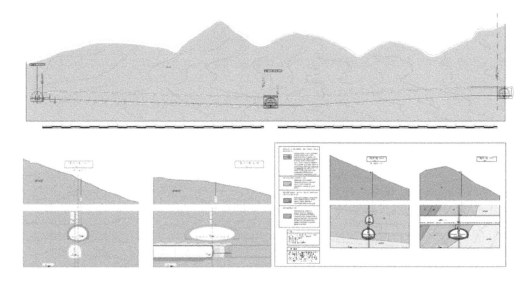

Figure 3. Rappresentative sections in Fegino worksite tunnels.

3 FEGINO JOBSITE – CAMPASSO TUNNEL, SOUTH PORTAL

The site is located in a morphologically depressed area, adjacent to the urban context of the city of Genoa and bordering on the east side with the existing railway line and with a residential area and the west side.

In order to reduce the overall time of the project, the mobilization of the Fegino site was simultaneously carried out with the excavation works. This choice added further complexity to the already not simple operations of the project.

From the previous picture it is possible to notice that the area identified for the Fegino site did not allow an easy access for the construction vehicles. For this reason it was decided to access the area with small-sized work vehicles, which allowed us in the first phase, to carry out the first activities for the mobilization of the site. We initially proceeded using exclusively internal earth movements for a volume equal to about 10,000 cubic meters, in order to realize the construction site plan necessary to build the bulkhead of the Galleria Campasso.

Figure 4.   Anteoperam area.

The operations inside the construction site also included the hydraulic adjustment of a small water stream interfering with the railway, which was provisionally channeled into an Armco-type pipeline.

Simultaneously with the construction of the south entrance of the Galleria Campasso, a box-like structure measuring 20 meters was completed. The purpose of the building was to create a very important suitable access to the site areas. The artifact was realized by pushing the box in the soil 80 centimeters below the railway embankment, using some hydraulic jacks. It is important to notice that the normal railway line activity was always guaranteed thanks to the realization of an Essen bridge, useful to support the rails during the pushing phases.

The pushing phase lasted between 4 and 5 days, using 4 hydraulic jacks at a pressure of about 20.000 kN, for a length of 25 m.

In just 6 months (October 2013 - April 2014) it was possible, to create the necessary access for the excavation works of the tunnel, and at the same time it was completed the bulkhead for the entrance of the tunnel. In this way it was possible to enter the site with heavy equipment vehicles and to start the activities for the mobilization of the construction site.

The first phase of construction included the installation of the plant engineering and tooling necessary to start the tunnel excavation (injection system for ground consolidation, generator, cabins for electrical distribution and transformation, electrical system, plumbing systems, fire system, air compression system, ventilation system ...). This let us start the tunnel excavation while the work to complete the construction of the site was still under way.

After the completion of the Galleria Campasso, operations of enlargement of the construction area continued quadrupling the available area, from the 5000 square meters of the first phase, to the 20,000 square meters of the final configuration. Furthermore, all the necessary preparations for the good health of a construction site were installed: concrete plant, mechanical and electrical workshop, warehouse, offices and the area used for the storage of the material necessary for the construction of the work (steel ribs, steel rebars and mucking).

Figure 5. Earth moving (a, b, c), south portal of Campasso tunnel realization (c), pipejacking tunnel realization (d, e).

4 FEGINO JOBSITE – CAMPASSO TUNNEL, SOUTH PORTAL – VALICO TUNNEL, NORTH PORTAL

At the same time of the execution of Phase 1 of the area in front of the southern entrance of the Galleria Campasso, the construction of the site area was carried out at the north entrance of the Galleria Campasso and the south entrance of the Valico tunnel. This area once completed, had three important purposes: the connection of the Galleria Campasso to the Valico tunnel, to host the ventilation system for tunnel excavation and to host the injection plant for consolidation of the initial part of the Valico tunnels.

The realization of the construction site lasted 21 months, from July 2014 to March 2016, ending with the demolition of the provisional Armco pipeline of the Trasta stream.

The construction site between the two entrances is located within a very small and steep valley, divided by the stream Trasta, a small river that runs at the base of a gorge. For this

Figure 6.   Fegino jobsite – Phase 2 realization.

Figure 7.   Portals realization.

Figure 8.   Anteoperam and completed Jobsite area.

reason, the construction of the two tunnel entrances and the mobilization of the site area was extremely difficult. For this reason, it was necessary to build a track to access to one side of the valley.

Exploiting the knowledge of naturalistic engineering, it was possible to create an easy access to the south entrance of the Valico tunnel, using many palings and fascinations 12 meters high. This let us start to build the entrance of the tunnel, to cross the Trasta stream using ARMCO type pipeline (diameter 6000 mm) and to build the access track to the north entrance of the Galleria Campasso using other palisades.

Once the entrances to the tunnels and the Galleria Campasso were completed, the final hydraulic box was created progressively to allow the excavation of the tunnels and then the passage of the work vehicles.

5   FEGINO JOBSITE – A NEW BYPASS

As previously shown the Fegino yard is responsible for the two Valico tunnels and two Voltri interconnection tunnels (up track and down track). These two consisting of two single track tunnels, with a total development of approximately 6.000 m (2.200 m down and 3.800 m up), which are linked to the Valico tunnels through grafting chambers.

In order to speed up the realization of the Voltri interconnections, a bypass tunnel has been devised to intercept the two secondary lines in a middle position between the Galleria Campasso and the interconnections. This allowed us to excavate simultaneously on two opposite directions (north and south). This connection consists of a first branch connecting the high-speed line to the up tunnel and a second branch connecting the down and up inter-connections tunnels.

The first branch (between the high-speed line and the up interconnection), is a tunnel not included in the project scope of work and therefore is considered part of the mobilization of the Fegino construction site. The second branch instead, it develops on a pedestrian by-pass connection already included in the scope of work. It was necessary then to adapt the project in order to make the by-pass drivable with heavy equipment.

Originally, an alternative route had been proposed for the construction of the bypass between the high-speed line tunnels and the interconnection tunnels. In this initial idea, the entrance of the by-pass was close to the south entrance of the Valico tunnel (only 18 m from the entrance bulkhead of the pass tunnel). According to this first hypothesis, the layout of this solution (length of about 100 m and longitudinal slope 11.32%) would have been critical because of small lateral coverage in the entrance area and with low roof coverage. Furthermore, this first idea would have affected a geomorphologically very critical area, characterized by superficial landslides, and presence of houses located next to the area of grafting with the line.

For this reason, an additional solution was analyzed and subsequently applied. This last one envisaged the displacement of the bypass in south direction, linking to the Galleria Campasso at a longer distance from the northern entrance, in order to allow a proper balance between the need to excavate with suitable coverings. Moreover, it was possible to move away from landslide-interested areas, but not to overstretch the development of the connecting tunnel.

In the end the applied solution provided for the realization of the bypass started from the Campasso tunnel, with grafting at about 80 m from the north entrance, with a path about 235 m long and a slope of 3.35%. The tunnel passed through an area not characterized by particular geomechanical problems with average stiffness.

This solution, with respect to the original hypothesis, also entailed the following advantages:

– Separation between the access to the high-speed and interconnection tunnels, reducing drastic the interference between building sites;
– It has been possible to avoid all the supplies of the entire construction site (six simultaneous excavation sites) to pass through only one entrance;

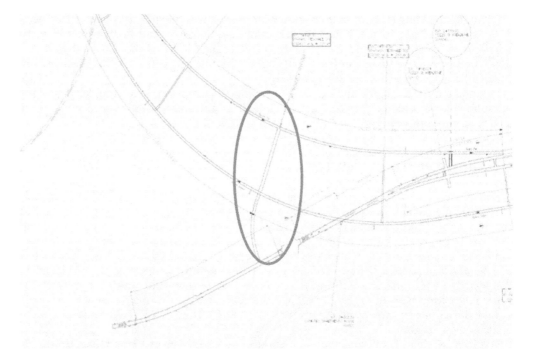

Figure 9. New bypass position.

– Direct connection between high-speed line and Voltri interconnection tunnels:
In this way all the interconnections are connected to the high-speed line by a autonomous path, allowing time saving for the excavation of the Interconnection tunnel (even track) and involving an importance reduction of the traffic load on a stretch of 200 m of the up track.
Moreover, it reduced the number of connections from 4 (line - 1st bypass, 1st bypass - BD interconnection, BD interconnection - 2nd bypass, 2nd bypass-BP nterconnection), to 3.
– Elimination of ventilation well during the mobilization phase:
The original hypothesis envisaged the realization of a ventilation well, which became necessary because of lacking of room in the southern entrance of the Valico tunnel, to host six ventilation pipes and their engines. The alternative solution proposed, allows the installation of four ventilation pipes inside the Campasso tunnel up to the engagement with the bypass connection.
– By-pass slope reduction (from 11.32% of the original solution to 3.35%):
– This significant reduction of slope, led to benefits in the quality and functionality of accessibility to interconnection sites.
– Reduction of the impact on the construction site area between the Galleria Campasso and the Valico tunnel because of reduction of the number of vehicle transits.

6 CONCLUSIONS

A good organization of the works, a careful preparation of the realization phases and a detailed study of the specific area permitted to achieve a great result, settling a functional construction site in a difficult area, during tunnels excavation.

In order to achieve this good result, two factors were essential. The first factor was to create a network between the people living near the jobsite and the construction site itself. In order to succeed in the settlement of the construction site, creating continuative neighborly relations is a very important factor, permitting to prevent many potential troubles. A good neighborly relation is also the right thing to do: the construction site is changing their daily life, by noise, dusts, vibrations, sight...

The second factor is connected to the monitoring of the area before the beginning of the works, and, in general, is related to the ante operam data, in order to highlight and prevent troubles associated with the tunnels excavations and works in general.

Tunnels and Underground Cities: Engineering and Innovation meet Archaeology,
Architecture and Art, Volume 2: Environment sustainability in
underground construction – Peila, Viggiani & Celestino (Eds)
© 2020 Taylor & Francis Group, London, ISBN 978-0-367-46579-7

Monitoring the water resources in the large railway projects: The Fortezza – Ponte Gardena case study, Southern Brenner Base Tunnel access

S. Rodani, G. Penna & F. Marchese
Italferr SpA, Rome, Italy

A. Scuri, F. Sciascia & D. Castioni
ENSER srl, Engineering company, Faenza (RA), Italy

L. Piccinini
Dipartimento di Geoscienze, University of Padova, Padova, Italy.

F. Cervi
DICAM, University of Bologna, Bologna, Italy

G. Benedetti
Enser s.r.l., Engineering company, Faenza (RA), Italy
DICAM, University of Bologna, Bologna, Italy

ABSTRACT: The Brenner Base Tunnel (BBT) is being built to connect Italy and Austria running from Fortezza to Innsbruck (55km). The southern access will link Ponte Gardena with Fortezza. As widely known, underground tunneling works may have significant impact on regional hydrogeological systems, causing a drawdown of water levels and a decrease of springs' discharge rates. Therefore, an extended monitoring of water resources can be crucial to define the possible interference between the tunneling and the surrounding environment. The analysis presented in this paper allowed to determine the main hydrogeological regime, characterizing this portion of the Isarco valley. The hydro-chemical analysis improved the knowledge of groundwater paths, mainly characterized by low-mineralized flows, running within the shallow portion of the rock mass or in correspondence of quaternary glacial deposits and not related to the deep-water network. Finally, isotopic correlations between rain and underground water have confirmed this hydrogeological layout.

1 INTRODUCTION

The case study presented in this paper refers to the southern access of the Brenner Base Tunnel (BBT), which is located in Bolzano (northern Italy). The designed tunnels will link Ponte Gardena to Fortezza (southern entrance of BBT) through two tunnels named Gardena and Scaleres, 6 km and 15 km long respectively.

The geological layout of the area is characterized by the presence of a metamorphic basement prevalently formed by phyllites with the occurrence of localized Permian granitic and dioritic intrusions. Thick quaternary deposits (fluvio-glacial and alluvial deposits) usually cover the slopes and the main valley in which river Isarco flows.

A comprehensive hydrogeological survey started in 2003 and is still ongoing. It consists of water sampling activities throughout an area larger than the one that will be involved in the tunnel excavation. This was necessary to identify areas in the vicinity of the tunnel that may be

Table 1. Water points surveyed in the area.

Springs	Wells	Rivers and river diversions	Piezometers	Rain water collectors
667	183	337	105	6

impacted by tunneling. As a result, a total of 1298 water points (namely: springs, wells, rivers, piezometers, precipitations) have been monitored in 400 Km^2. Water points resumed in Table 1.

Most of water points are represented by springs that can be considered as the most representatives of the hydrogeologic network.

2 SPRINGS DISCHARGE RATES

A preliminary analysis of springs' discharge rates (for 418 springs with at least 1 measure derived from different data sources such as Provincia Autonoma di Bolzano and Italferr database) have been performed.

The results are shown in Figure 1. They show that springs with average discharge rates lower than 1 l/s are about the 85%. This means that groundwater network is characterized by diffuse shallow water flows, which are not able to produce significant water discharge points (except for some localized area developed along tectonic alignments, such as Scaleres and Spelonca valleys).

3 MONITORING PROGRAM

A reduced number of water points were selected for further hydrochemical monitoring because of relevant flow rates magnitude and proximity to the tunnel (Table 2).

A total number of 198 points have been monitored in last 2017 campaign.

In Figure 2 the distribution of monitored water points is shown in relation to the railway track alignment and drainage basin subdivision, useful to determine the presence of water points clusters with different chemical characteristics and origin.

To define the main hydrochemical characteristics of water points a complete set of analysis, consisting of physico-chemical parameters (discharge rate, water depth, temperature,

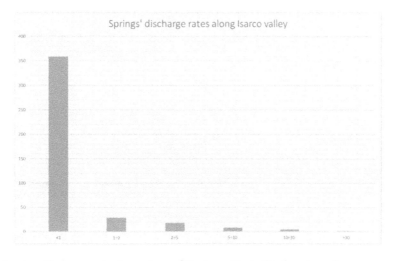

Figure 1. Average discharge rates for springs of Fortezza-Ponte Gardena tunnel.

Table 2.   Water points monitored in 2017/18 campaign.

Springs	Wells	Rivers	Piezometers	Rain water collectors
96	14	15	67	6

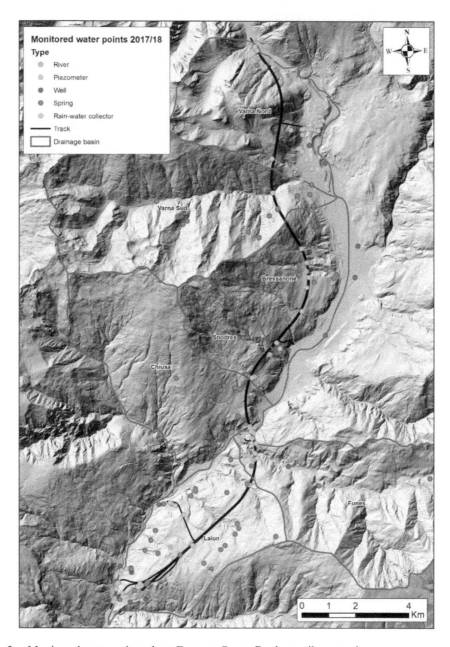

Figure 2.   Monitored water points along Fortezza-Ponte Gardena railway track.

conductivity, pH, oxygen, ORP and turbidity), major chemical elements (bicarbonate, calcium, chloride. magnesium, nitrate, potassium, silica, sodium and sulphate), trace elements and water isotopes ($\delta^{18}O$, δ^2H, ^3H), has been performed.

4 PHYSICAL AND CHEMICAL PARAMETERS OF SPRING WATER

4.1 *Specific Conductivity(25°C) SC*

This parameter expresses the quantity of dissolved mineral in water and provides useful information about the degree of mineralization.

In fact, the analysis is useful to differentiate potential shallow water flows (lower values of SC) from those of deeper origin (higher values of SC). Based on the homogeneous lithologic characteristics of the project area (manly phyllites, granites, diorites) the investigation can be considered reliable for the scope of work.

Data presented in Figure 3 show that about 55% of spring water have a SC lower than 200 µS/cm indicating an overall low degree of mineralization. Anyway, an increase in SC develops from the top of the mountains (right part of Figure 3) to the valley (left part of Figure 3) following an inverse exponential relationship. This means that shallow waters progressively increase their mineralization as a result of the interaction with the rock bulk.

A reduced number of points present unusually higher or lower SC values that deviate from the exponential relationship. These points have been reported with their identification code in Figure 3 and further analyzed with water isotopes to estimate their recharge areas (see 6.1).

4.2 TEMPERATURE

A similar relationship than that of SC has been found for water temperature. In fact, springs at lower altitude exhibit high average temperatures (up to 16°C) while points located at higher elevation have low average temperatures (up to 4°C). A linear relationship between springs temperatures and altitudes have been found (Figure 4). As in the case of SC, monitored water points deviating from the linear relationship have been reported with their identification code and further analyzed with water isotopes to estimate their recharge areas (see 6.1).

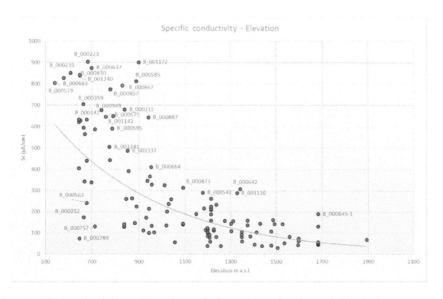

Figure 3.   Specific Conductivity vs Elevation ratio for monitored spring point in study area.

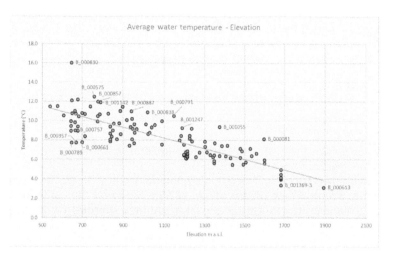

Figure 4. Temperature vs elevation ratio for monitored springs in project area.

5 GEOCHEMICAL FACIES

To determine the geochemical characteristics of water a Piper plot diagram is used (Figure 5). It is a ternary diagram where cations and anions are plotted to define the chemical characteristics of water.

Most of the spring water belong to the Calcium-Bicarbonate hydro-facies (78%). These waters are the results of dissolution processes taking place within quaternary deposits that cover the mountain slopes. Even very low quantities of calcite, due to its high solubility, determine a significant increase in bicarbonate content.

The other main hydro-facies is that of calcium-sulfate water (22%). The sulfate ion is due to dissolution processes involving Sulphur mineralization within phyllite rocks.

Due to the high number of water samples (around 400), Varna Sud and Bressanone drainage basin are reported separately (Figure 6). As represented in Figure 7, well and river water follow an analogous distribution, in terms of geochemical facies, as springs, confirming the similar origin of the water flows.

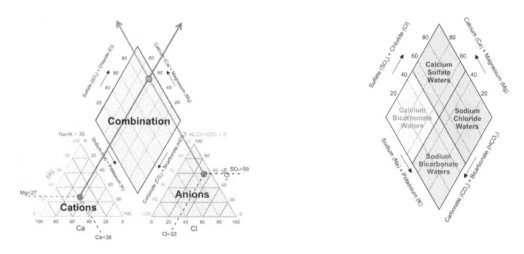

Figure 5. Piper diagram example with different geochemical facies obtained in final diamond plot.

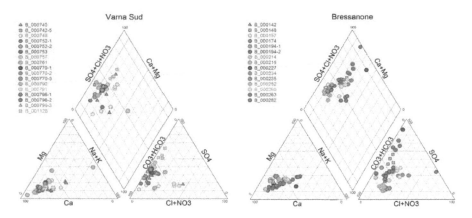

Figure 6.   Piper plot for springs in Varna Sud and Bressanone drainage basins.

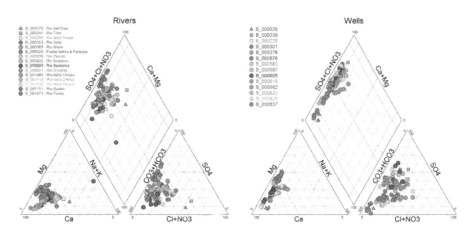

Figure 7.   Piper plot for Wells and Rivers across all the area.

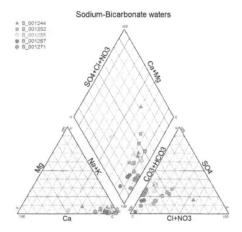

Figure 8.   Piper plot for Sodium-Bicarbonate water from some of deep piezometers water.

Geochemical analysis of piezometers' water highlights a distribution like that already seen for springs. The only difference is represented by the presence of 5 water points (Figure 8) falling into the sodium-bicarbonate facies. Higher contents in Na+ and/or K+, coupled with high HCO3 and variable SO_4^{2-} concentrations is likely attributable to dissolution of alkali minerals such as aluminosilicates (sodium and potassium plagioclase), supported by the conversion of aqueous CO2 into bicarbonate ion and, partially, by oxidative dissolution of pyrite. These samples can be considered as the expression of a deeper and older water circulation.

6 ISOTOPES

Water isotopes provide significant information concerning groundwater flow paths, with emphasis to recharge altitudes and residence time within the aquifer.

Four water collectors distributed in southern and northern portion of the area (as shown in Figure 2) have been used starting from 2014 and have been then supported by two additional PALMEX rain samplers, specifically built for this scope.

This was necessary to compare the isotopic characteristics of rain water with those of groundwater and to define the variations, in isotopic content, occurred from infiltration to discharge at single monitored water point.

Analyzed isotopes are represented by ^{18}O (oxygen), ^2H (deuterium) and ^3H (tritium). The first two isotopes are stable isotopes, which remain unchanged after their formation (rain drops), while the third one (tritium) is an unstable isotope that undergoes a radioactive decay which generates a new chemical element ^3He (Helium) and a decay product β^-.

6.1 STABLE ISOTOPES ($\delta\,^{18}O$, δ^2H)

The concentration of these isotopes is expressed as a relative difference δ (‰) between the isotopic ratio of the sample (e.g.: $^{18}O/^{16}O$ o $^2H/^1H$) and that of a reference standard (RV-SMOW where V-SMOW is the acronyms of Vienna Standard Mean Ocean Water) as shown in Equation 1:

$$\delta(‰) = \left(\frac{R_{SAMPLE}}{R_{V-SMOW}} - 1\right) \cdot 1000 \tag{1}$$

Where R_{SAMPLE} = Isotopic content of the sample; $^R{}_{V-SMOW}$ = Isotopic content of a standard sample.

During the monitoring time (June 2014 to July 2017) rain water showed an average $\delta^{18}O$ value of -8.5‰, while springs -10.2‰, wells -10.6‰ e piezometers -10.3‰. Minimum values of -12.0‰ (rain water), -12.20‰ (springs), -12.3‰ (wells) e -12.1‰ (piezometers); maximum values of -0.5‰ (rain water), -8.9‰ (springs), -8.45‰ (wells) e -8.7‰ (piezometers).

These data indicate a depletion of $\delta^{18}O$ in groundwater compared to rain water. Furthermore, a low variability of the isotopic data emerges in groundwater (springs: $1\sigma = \pm 0.6$‰; wells: $1\sigma = \pm 0.9$‰; piezometers: $1\sigma = \pm 0.7$‰) when compared to rain water ($1\sigma = \pm 2.4$‰).

The higher depletion that distinguishes groundwater, combined with lower variability of isotopic data, suggests both a recharge occurring at higher altitudes and a homogenization role played by the aquifers. This process tends to smooth the isotopic variability of groundwater when compared to rain water precipitating in different periods of the year and at different altitudes. In fact, at lower temperatures (winter) depleted values of $\delta^{18}O$ are expected while at higher temperature (summer) enriched values of $\delta^{18}O$ are found.

Moreover, and due to the "altitude effect" (the progressive decrease in air temperature with the elevation), an isotopic gradient can be found for rain water precipitated at different altitudes. This makes possible to calculate the mean altitude at which rain water infiltrated and thus estimate the groundwater flow paths length (Mazor, 1997). An average decrease of

-0.3‰/100 m for $\delta^{18}O$ and of -2.5‰/100 m for δ^2H is reported for the entire Italian territory by Celico,1986. Other authors found different gradients (Longinelli, 2003; Penna, 2014) in areas close to the Isarco valley.

Evaluating available isotopic data from rain collectors (2014–2017), it has been possible to calculate the $\delta^{18}O$ gradient for the area. A decrease of – 0.25‰/100m has been defined, in line with those presented by the abovementioned Authors.

Data for monitored water points have been compared to the $\delta^{18}O$ gradient line from rain col- lectors as shown, for springs, in Figure 9.

Points falling closer to the line, as B_000234 located at 1600 m a.s.l., are supplied by small recharge basins (between 1780 and 2040 m a.s.l.). On the contrary, points away from the line, as B_000174 located at 880 m a.s.l., are supplied by larger hydrogeological basins where rain

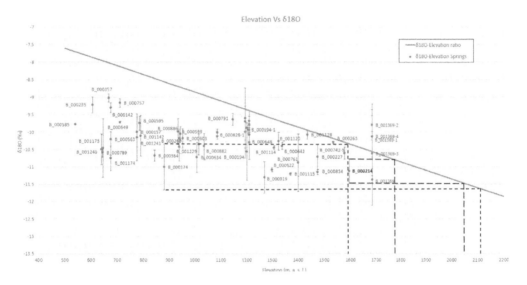

Figure 9. Relation between isotopic values recorded at different springs (std. dev. ±1σ) and δ18O gradient line from rain water.

Table 3. Estimated infiltration elevation for wells and piezometers.

ID	Type	Water table elevation	Average $\delta^{18}O$	Estimated infiltration elevation	Elevation difference
		m. a.s.l.	‰	m. a.s.l.	m. a.s.l.
B_000228	Well	555	-11.5	2060	1505
B_000376	Well	714	-11.0	1860	1146
B_000578	Well	662	-9.6	1300	638
B_000580	Well	774	-10.0	1460	686
B_000587	Well	483	-11.5	2060	1577
B_000605	Well	481	-11.2	1930	1449
B_000618	Well	680	-9.6	1300	620
B_000662	Well	522	-10.2	1540	1018
B_000837	Well	619	-10.0	1460	841
B_001244	Piezometer	941	-11.2	1940	999
B_001245	Piezometer	862	-10.3	1580	718
B_001252	Piezometer	758	-10.3	1580	822
B_001255	Piezometer	802	-11.2	1940	1138
B_001268	Piezometer	683	-9.7	1340	657
B_001270	Piezometer	681	-10.1	1500	819
B_001271	Piezometer	878	-10.6	1700	822
B_001353	Piezometer	614	-10.2	1540	926

water infiltrates at higher altitude respect to the spring point elevation (between 1600 and 2100 m a.s.l.). The same relation can be extended to wells and piezometers. In this case the difference is rep- resented by a general higher difference in elevation (at least 600 m) between the sampled points and their estimated recharge basins. This fact testifies, for piezometers, the usually deeper and more developed groundwater flow paths. In case of wells, which are always drilled through qua-ternary deposits along the river valleys, which are in turn fed from higher elevation (Table 3).

6.2 UNNSTABLE ISOTOPES (^{3}H)

Tritium is a hydrogen isotope that forms naturally in atmosphere. From 1960's nuclear tests significantly increased its concentration, so that it could have been used, due to its radioactive nature (half-time $T_{1/2}$ = 12.4 years), as a tracer in hydrogeology to estimate the time of water from their infiltration to the discharge point.

It is possible to calculate the time of residence of water within the aquifer based on the abundance of tritium with the following Equation 2:

$$t = \left(\frac{T_{1/2}}{0.693}\right) \cdot \left(lnC_p - lnC_s\right) \tag{2}$$

Where Cp = quantity of tritium in rain water (rain collectors) and Cs = quantity of tritium in underground water.

Water with high concentration of Tritium infiltrated in recent times, while water with very low quantities of this isotope can be considered older than 50–60 years, infiltrated before the beginning of nuclear tests.

Variable concentrations of Tritium during different survey campaigns have been recorded. Therefore, the residence time is calculated for every survey. In following table an overall aver-age value is reported. This has been possible just for those water points which show a low variability among different surveys. In fact, the maximum variation threshold for which data can be considered as reliable, has been based on the highest instrumental error among the sur-veys (2.5 TU, Tritium Units).

Table 4.   Estimation of groundwater age.

ID	Type	Aver. ^{3}H TU	Aver. age Years	ID	Type	Aver. ^{3}H TU	Aver. age Years
B_001244	Piezometer	<LD	>50	B_000649	Spring	5	8
B_001245	Piezometer	7	3	B_000745	Spring	5	8
B_001248	Piezometer	6	5	B_000752	Spring	6	4
B_001252	Piezometer	6	5	B_000757	Spring	5	8
B_001255	Piezometer	<LD	>50	B_000761	Spring	6	5
B_001268	Piezometer	8	2	B_000828	Spring	5	11
B_001270	Piezometer	7	2	B_000834	Spring	5	8
B_001271	Piezometer	<LD	>50	B_000886	Spring	5	7
B_001353	Piezometer	7	3	B_001114	Spring	6	6
B_001372	Piezometer	2	23	B_001120	Spring	5	8
B_001382	Piezometer	5	8	B_001128	Spring	5	9
B_000227	Spring	5	9	B_001174	Spring	7	3
B_000235	Spring	5	8	B_000376	Well	6	6
B_000260	Spring	7	4	B_000578	Well	-6	11
B_000263	Spring	6	6	B_000580	Well	-4	10
B_000508	Spring	5	7	B_000587	Well	4	6
B_000585	Spring	5	9	B_000618	Well	13	4
B_000642	Spring	7	3	B_000662	Well	4	7

7 CONCLUSIONS

Most of spring water show low mineralization degrees which allow to consider them as originated by short groundwater flow paths: the latter are mainly developed within the altered and fractured phyllite rock mass and the overlying quaternary fluvio-glacial deposits.

Only some areas are affected by more developed groundwater flows: the southern Laion-Ponte Gardena area and the confluence between the Isarco and the Funes valleys.

The latter is characterized by the presence of a remarkable tectonic structure (Funes thrust) that allows groundwater-mixing processes between shallow and deeper flow paths.

In Ponte Gardena area, water flows show higher mineralization due to the more extended fracturing degree of the bedrock. Thanks to isotopic data, it has been possible to highlight the presence of localized water flow-paths fed by higher altitudes. Nevertheless, low discharge rates justify shallow groundwater flow paths.

Concerning piezometers, they usually show similar characteristics to those seen for springs. Anyway, exceptions were found. Some of deeper water fall into the bicarbonate-alkaline hydrofacies. Thanks to the isotopic analysis, these waters can be considered as the expression of deeper groundwater flow paths developing within the rock mass.

However, it is important to highlight that no bicarbonate-alkaline water was found in springs, wells and rivers which represent the final outlet of the groundwater from the area. This is the focal point of the analysis due to their importance in terms of water resources exploitation.

Therefore, for all the geological, hydrogeological and geochemical reasons explained above, the forecasted impact of tunneling construction on water resources is expected to be very limited.

REFERENCES

Celico, P. 1986. *Prospezioni idrogeologiche Vol. 1*. Napoli: Liguori Editore.

Civita, M. 2005. *Idrogeologia applicata e ambientale*. Milano: CEA.

Longinelli, A. & Selmo E. 2003. Isotopic composition of precipitation in Italy: a first overall map. *Journal of Hydrology* 270: 75–88.

Mazor, E. 1997. *Chemical and Isotopic Groundwater Hydrology*. New York: Marcel Dekker Inc.

Penna, D., Engel, M., Dell'Agnese A. & Bertoldi, G. 2014. Tracer-based analysis of spatial and temporal variations of water sources in a glacierized catchment. *Hydrol. Earth Syst. Sci.* 18: 5271–5288.

Tunnels and Underground Cities: Engineering and Innovation meet Archaeology,
Architecture and Art, Volume 2: Environment sustainability in
underground construction – Peila, Viggiani & Celestino (Eds)
© 2020 Taylor & Francis Group, London, ISBN 978-0-367-46579-7

Muck management in infrastructure projects

A. Sharma
Afcons Infrastructure Ltd, Kolkata, India

ABSTRACT: Growing urbanization has caused rapid growth in urban infrastructure. Future prospects indicate exponential infrastructure growth in the underground infrastructure. Construction process of tunnels and associated stations and shafts consists of excavation of tremendous quantities of earth. Steep growth in construction of underground tunnels will cause a tandem rise in excavation quantities of underground spoils of varying nature. This calls for efficient management of the massive quantities of excavated earthen spoils. Current study investigates possible applications of excavated earthen materials based on their material type. Applications range from their industrial applications to large scale utilization of muck in infrastructural aspects such as embankments, dams and landfills. There lies an immediate need for channelizing the demand using relevant earthen spoil banks. This paper studies the geotechnical, ecological and economical aspects associated with channelizing of the muck management operations.

1 INTRODUCTION

Muck can basically be defined as the Earth produced as a by-product of an infrastructure/construction activity mainly involving excavation activity. In developing country like India, urbanization is on a constant rise. This continually increases the population per square km and hence increases the impact on the dimensionally limited existing infrastructure. This has led to a rapid rise in the construction of underground infrastructure. The advancement in boring technology has now enabled us to efficiently construct simple roads and path ways through the hilly and mountainous terrains thereby effectively reducing the rate of accidents. Similar increasing trends in urban needs have also led us to create hydro power units at a larger scale than ever. The common technology in the abovementioned infrastructure operations is advancement of technology in construction of tunnels using mechanized. Rising underground infrastructure has led to rising.

2 HISTORIC USES, LITERATURE REVIEW & CURRENT SCENARIO

Construction of tunnels can be dated back to 2000BC. Initial application of tunnels were restricted to construction of pipelines for sewerage, irrigation etc. Since then, tunnels have found their application in transportation like roadways, railways etc. and thereby increasing the quantity of muck produced.

Since ancient times, earth has been a major source of raw material. Items ranging from bricks to pots were made using the Earth. Earth, be it in form of rocks, stones or soil, has found its application since ancient times and they can be dated as back as 4000BC.

Even in construction, earth finds its special significance. Earth finds it's usage in some or the other form in most of the construction material. Major materials such as cement, bricks, concrete, ballast, subgrade etc. find their origin from earth i.e. either soil/clay/rocks/gravels etc.

There have been several instances where the excavated muck has been utilized for creating a sustainable environment. Construction of London Underground can be traced as one of the earliest instances where the excavated muck was re-utilized at the source site itself. In 1817, first underwater tunnel was being constructed under River Thames. The geology through which tunnel was being constructed was primarily clayey in nature. The excavated muck was re-utilized for production of bricks out of it. These bricks were eventually used to provide a lining for the tunnel. There have been similar instances where the excavated muck of rocks has been re-utilized for serving as a source of aggregates for shotcrete etc.

With recent advancements, the technology has enabled us to produce the required items such as bricks, dressing tools etc. in a very mobile and site-specific way, yet, the efficient management and the potential re-use of muck are usually not planned, designed or sought upon.

3 GEOTECHNICAL CHARACTERIZATION OF MUCK

Underground projects being undertaken are constructed in different parts in complex varying geology. Standard practices for construction includes a detailed soil/geotechnical investigation before construction of underground works as the design relies heavily on it. Geotechnically, the soil or the muck can be broadly classified as:

a. *Rocks*

Rocks can be further classified geologically as:

Igneous Rocks: Mainly formed due to cooling & hardening of molten magma (molten rocks deep within earth surface). These forms either within the earth or on the surface of earth (cooling of lava, i.e. molten eruption of through volcanoes). Examples of Igneous rocks include basalt etc.

Sedimentary Rock: Sedimentary rocks are formed due to sedimentation or deposition of sand, shells, pebbles etc. over each other over a long period of time into a hard rock deposition. This hard rock deposition is called Sedimentary rock. Examples include sandstones, limestones etc. and are generally found to comprise of fossils.

Metamorphic Rocks: Metamorphic rocks are formed due to metamorphosis or change of igneous and sedimentary rocks due to heat and pressure over a period of time. They generally have a ribbon like structure. Examples include marble and gneiss.

Figure 1 below shows the further classification of rocks.

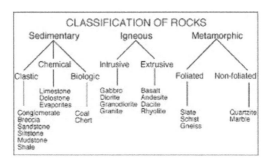

Figure 1. Classification of Rocks based on their origin.

b. *Soils*

Almost all the soils are formed from weathering of rocks. Weathering may be either physical or chemical agents. Depending upon the formation of soil they may be geologically classified into several types such as alluvial, marine, lacustrine, loam etc. Based on particle size distribution, the soil can be classified as; Coarse grained and fine-grained soils. Figure 2 depicts the classification of coarse grained & fine-grained soils. Since mainly soils are formed by physical & chemical weathering and are deposited over each other for millions of years, they occur in heterogeneous conditions. For detailed characterization of soils, they can be further classified based on their descriptive & engineering properties. Although it is a standard practice to

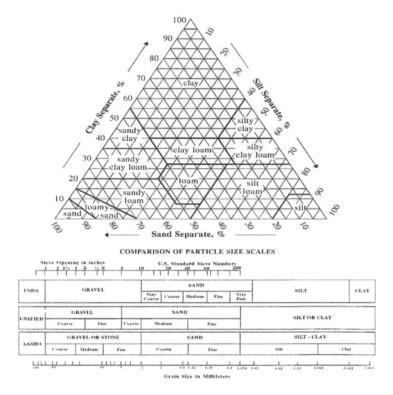

COMPARISON OF PARTICLE SIZE SCALES

USDA	GRAVEL		SAND				SILT	CLAY
			Very Coarse	Coarse	Medium	Fine	Very Fine	

UNIFIED	GRAVEL		SAND			SILT OR CLAY
	Coarse	Fine	Coarse	Medium	Fine	

AASHO	GRAVEL OR STONE			SAND		SILT - CLAY	
	Coarse	Medium	Fine	Coarse	Fine	Silt	Clay

Grain Size in Millileters

Figure 2. Classification of Soil based on Particle size.

investigate the ground conditions and properties, yet since the obtained muck is often highly disturbed composition of soil, hence post muck collection the properties of muck should be investigated for ensuring its proper re-utilization. Following are some engineering properties that should be considered for post muck testing.

Engineering Properties such as mentioned below must be considered post muck collection.

Particle Size Analysis
Density/Specific Gravity
Strength/Deformability
Compaction Properties
Hydraulic Conductivity

4 APPLICATIONS/UTILIZATION OF MUCK

Since investigation of ground conditions and soil engineering properties are done prior to commencement of excavation works, hence based on the envisaged quantity of muck being collected, the possible re-utilization or application of retrieved muck can be well planned in advance. Following are some potential applications of muck reuse:

4.1 *Utilization at Source Project*

Major infrastructure project producing massive muck quantities are:

Underground Metro Railway Projects
Underground Roadways/Subways
Hydro Power Projects

The former is mainly located in urban locations while latter in non-urban locations. Either being the case, excavated muck can be potentially re-utilized at the source site itself. There have been several case studies, where the excavated muck has been re-utilized at the source site itself. An underground project essentially comprises of several construction activities such as roads, buildings etc. where the excavated muck can be re-utilized. Proper re-utilization of muck would make the project partly self-sustainable. Depending upon the geotechnical characterization of the soil/rock type available, the muck could be re-utilized in following construction activities:

1. Backfilling of Shafts/Cut and Cover boxes
2. Sub-grade for diversion or service roads
3. Raw material for concrete/shotcrete i.e. coarse or fine aggregate
4. Backfill for embankments/railway over bridges

Following steps may be undertaken for selecting potential re-utilization of muck at source site:

1. Channel the muck and geotechnically classify collected muck
2. Check engineering & chemical properties of collected muck
3. Plan for potential re-utilization referring the table below for possible re-utilization based on geotechnical characterization

4.2 Urban & Non-Urban Infrastructure Projects

Depending upon whether the source being urban or non-urban, the excavated may preferably be used at source site, but in absence of any potential re-utilization at source site, the excavated muck can be reused at relevant infrastructural projects. Depending upon the source being urban or non-urban, the muck can be reutilized in ongoing or proposed or planned infrastructural project. Soil is a major construction material and often major infrastructural projects have constructional value & hence require specified soil types. Often the required material may not be available at the source site or in close proximity. Even in case of availability, it would call for unnecessary cuttings and excavations and additional resources. Different infrastructure projects follow difference specifications and have specified engineering requisites. Excavated spoils or muck being produced can be planned for possible re-utilizations in such cases as it would lead to

4.2.1 Construction or maintenance of roads

Generally, the desirable properties of highway or road material include incompressibility, durability and stability in adverse weather, minimal volumetric changes, ease in compaction and good drainage characteristics. In case if the excavated material is soil, based on the index & engineering properties such as grain size analysis, Atterberg limits, field moisture content of excavated muck, potential application for construction of roads in nearby locations can be ensured. In case of availability of range of soil types, they can further be physically mixed with each other to form well graded & stabilized soil.

In case of rocks or gravel muck, they can be potentially used as aggregates. Aggregates form major portion of structure as they bear stresses occurring due to wheel loads on pavement and on surface course. Aggregates are often used as granular sub base course and base course. The excavated muck can be further dressed down using tools to meet the required purpose. Desirable properties of road aggregates are mainly strength, hardness, toughness, durability, shape of aggregates.

Excessive quarrying leads to ecological imbalance, hence it is important that the excavated muck be undertaken for proper use.

Following are some Tests that can be undertaken on the excavated material post dressing.

Crushing Test
Abrasion Test
Impact Test
Soundness
Shape Test

As discussed in previous section, the construction of tunnel could occur in urban or non-urban areas. Hence, depending upon the location of underground excavation the much could be re-utilized in urban roads, rural roads or non-urban roads. In case of non-urban roads, as they are mostly located in less populated areas hence, transportation & storage of muck becomes a logistic issue.

4.2.2 *Embankments*

Existing topography often forces infrastructure works to create embankments by filling soil to form a route to connect places. Often the required soil type again has to be acquired from some place and thereby creating a void in the land topography. Construction projects such as nuclear power plants, dams, thermal power plants etc. usually require such creation of embankments. The stability of embankments predominantly depends upon the type of fill soil used which should have inherently freely draining properties. Hence, the required soil type could be acquired from the excavated muck. Often reasons of embankment failures are attributed to the type of fill material being used. For procuring the required fill material in the required quantity, the region has to further excavated and this creates a lowland which further leads to ecological imbalance. Use of excavated muck could thereby solve the constraint related to construction of embankment as it could serve as a ready source for the fill material.

4.2.3 *Embankment/Earthen Dams/Dykes or Rock Fill Dams*

Construction of Embankment/Earthen Dams/Dykes. In case of muck comprising of fine grained soils such as clay or silt, they find their potential application during construction of central cores of embankment/earthen dams. Constructions of Earthen Dams or Dykes mainly require retention of water bodies using soil and additives. In such construction activities, fine grained soils with low permeability or impermeable soils are required. Use of fine grained part of the muck can be well facilitated in such projects. Such projects are usually located in non-urban areas, hence again the associated transportation of muck becomes a logistic issue.

4.2.4 *Solid Waste Landfill Embankments*

One of the methods of disposing off the city waste is to dump it in the dumping yard. This creates an unpleasant and unhygienic condition of the portion of the land. A more recent optimized method is to use the waste as a fill material for construction over them usually called as Solid Waste Landfills. Figure 4 shows a schematic of use of Solid waste landfills. The usual disposal of excavated muck from tunneling operations is creation of landfills. Unplanned landfills can cause ecological and environmental disruption as the creation could serve as a barrier to the natural precipitation run off and cause water logging. Considering these notions, it can be established that unplanned disposing of the muck should be avoided as they can pose long term threats. Often it may be the case that the muck being produced may not be a proper fit for the required soil type and hence in such cases, physical or chemical stabilization can be undertaken for ensuring the required specifications and engineering properties.

4.2.5 *Source of course/fine aggregate*

Concrete mainly comprises of cement, water with coarse & fine aggregates. A similar composition is grout or shotcrete which comprises of same elements with just fine & coarse aggregates

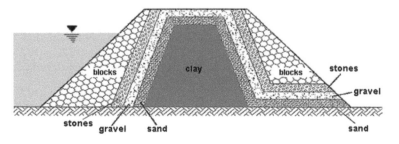

Figure 3.   Typical Schematic of an Earthen Dam.

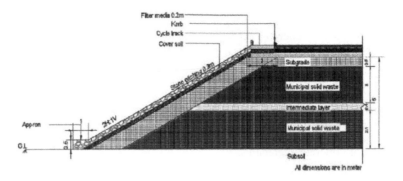

Figure 4. Typical details of MSW Embankment.

respectively. Concrete is a major construction element and is nowadays used in almost all construction works. Regular application of excavated muck for production of concrete, shotcrete or grout could lead to a sustainable construction environment. Conventionally, fine & coarse aggregates are produced from the quarries. With advent of growth in infrastructure, the excessive quarrying for material production would eventually lead to ecological imbalance. Hence, utilization of excavated muck as a source of course & fine aggregates should be a regular practice to create an ecologically balanced environment.

4.3 *Industrial Applications*

Earth is a major source of raw material. The excavated muck, may it be of any type, any geotechnical characterization, can serve as a source of raw material for industrial production. Historically, soil has been used for producing numerous appliances useful for mankind. May it be an earthen pot or use of soil in agriculture, soil has endless arrays of uses in our day to day life. Some of the major industrial applications have been listed below:

4.3.1 *Bricks/Mud blocks*
Earliest instances of use of soil for construction works can be traced back to 4000 BC. Brick is one of the major building materials that have been in use since over thousands of years. Brick is usually composed of clay. A brick can be composed of clay-bearing soil, sand, and lime, or concrete materials. Bricks are produced in numerous classes, types, materials, and sizes which vary with region and time period, and are produced in bulk quantities. Two basic categories of bricks are fired and non-fired bricks. Block is a similar term referring to a rectangular building unit composed of similar materials but is usually larger than a brick. Lightweight bricks (also called lightweight blocks) are made from expanded clay aggregate.

4.3.2 *Aggregate*
Aggregates are naturally occurring inert granular materials such as sand, stone, gravel etc. used in production of concrete. They form one of the major constituents of concrete which are used to provide bulk to concrete. They form the inert matrix in the concrete and form 75% of the body. Aggregates are used in concrete in two forms i.e. coarse and fine. Coarse aggregate are mainly stones, gravels etc. and mainly impart the stability and durability to concrete while fine aggregate assist in workability and help cement paste to hold coarse aggregates. Since aggregates are required to impact strength and durability to concrete, hence they are required to meet certain specifications such as shape (cubical or rounded), clean, hard and well graded. They can be classified as coarse or fine based on their particle size or grain size analysis. Particle sizes ranging from 40mm–4.75 mm are classified as coarse-grained aggregates whereas particle sizes from 4.75mm to .07mm are classified as sand. In other words, as described in previous section, gravel and boulders can be classified as coarse-grained aggregates whereas sand can be classified as fine aggregates. Generally, they are obtained from naturally occurring deposits such as quarries or river beds etc. and further modified to the required

dimension. Muck excavated from tunneling can serve as a good source for production of coarse or fine aggregates. Depending upon the existing geology of tunneling, the muck can be utilized for production of concrete and can utilized in several purposes. Geologically, Igneous rocks are considered to be a good source because of the hard, tough and durable nature. They have crystalline/glassy texture. Sedimentary rocks such as limestones and siliceous sand stones have proved to be good source of coarse aggregates. Similarly, metamorphic rocks such as quartzite or gneiss have proven to be good source. In both, metamorphic and sedimentary rocks, the foliated nature may reduce the strength of material and hence must be considered prior to selecting the material. Following properties/tests should be considered prior to selecting the material for concrete aggregates:

Shape/Angularity
Density
Strength
Particle shape & Texture
Specific Gravity
Porosity
Moisture content
Chemical & Thermal properties

4.3.3 *Ceramics*

Fine grained soils such as clays have tendency to exhibit plasticity upon mixing with water up to a certain limit. Upon drying, it becomes firm. When thermal changes are induced in clay, permanent physical and chemical changes occur within the matrix of clay called Ceramics. These properties enable application of clay into several works such as pottery, earthen ware, floor tiles etc. These products find their extensive application in day-to-day industrial works. Different thermal changes induce different effects in clay which finds its application in like porcelain, stone ware etc. which are used in decorative and art works. Muck excavated with predominant fine-grained soil can be particularly used for these purposes. These can be trace back to ancient times and use of such products can be dated as back as 14000 BC. Cooking pots, art works, dishware even musical instruments are the products of clay. It is also used in many industrial processes, such as paper making, production of cement etc.

4.3.4 *Agricultural Uses*

India is an agrarian country. Historically agriculture has been primarily contributing to the country's economy. Muck being produced may be well utilized in agriculture based on the type of muck. Muck can be treated or mixed in suitable manner to suit the requirements for agricultural purposes. Based on geotechnical characterization and chemical properties of soil, the type of treatment required for use of muck in agricultural purposes can be ascertained.

4.4 *Soil Banks*

Increase in tunneling projects has caused increase in muck quantities and are expected to increase further. This would call for more efficient & planned use of excavated muck. Continuous dumping/deposition of excavated muck could potentially cause ecological imbalance as it could alter the existing topography and impact the hydrological run-off course. Hence it can be established that it is economically and environmentally in best interest to utilize the excavated muck in above mentioned potentially productive applications. The construction projects usually have large time frame, ranging from a few months to several years and are often associated with unforeseen delays.

In order to enable utilization of excavated muck, it is important that the soil be in required proper form and free from unwanted chemicals & contamination. In tunneling projects, since the excavated soil/rocks are obtained in highly disturbed &chemically mixed condition, hence cannot be utilized directly. Hence, it is necessary that the excavated soil undergo required treatment and remediation. The excavated soil should hence be dried & pulverized for further use. Depending upon the application of soil, the soil should be further treated accordingly.

The idea of soil banks basically comprises of installation of a facility within or nearby the source site whereby the soil can be directly stored, treated and prepared for potential utilization

of muck. Generally, the muck excavated from tunnels are transported using conveyor belts to the shaft and then to muck pit. The Soil Bank can be further directly connected to the muck pit for large storage of soil. These can be created underground connected to the muck pit, particularly in urban areas where space is a constraint. Based upon the anticipated quantity to be excavated, the size of the storage can be planned.

The Soil Bank should comprise of following elements for their functions.

De-Sander: to separate the sand particles from the excavated muck
Deflocullater: to make the clay slurry workable to allow separation of clay and silt
Sedimentation Tanks:
Large Scale sieves of relevant mesh sizes: for batching the soil particles according to their grain sizes
Large capacity storage tanks
Procedure:

The excavated soil would be dumped to the muck pit. From muck pit, through the interconnected pipes or though suitable transportation channels, the muck would be passed transferred to the storage compartment or tanks where it can be thoroughly washed. Before the muck enters, the soil would be passed though array of sieves to differentiate the soil as per their grain sizes. The excess water can be dewatered. Since the fine-grained particles have tendency to flocculate, they will be deflocculated & separated. Figure 5 shows the mechanism of working of Soil Banks.

With this we will be able to obtain the soil types according to the grain sizes.

Set up of Soil banks may comprise of above elements based upon the anticipated muck to excavated and required treatment/remediation of the muck based on its further use to ensure the soil be free from contaminants.

In case of excavated muck comprising of rocks, dressing of the rocks/boulders can be done to suit the requirements.

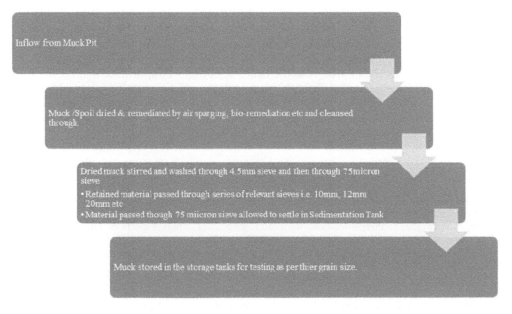

Figure 5.   Procedure for workings of Soil Banks.

5 ECOLOGICAL & ECONOMIC ASPECTS

5.1 Ecological Aspects

Excessive mining, quarrying or creation of unplanned landfills by unplanned dumping of earth has led to ecological imbalance and has had disastrous results. Rampant stone quarrying and excessive earth digging for procurement of earth for industrial production and construction has caused several landslides and landslips. Same coupled with massive rains has also been reason for massive floods.

Similarly, unplanned disposal of excavated earth causes unprecedented terrain changes. These changes impact the natural stream flow and can be hazardous during heavy rainfalls.

Utilization of excavated earth from tunneling projects for construction works or for industrial purposes can serve as a tool to create an ecological balance. This potential utilization of earth could serve as a win-win situation on all fronts as the construction, industry and environment could benefit from it.

5.2 Economic Aspects

Utilization of excavated muck from tunneling projects situated in urban or non-urban locale would on a first look call for a lot of resources and expenditures. Installation of plants for Soil Banks as detailed in previous section, Testing, Planning and Design and related services would add up the economic front. Land, storage places, pulverization, treatment of muck such as bio-remediation or air-sparging, associated logistics, transportation, storage etc. would add up to the total cost of the project and may not seem productive as it would not draw any immediate revenue.

Looking the same from a larger perspective, the utilization of muck could draw revenue as it could serve as a source of raw material for both construction & industrial purposes. Re-utilization of excavated muck in the source site of production would lead the site to increase its self-sustainability index. Considering all above points, there seems to be a supply-demand scenario which can be actively worked up considering the above points. Ecological benefits with added sustainability will in long term serve the economic purposes as well. With these points in view, it can be said that in long term, the same would be an economically feasible solution.

6 CONCLUSION

With growing trend in urbanization, the parallel need for efficient infrastructure would call further rise in tunneling in urban and non-urban locations. This would cause a steep growth in the quantity of excavated Earth. It would hence be of utmost importance that the increasing amount of excavated muck be efficiently managed. The efficient use of excavated muck in industrial and construction works would enhance the industrial sustainability and lead ecological balance. Use of large storage and muck remediation tanks called Soil Banks at Source site of Muck Production would create a channel for its usage.

REFERENCES

Bellopede, R & Marini, P. 2011. Aggregates from tunnel muck treatments, Properties and uses in Physicochem. Probl. Miner. Process. 47: 259–266

Bellopede, R. Brusco, F. Pierpaolo, O & Maurizio Pepino. 2011. Main Aspects of Tunnel Muck Recycling. American Journal of Environmental Sciences 7 (4): 338–347

Oggeri, C., Fenoglio, T.M. & Vinai, R. Muck Classification: Raw Material Or Waste In Tunneling Operation

Oggeri, C., Fenoglio, T.M. & Vinai, R. 2017. Tunneling Muck Classification: Definition and Application in World tunnel Congress Proceedings 2017.

US Department of Transportation Urban Mass Transportation Administration Office Of Technology Development And Deployment Washington Dc 20590, Report No. Umta-Ma -06-0025-77-15. Muck Utilization In The Urban Transportation Tunneling Process

Tunnels and Underground Cities: Engineering and Innovation meet Archaeology,
Architecture and Art, Volume 2: Environment sustainability in
underground construction – Peila, Viggiani & Celestino (Eds)
© 2020 Taylor & Francis Group, London, ISBN 978-0-367-46579-7

Influence of bacteria inoculum and organic concentration on the biodegradation of soil conditioning agents in aqueous solutions

G. Vilardi, I. Bavasso, D. Sebastiani, S. Miliziano & L. Di Palma
Università degli Studi di Roma "La Sapienza", Rome, Italy

M. Pirone, F. Carriero & R. Sorge
Astaldi, Rome, Italy

ABSTRACT: Earth Pressure Balance (EPB) technology is currently the most widely used technique in mechanized tunnelling with Tunnel Boring Machines (TBMs) and particularly in urban environments, because of several advantages, as the possibility to excavate in different geological conditions and effectively controlling the induced effects on the pre-existing structures. EPB technology requires the continuous injection of chemicals during the advancement: this may induce the accumulation of xenobiotic compounds in the excavated debris. The present work aims at studying the role of several parameter, as dosage of conditioning agents and environmental conditions, in the biodegradation process through the development of an experimental activity involving the use of commercial soil conditioning agents. Results provide elements useful to improve the knowledge of the interaction phenomena between the soil and the chemicals, the management of excavated soil and the design of the environmental monitoring activity.

1 INTRODUCTION

In mechanized tunnelling with TBM-EPB technology the management of the soil conditioning process plays a key role in the success of the excavation; different conditioning parameters are constantly monitored and modified step by step to accommodate the variability of the conditions that the TBM faces.

Nature and chemical composition of the commercial products used during the excavation and the management of several conditioning parameters determine the amount of single compounds injected into the soil and, consequently, their environmental impact.

In recent times is visible in Europe an increase in the attention reserved to the theme of the reuse of non-renewable natural resources and of course the management of large volumes of soil produced by the tunnels and deep excavations is fully integrated into this virtuous path.

The desire to reuse as much as possible the soil produced by the excavation and the increase in excavation performance (and consequently the quantity of chemicals injected) have led to the need to develop studies of the interaction between these chemicals and the soil. These studies are based on two main elements: the environmental impact of the compounds present in the commercial products and the biodegradation process; from the combination of these two independent factors depends the possibility and the methods to safely reuse of the soil.

In-depth studies and specific laboratory activities can be carried out to analyze the environmental impact of a specific chemical substance. These activities are conceptually developed by administering to a series of target micro-organisms increasing doses of the chemical involved and monitoring a series of parameters such as mortality and growth.

The biological degradability or "biodegradability" is the process for which the organic compounds (as surfactants present in the foaming agents injected during the excavation) are reduced as simple molecules by microorganisms. In particular, the development of the biodegradation process in aerobic conditions may lead to the complete oxidation (mineralization) of the organic substance (with the formation of CO_2 and H_2O) and, where nitrogenous species or compounds rich in phosphorus are present, also to the formation of nitrates and phosphates which are used as fertilizers for the growth of algae and other organisms. In particular case, the biodegradation could also be partial, leading to the formation of simpler organic compounds. In anaerobic conditions, the main product is biogas, made of methane and other reduced gaseous compounds [Garg, 2017].

As can be easily understood, the study of biodegradation is influenced by numerous factors, some of which are related to the characteristics and the dosage of the chemical products, while others related to the environmental conditions in which the biodegradation process takes place.

For the description of the biodegradation process different parameters are required: the Biochemical Oxygen Demand (BOD) that identifies the amount of oxygen necessary for the biological stabilization of the organic compounds, the Chemical Oxygen Demand (COD) that identifies the oxygen required for the chemical oxidation of organic compounds and the organic carbon content in a medium, defined by Total Organic Carbon (TOC) parameter.

Respirometric tests are performed for the determination of BOD value on time; if the oxidable compounds present in the medium are all biodegradable the initial COD value can be consider as possible BOD ultimate value. In any case, all three parameters are required for a complete investigation of the biodegradation process.

A joint research activity between Astaldi and Sapienza University of Rome to study the biodegradation of the most commonly used chemicals has been developed, and in the present work several laboratory biodegradation tests performed using selected conditioning agents are reported.

The presented results will allow to deep understand the effect of the chemical composition and dosage of the commercial product as well as the environmental conditions on the biodegradation process.

Results provide elements useful to improve the knowledge of the interaction phenomena between the soil and the chemicals during the biodegradation process. In particular, the main factors affecting the biodegradation process of the chemicals used in tunnel excavation with TBM-EPB were highlighted, thus allowing to share some suggestions on the most effective methodologies for laboratory tests and, finally, to offer some interesting insights for constructive discussions and future studies.

2 EXPERIMENTAL PROCEDURE

Different solutions were prepared by the dilution of chemicals in distilled water. The set-up of the system used for the biodegradation tests was developed by using synthetic solutions containing a biodegradable compound such as sodium acetate at 1 g/L [Kuokkanen et al., 2004]. The biodegradation of sodium acetate solution was compared with the biodegradation of a solution containing fumaric acid at the same concentration and at the same organic carbon content.

Biodegradation tests of conditioning agents, identified by capital letters A, B and C, were performed by preparing different solutions at different dosages of foam agents (0.10% and 0.20%).

For the BOD tests, a volume of 400 mL of synthetic solution was filled in a Closed Bottle system with the addition of a mix of salts according to the procedure 301D OCDE/OECD [APAT-IRSA/CNR] and 2 mL of a microorganism source. In case of specific microorganisms, *Bacillus Clausii* (BACT1) was used as bacterial inoculum while in the other tests the inoculum source derives from a soil humus (BACT2), obtained washing in water a proper soil amount (100 g per 1 L of water) and then acclimating in batch aerated reactor the supernatant before using [APAT-IRSA/CNR].

BOD value at 5, 10 and 28 days were recorded and to complete the evaluation of the biodegradation rate TOC (Shimadzu TOC-L CSH/CSN analyzer) and COD were measured according to the standard procedures.

The removal efficiency of TOC that suggests the mineralization of the organic carbon into CO_2 and H_2O was calculated as $R(\%) = 100(TOC_0 - TOC_f)/TOC_0$, where TOC_0 identify the initial value of the specific parameter and TOC_f the final value of the same.

The initial TOC and COD value are summarized in Table 1 reported below.

Table 1. Total Organic Carbon and Chemical Oxygen Demand of foam agent solutions at different initial concentration.

Product	TOC [mg/L]		COD [mg/L]	
	0.10%	0.20%	0.10%	0.20%
A	140.55	297.33	160.23	310.42
B	67.94	134.32	78.17	175.16
C	75.60	146.02	78.35	147.30

All tests were conducted at room temperature and neutral pH conditions (the pH was measured using a Crison GLP 421).

3 RESULTS AND DISCUSSION

3.1 *Set-up test*

The biodegradation curves of sodium acetate and fumaric acid as standard solutions are reported in Figure 1 where the biodegradability is represented by the ratio between BOD value at time t and the initial value of COD.

In Figure 1 it is possible to observe a classical BOD curve trend: it is known that sodium acetate is an easy biodegradable compound and its biodegradation curve showed high BOD values according to the literature [Baker et al., 2000]. When fumaric acid is used at the same concentration of 1 g/L the BOD values recorded were lower that the corresponding value of sodium acetate solution. This because of fumaric acid TOC content was higher than sodium acetate: both are linear organic molecules but the carbon content of fumaric acid is higher than that of sodium acetate and required more time for its removal as a consequence of a possible inhibitory effect by excess of substrate. At the same TOC initial concentration, the BOD value of a solution of 0.71 g/L of fumaric acid increased but the BOD curve maintained a lower profile than those of sodium acetate. This confirm the hypothesis of excess substrate and the possibility to appreciate these differences with known biodegradable matrices make the system ideal for our purpose. Considering the commercial conditioning agents, no details in term of chemical composition are

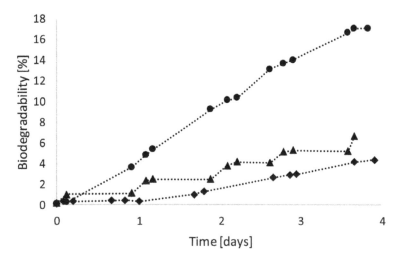

Figure 1. BOD curves for Sodium Acetate and Fumaric Acid solutions. Conditions: (●) 1 g/L sodium acetate solution, (▲) 0.71 g/L and (♦) 1 g/L of Fumaric Acid solutions.

known and only the overall index (TOC and COD) were defined (Table 1). However, the system allows to detect similarities and/or differences with the above reported completely biodegradable substance. The performed preliminary tests were carried 5 times for each chemical compound, showing a remarkable reproducibility of the tests with a percentage of 5% as deviation standard.

3.2 *Biodegradation: effect of conditioning agents initial concentration*

Three soil conditioning agent (A, B and C) solutions at 0.20% as concentration were filled in the Closed Bottles system and the results about the BOD trends up to 28 days are reported in Figure 2.

How it is possible to observe in Figure 2, all BOD curves described similar behavior: after a delay of about 3 days, the BOD value exhibited during the test with A solution reached the same value of B and C. This delay suggested that A is a mixture characterized by a slower kinetic of biodegradation in comparison with B and C, but after 28 days the BOD values recorded were similar: 74.00 mgO$_2$/L, 64.00 mgO$_2$/L and 62.00 mgO$_2$/L for A, B and C respectively.

In case of complete biodegradation of the oxidable compounds for each solution the BOD ultimate has to be close to the COD initial value (Table 1). After 28 days only a 23.84%, 36.54% and 42.10% of the oxidable compounds respectively for A, B and C was removed in comparison to their initial COD values and an additional time is required to appreciate high removals. In case of TOC removal, as reported in Figure 3, the 18.07%, 43.21% and 57.36% for A, B and C was measured. With a summary of these results C and B solutions can be considered more ready biodegradable mixtures in comparison to A.

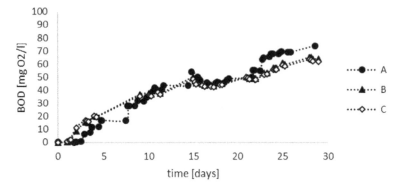

Figure 2.   BOD curve for A, B and C solutions at 0.20% as concentration.

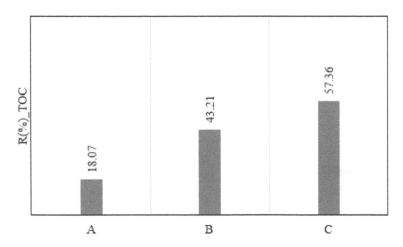

Figure 3.   TOC/TOC$_0$ value for mixtures A, B and C after 28 days of biodegradation process.

To investigate if a possible inhibitory effect by excess substrate might be a cause of the results previously reported similar tests were conducted with the same mixtures at lower concentration by providing a dilution with a dosage of 0.10%. After 10 days the BOD values of 91.50 mgO$_2$/L, 101.00 mgO$_2$/L and 112.67 mgO$_2$/L for A, B and C respectively were recorded.

The high concentration acted as inhibitor of biodegradation process and for this reason a general delay of the process was observed in Figure 2. The same effect on TOC removal was observed and a 59.95%, 81.33% and 78.99% for A, B and C was calculated. Considering the ratio between BOD ultimate and initial COD it was possible to observe that in case of B and C solution this ratio, after 10 days was up to 1. This result suggested the presence of compounds that can be oxidized only through a biological process and were not chemically oxidable such as nitrogenous compounds by nitrification process [Fenchel et al., 1977].

In conclusion, all mixture showed good biodegradability as reported among the final remarks of several similar studies [Barra Caracciolo et al., 2017] but the high organic carbon content of A, defined by the TOC initial value, resulted in a slow biodegradation process. Conversely, B and C showed a faster biodegradation and this result gives the possibility to detect after 10 days the presence of possible non-carbonaceous and not chemically oxidable compounds.

The initial concentration of conditioning agent in the soil sample is important as directly related with the time in case of preliminary study devoted to establish the biodegradation behavior of chemicals adopted during the excavation process.

3.3 Biodegradation: effect of bacteria inoculum

As previously assessed, the biodegradation process of a single chemical product may be affected by the environmental conditions in which the conditioned soil is placed. To quantify the difference in the biodegradation curves, BOD tests at 0.20% of soil conditioning agent's concentration were repeated by using soil bacteria source (BACT2) taken directly from a soil sample and consequently containing several different bacteria families frequent in natural environments. The results are reported in Figure 4.

The use of a mixed consortium enhanced the biodegradability and high values of BOD after 28 days were recorded: in this case, no delay was detected and the BOD curve of mixture A reached higher values than B and C. The inhibitory effect by excess substrate was overcome by changing the kind of inoculum.

In order to define the advantages on using a mixed consortium, close to the real conditions on site, it is possible to compare the BOD value after 10 days recorded in test with *Bacillus Clausii* (BACT1) and with the mixed consortia (BACT2). The results are reported in Table 2.

The adoption of a mixed culture enhanced the speed of biodegradation process: the differences in term of BOD were extremely different even at 10 days (one order of magnitude). This time can be considered sufficient to evaluate if the biodegradation occurs and the oxidation degree directly

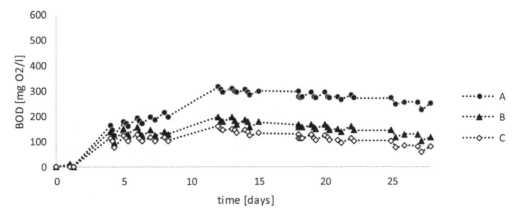

Figure 4.   BOD curve for A, B and C solutions at 0.20% as concentration with BACT2 as inoculum.

Table 2. BOD removal after 10 days with a mixed inoculum at 0.2% foam agent dosage.

Foaming Agent	BOD$_{10}$ [mgO$_2$/L]	
	BACT1	BACT2
A	32.90	316.00
B	32.99	198.00
C	36.73	123.00

by using a mixed inoculum. Indeed, the development of test with a real inoculum is correlated directly with the excavation site and is a description of a real biological process, but the use of a specific inoculum guarantees the achievement of repeatable results, fundamental during a comparative study.

4 CONCLUSIONS

The present study is an investigation about the biodegradability of soil conditioning agent solutions used in TBM-EPB technology. Experimental tests at different dosages and bacteria sources were performed.

Results allow to deep understanding the key role played, from one hand by the chemical composition and dosages of the foaming agents and, on the other, the role played by the environmental conditions in which the biodegradation process take place.

Particularly, the BOD test is able to detect the difference in the biodegradation of different products at the same dosage, underlining the importance of the chemical composition in the biodegradation time. Increasing the dosage of the injected chemical can be induced an inhibition effect in the microorganism responsible for the biodegradation of the product and finally leading to a higher biodegradation time.

Finally, the kind and composition of the bacteria inoculum selected for the laboratory test and, on site, the microorganism consortium and the environmental conditions may have a fundamental impact in the biodegradation process.

These factors, all affecting the biodegradation process, should be seriously taken into consideration in the experimental activities realized to foreseen the features of the excavation debris during time and finally to design the management of the excavated soil.

Even if the surfactants mainly present in the foaming products used in the tunnel excavation with TBM-EPB are well known and studied, there are still missing information on the environmental impact and on the effects on the biodegradation process of several additives and polymers often added to the chemical formulation or directly injected during the excavation

REFERENCES

APAT-IRSA/CNR. 2003. Manuals and guidelines. ISPRA.

Baker I.J.A., Matthews B., Suares H., Krodkiewska I., Furlong D.N., Grieser F. & Drummond C.I. 2000. Sugar fatty acid ester surfactants: structure and ultimate aerobic biodegradability, *Journal of Surfactants Detergent* 3: 1–11.

Barra Caracciolo A., Cardoni M., Pescatore T. & Patrolecco L. 2017. Characteristics and environmental fate of the anionic surfactant sodium lauryl ether sulphate (SLES) used as the main component in foaming agents for mechanized tunneling. *Environmental Pollution* 226: 94–103.

Fenchel T.M., & Jørgensen B.B. 1977. Detritus Food Chains of Aquatic Ecosystems: The Role of Bacteria. In: Alexander M. (eds), *Advances in Microbial Ecology*. Springer, Boston, MA.

Garg S. 2017. Bioremediation of agricultural, municipal, and industrial wastes. *Handbook of Research on Inventive Bioremediation Techniques*.

Kuokkanen, T., Vähäoja, P., Välimäki, I. & Lauhanen, R. 2004. Suitability of the respirometric BOD Oxitop method for determining the biodegradability of oils in ground water using forestry hydraulic oils as model compounds. *International Journal of Environmental Analytical Chemistry*, 84 (9): 677–689.

Tunnels and Underground Cities: Engineering and Innovation meet Archaeology,
Architecture and Art, Volume 2: Environment sustainability in
underground construction – Peila, Viggiani & Celestino (Eds)
© 2020 Taylor & Francis Group, London, ISBN 978-0-367-46579-7

Tunnel dust control project

T.W Warden & C.M. Warden
Englo Inc., Bluefield, WV, USA

ABSTRACT: The case study describes implementation of a new Wet Type Dust Extraction System designed to control fugitive dust generated at a large tunnel project in British Columbia, Canada. Tunnel specifications called for a tunnel 1000m in length and approximately 14m diameter. Regulators found unacceptable levels of dust and required mitigation devices to be implemented before the project could commence again. The paper briefly describes expedient implementation and operation of a highly efficient exhaust ventilation and dust control system that would remove particulates to meet both emission and ventilation parameters for the construction area. The discussion includes the implementation of the exhaust ventilation and dust control device and the corresponding test data to illustrate dust levels before and after implementation at critical tunnel locations in the construction area.

1 INTRODUCTION

This paper presents a case study of a Road-header tunneling operation that regulators had noted experienced dust levels which contributed to poor air quality within their operations. The ventilation plan for the tunneling operation initially called for a blowing ventilation system. This is where the air is blown in from outside longitudinally along the tunnel up to the face. Blowing ventilation induced several issues, including poor air quality and visibility within the tunnel. Once the regulators tested the air quality and saw the high dust levels, the tunnel construction was immediately halted until a remedy could be found. The tunnel construction contractor contacted our company to provide some assistance with reducing the dust level emissions. Our company manufactures wet de-dusting equipment which can provide ventilation and air cleaning simultaneously. Several mitigation options were presented to the client in order address the dust issues and to resume the construction project. In working with the ventilation construction company's ventilation engineer several concepts were discussed concerning best approaches and installation arrangement for the de-dusting unit. In essence the Wet De-Duster in a tunnel ventilation application can be affixed in three different arrangements.

- Equipment Mounted (Figure 1)
- Trailing unit (Figure 2 and Figure 3)
- Fixed Mounted (Figure 4)

1.1 *Equipment mounted. Advantages and Disadvantages...*

In this mounting arrangement the de-duster is mounted directly on the Road-header as close to the cutting head as possible. As you can see from Figure 1 the inlet to the de-duster extends as close to the cutting head. It can be either left or right mounted, depending on the space available on the Road-header, and/or the preferred air direction across the face. The advantage to this method is that it is always optimal to collect and control dust as close to the source as possible. This provides close proximity collection and exhausts clean air back into the tunnel. One typical disadvantage is that space constraints require that only smaller units be fitted. The inherent problem is that many times larger units are needed in order to maintain

Figure 1. Equipment Mounted De-Duster.

proper air volumes that are sized for the tunnel cross-sectional area and the various equipment needs within the tunnel. The other advantage with this method is the water discharge from the Wet De-duster can be gravity fed onto the material handling conveyor wetting it down so that there can be additional dust control in the material handling process.

1.2 *Trailing unit*

A trailing unit is where the De-duster is positioned behind the Road-header or TBM and is essentially "parked" in an area out of the way of the operating equipment. In this arrangement inlet ducting is added to the inlet to the De-Duster so that the proximity of the inlet suction of the De-Duster is closer to the dust source which yields better control of air movement and air quality. Typically the ductwork travels 30-60 meters from the parked de-duster to as close to the cutting area as possible. The air is suctioned from the inlet of the ductwork and cleaned by the De-duster; the clean air exhaust from the de-duster exists into the tunnel, or another section of positive pressure duct is used to channel the clean air exhaust outside of the tunnel. The advantage to this method is that you can use a larger capacity machine, exhaust ventilation and dust control remain close to the source and the mobility of the unit provides some flexibility to the operations. The unit can be operated without the inlet duct as shown in Figures 2 and 3. Note that overall

Figure 2. Trailing De-Duster.   Figure 3. Trailing De-Duster.

effectiveness of dust capture is reduced due to the expansion of the dust particles within the air stream. The downside is that the unit must be moved periodically in order to maintain the proper air flow/static pressure ratio and inlet ducting and usually exhaust ducting is needed. Also the parking location must be planned or a cut out made along the tunnel in order to locate the unit.

1.3 *Fixed mounted*

A third option for locating the De-Duster is a fixed mounted machine outside of the tunnel as shown in Figure 4. In this arrangement the unit sits on a permanent stand and inlet ducting as shown in Figure 4. The unit is extended into the entrance to the tunnel and the inlet is placed as close to the cutting area as possible. This method again assumes that "exhaust" ventilation is the preferred approach and the dirty air is suctioned from the face area and cleaned by the De-Duster before it exhausts to atmosphere. In this arrangement the fixed mounted unit can be easily accessed for maintenance; it is out of the tunnel construction area so no need for a parking space, and the air volume can be as much as needed to ventilate and clean the tunnel. The downside is that more horsepower is needed in order to overcome the static losses for the full length of the tunnel, so multiple units may be needed, or a booster in order to maintain air flow as the tunnel advances. All of the ducting must able to handle a high static pressure in order which tends to add to the overall cost of the system.

Recapitulating, the most common ventilation methods are blowing ventilation and exhaust ventilation.

1.4 *Blowing*

Air entering the tunnel construction area is brought in under positive pressure using ventilation fans from outside the tunnel; this is called "blowing" ventilation. The concept is to use clean air under positive pressure to move air through the tunnel and across the tunnel face area exiting the other side of the tunnel. The disadvantage with this method is that it is very difficult to control the air movement unless ducting is carried forward to the face. In many cases this positive pressure creates more dusting issues than it solves. The air still passes over all the diesel equipment in order to maintain breathable air quality for the workers. Note that this often this picks up dust under pressure and carries it throughout the tunnel. This can be used in conjunction with exhaust ventilation techniques, increasing effectiveness.

1.5 *Exhaust*

Exhaust ventilation occurs when the air is pulled from the face area using a series of exhaust fans or with a combination of fans and/or De-Dusting equipment. The negative pressure

Figure 4.   Fixed Mounted De-duster.

Figure 5.   Ducting Inside of Tunnel.

created from the exhaust method pulls air from the outside to sweep over the mobile equipment thereby creating a continuous stream of clean air which gradually moves across the tunnel face area and subsequently exiting the tunnel.

2 SITUATION AT TUNNEL WITH BLOWING

In this case study the tunnel construction project employed a blowing ventilation system in the tunnel. As previously mentioned this caused severe dusting issues in nearly every location of the tunnel, which led to regulators shutting down the project until an effective remedy could be found.

2.1 *Regulators shut project down due to dusting issues*

According to the regional regulatory authority, Work Safe BC, the following dust control requirements must be met for underground mining or tunneling projects:

a) Mechanical excavating devices, such as tunnel boring machines and road headers, must have an effective dust control and ventilation system which maintains workers' exposure to dust below the applicable exposure limits in this Regulation.
b) Such systems must be maintained in good working order and must be operational whenever the mechanical excavating device is working (Work Safe BC, Controlling Exposure Policy and Procedure, 2015).

Dust containing silica has specific silica limits as outlined by Work Safe BC with control method suggestions. These control methods are framed via a series of questions (Work Safe BC, Underground Workings Policy and Procedure, 2003):

a) Can local exhaust ventilation be used on all equipment that generates silica dust?
b) Can water be used to prevent dust from becoming airborne?
c) Can the areas that generate large amounts of dust be enclosed, and have proper ventilation to clean the air?

This paper is not meant to detail all the regulatory compliance details for the various types of silica other than to point out that the observed limits for Amorphous Silica exceeded the 1.5 mg/m3 limit in the worker area.

In addition to concerns over silica dust exposure, Work Safe BC also states total dust limits as described below:

The Board categorizes particulates that are insoluble or poorly soluble in water and do not cause toxic effects other than by inflammation or the mechanism of "lung overload", as "nuisance dusts". A "nuisance dust" will have an exposure limit or TLV of 10 mg/m^3 for total particulate. It is recognized that the respirable fraction of "nuisance dusts" may also be measured. The equivalent exposure limit for respirable particulate is 3 mg/m^3. Respirable particulate refers to the fraction of inhaled dust that is capable of passing through the upper respiratory tract to the gas exchange region of the lung. Total particulate refers to a wide range of particle sizes capable of being deposited in the various regions of the respiratory tract (Work Safe BC, Occupational Exposure Limits – Chemical and Biological Substances, 2004).

3 DE-DUSTER SELECTION

There are basically two main criteria for the selection of the proper De-Duster for an exhaust ventilation system; 1) Air Volume and 2) Static Pressure. The entire selection process boils down to those two main parameters. There are extensive calculations for determining the air volume requirements as discussed in Section 4 below which are for this discussion based on the tunnel dimensional parameters as listed in Table 1. The length of the tunnel is then used

Table 1. Tunnel Dimensions.

Initial Tunnel	Design Criteria	
	m	feet
Tunnel Length	1012	3320.2
Fan Dist. to Portal	15	49.2
Duct Dist to Face	18	59.1
Total Duct Length	1045	3428.5

along with the desired duct carrying velocity for the specific type of dust particles to in turn provide the criteria to calculate the duct friction loss factors for total static pressure that the fan must overcome.

4 AIR FLOW REQUIREMENT CALCULATIONS AND DEDUSTER PRINCIPLES OF OPERATION

The first step in determining the air flow requirements is to calculate the amount of air needed for each piece of diesel equipment. The calculation should also take into account the recommended amount of air so to provide clean air at the face. It should be noted that the diesel equipment requirements are determined by the size of each piece, taking into account empirical standards on how much air flow is needed to clear away the exhaust fumes from each piece as shown in Table 2 below. Dust level control requirements are based on the cross sectional area of the face and the velocity needed across the face in order to move the dust in a controlled manner. The larger of the two numbers is used for design purposes.

Table 2. Air Flow Requirements.

	m3/sec	cfm
Equipment Airflow	20.58	43,600
Losses (10%)	2.06	4,360
Total Air for Diesel Loading*	22.63	47,960
Total Air for Dust Control	21.24	45,000

* Note Total air flow for diesel loading to be used for design purposes.

4.1 Machinery & Dust Control Needs

Secondly once the total air-flow requirements are determined the static pressure of the system is calculated. This is done using standard principles from the Industrial Ventilation Handbook. The handbook combines the actual duct length with corresponding duct length factors such as exit losses and coupling losses. This will provide an effective duct length as shown in Table 3.

Table 3. Effective Duct Length Calculations.

Duct Length and Equivalent Length Calculations			
	m	Feet	NOTES
Duct Length	1009	3310	
Exit Loss	30.5	100	Equivalent Length
Coupling Losses	111	364	Equivalent Length
Total Duct Length	1150	3774	Includes Other Losses

Table 4. Static Pressure Calculations.

Criteria	Static Pressure	Dimensions	
	Inches WG	m	inches
Diameter of Duct		1.12	44
Static Resistance Total	18.87		
Safety Factor	3.77		
Entrance Losses	0.85		
Elbow Loss	0.65		
Velocity Loss	1.30		
Total Ventilation Losses	25.44		

Using Industrial Ventilation Handbook principles the static pressure is then calculated based on the "effective duct length", total air-flow requirements, duct diameter, and the calculated air velocity inside of the ductwork. Other losses in turn add to the total losses, such as the entrance losses on the duct inlet, elbows, and a velocity loss due to the air velocity in the ductwork. Table 4 shows the anticipated breakdown of these friction losses and the total design loss to be used in the De-Duster selection.

4.2 *Principles of the wet De-Duster operation*

In Figure 6 the inlet ducting is visible to the right side of the picture. This ducting carries the dirty air from the tunnel into the De-Duster inlet. The integrated impeller, Figure 7, mixes the dust and water through a very rapid pressure differential brought about by its dynamic action. This integrated impeller is shown in Figure 7.

After the rapid mixing of dust with the water, the dust-laden water travels around the motor in a sealed compartment so to keep the motor operating in clean air as illustrated in Figure 8. After leaving this bifurcated section it is again sprayed with water using an internal water spray system as shown in Figure 9. To further saturate the dust particles and to provide better air cleaning the next step is to separate the dust-laden water from the air stream. This is accomplished in the rear rectangular section of the machine using a unique series of mist eliminators and impingement panels.

4.3 *Utility needs*

Utility need requirements must be evaluated and considered for each project. In this particular case the amount of horsepower (kw). To power the De-dusting unit's integrated impeller, the amount of supply water, and the drainage location all had to be evaluated prior to installing the unit on a permanent basis.

Figure 6. De-Duster Ducting.

Figure 7. Integrated impeller.

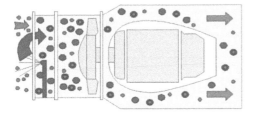

Figure 8.   Dust Capture Principles.

Figure 9.   Internal Water Sprays.

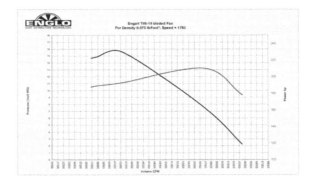

Figure 10.   Fan curve illustration.

4.4   Fan curve and power consumption

Each De-duster has an integrated impeller with a characteristic fan curve that relates air volume to static pressure of the system as illustrated in Figure 10. The curve also estimates power consumption based on the air volume against its relative static pressure which is shown on the red line in Figure 10 as well. Frequently, in order to achieve and maintain the total static pressure of the tunnel as it advances, a booster fan is placed in series with the De-duster, as needed. When placed in series the total static pressure is increased by the sum of the two units, while the volume curve remains. In effect as the tunnel advances and the duct length increases the additional static pressure is compensated with the addition of a booster.

In this application a VFD (Variable Frequency Drive) was used to vary the running speed of the impeller and in effect control the air volume more accurately. Each fan motor for the De-duster and the booster was equipped with a VFD. This allows the running speed to be increased as the tunnel advances and therefore provides an additional control mechanism for the ventilation system. The fan performance and static pressure capability increase based on standard fan laws as the running speed increases.

5   IN SITU TESTING

Final results of the system proved very successful for total dust control. Dust level readings were taken at various locations throughout the tunnel to determine if the system met the regulators' approval. The system was approved and tunnel construction was allowed to proceed.

Airflow readings were taken with a Fluke digital anemometer and the results met the required minimum design values. These results are shown in Table 5 below. In-situ dust level readings were measured in milligram per cubic meter and are shown below in Table 6 below.

Table 5. In-Situ Air Flow Results.

	Design Criteria m3/sec	cfm
Airflow - In Duct (avg)	20.501	52,400
Airflow Above Roadheader	25.34	53,700

Table 6. In-Situ Dust Testing Results.

Dust Results	mg/m3 Before De-duster	mg/m3 After De-duster
Roadheader (avg)	7.325	0.344
Scoop #1	3.321	0.030
Scoop #2	2.15	0.02

6 CONCLUSIONS

In effect, the wet extraction system met or exceeded all of the safety and regulatory requirements for dust levels and airflow. Subsequently tunneling operations were allowed to proceed, allowing construction to continue and thus saving the project. Furthermore, the system was approved for use on a permanent basis and for all future tunneling projects.

REFERENCES

Occupational Health and Safety Regulations, Part 22, Underground Workings, 2003 Oct 29. Vancouver, B.C., Work Safe BC; [accessed 2018 Sep 1] https://www.worksafebc.com/.
Occupational Health and Safety Regulations, Policies Part 05, Controlling Exposure, 2010 Sep 15. Vancouver, B.C. Work Safe BC; [accessed 2018 Sep 1] https://www.worksafebc.com/.
Occupational Exposure Limits – Chemical and Biological Substances, Discussion Paper, 2004. Vancouver, B.C., Work Safe BC; [accessed 2018 Sep 1] https://www.worksafebc.com/.

Tunnels and Underground Cities: Engineering and Innovation meet Archaeology, Architecture and Art, Volume 2: Environment sustainability in underground construction – Peila, Viggiani & Celestino (Eds)
© 2020 Taylor & Francis Group, London, ISBN 978-0-367-46579-7

Author Index